DYNAMICAL METHODS IN SOIL AND ROCK MECHANICS
1: DYNAMIC RESPONSE AND WAVE PROPAGATION IN SOILS

DYNAMICAL METHODS IN SOIL AND ROCK MECHANICS

1. Dynamic response and wave propagation in soils
2. Plastic and long-term effects in soils
3. Rock dynamics and geophysical aspects

DYNAMICAL METHODS IN SOIL AND ROCK MECHANICS
PROCEEDINGS / KARLSRUHE / 5–16 SEPTEMBER 1977

1

Dynamic response
and wave propagation
in soils

edited by
B. PRANGE
Institute of Soil and Rock Mechanics, Karlsruhe

A. A. BALKEMA / ROTTERDAM / 1978

The text of the various papers in this volume were set individually
by typists under the supervision of each of the authors concerned.

For the complete set of three volumes, ISBN 90 6191 023 4
For volume 1, ISBN 90 6191 024 2
For volume 2, ISBN 90 6191 025 0
For volume 3, ISBN 90 6191 026 9

©1978 A. A. Balkema, P. O. Box 1675, Rotterdam

Printed in the Netherlands

Contents

Preface

Dynamical Methods in Soil and Rock Mechanics was the topic of
an International Symposium and a Nato Advanced Study Institute,
held at the University of Karlsruhe, Germany, on September
5 - 16, 1977, organized by the Institute of Soil and Rock
Mechanics. It was aimed at the exchange of information on the
state of the art and the presentation of results of basic research and practical experience in Soil and Rock Mechanics.
The lectures of the main sessions and the discussion contributions of the accompanying workshops are comprised in three
volumes of the proceedings:
    I   DYNAMIC RESPONSE AND WAVE PROPAGATION IN SOILS
    II   PLASTIC AND LONG-TERM EFFECTS IN SOILS
    III ROCK DYNAMICS AND GEOPHYSICAL ASPECTS
The material presented is of interest to those involved in research as well as to the practising engineer dealing with dynamically and cyclically loaded structures and with environmental vibration problems.
Volume I deals with problems related to the momentary response
of soils and soil-structure systems to dynamic loads.
Consequently, the contributions cover:
    The investigation of the dynamic soil properties by experi-
    mental field and laboratory methods
    The analysis of soil-structure interaction and structure-
    soil-structure interaction with numerical and analytical
    methods
    The mathematical and experimental treatment of wave propaga-
    tion and wavefields
    The problems of measuring vibration data of structures and
    wavefields as well as the dynamic excitation
The 21 articles of volume I may be considered to be a brief
survey of the theoretical and experimental methods available at
the present time to deal with problems of momentary dynamics
in Soil and Rock Mechanics.
A short introduction relates the articles to the respective
topics (page numbers in paranthesis):
The determination of the elastic and damping properties of soils
in the field and in the laboratory is covered by the principle
contribution of Richart (3), the particular parameters affecting
such properties and their relevance are investigated by Woods
(37) and Prange (61).

Mechanical analoga of dynamic soil-structure systems are described as lumped parameter models by Woods (79) assuming the subsoil to be an ideal elastic halfspace and the structure at the interface to be rigid.

If the superstructure in contact with the soil is flexible, the deformability of the interface has to be investigated by a more rigorous analysis. Holzlöhner (103) and Gaul (167) apply both the halfspace theory and finite-element methods to the problem of soil-structure interaction, while the dynamic interaction of adjacent structures through the coupling soil is investigated by Roesset and Gonzalez (127).

In connection with pile problems, Novak (185) presents impedance functions of embedded piles, gained by analytical methods, to describe the response of piles to dynamic loading.

The importance of soil-structure interaction is reflected by the fact, that four discussion contributions to the workshop "Lumped Parameter Models" deal with this problem, Stokoe (201), Werkle (205), Kausel (213) and Prange (217).

Based on the assumption of an ideal elastic halfspace, the analytical methods available for the computation of wavefields and soil-structure interaction are discussed in detail in a principle contribution by Savidis (225). Haupt (255) treats the problem of wavefields with numerical, namely finite-element methods applying a new boundary condition to varify the halfspace at the grid boundary. Interferences in the nearfield and due to subsoil anomalies are studied by Prange (281).

Cross-Correlation methods are a powerful tool for the investigation of different wave-types, effects of anisotropy of polarized waves and for measurements at extremely low signal/noise ratios. Roesler (309) introduces this technology to Soil Dynamics.

From the point of view of screening of waves, Haupt (335) deals with a specific soil-structure interaction problem applying finite-element methods with the already introduced boundary conditions.

Some basic problems of the measurement of vibrations are discussed by Prange (369) with emphasis on inertia effects of vibration transducers.

The method of holographic interferometry presented by Woods (391) enables one to actually see wavefields and study directly the screening effects of trenches and barriers.

Rücker (407) investigates the dynamic response and the input excitation of systems by the measurement of random vibrations caused by traffic. A method to decouple the vibration modes of electro-dynamic exciters is described in a discussion contribution by Prange (423).

Not all the problems of the momentary dynamic response of soils and soil-structure systems can be covered in detail in this volume. The reader will, however, find some useful information immediately applicable to practical dynamical problems as well as references where further experimental and analytical research is necessary.

The editor wishes to express his appreciation to all the contributors and to all who assisted in the preparation of the manuscripts.

B. Prange

Field and laboratory measurements
of dynamic soil properties

F. E. RICHART, Jr.
University of Michigan, Ann Arbor, Mich., USA

ABSTRACT

The shear modulus and value of material damping are the two dynamic soil properties needed for evaluation of dynamic wave propagation and dynamic response of foundations.

Field methods are described for determining the shear wave velocity, from which the low-amplitude shear modulus may be calculated. Laboratory methods, particularly the resonant column method, provide correlations between laboratory and field tests for the low-amplitude shear modulus for undisturbed samples, when the secondary time effects are included. Laboratory tests permit study of factors which influence the shear modulus values. A factor of particular influence is the shearing strain amplitude, which may be varied in resonant column tests up to about one percent strain in the high amplitude machines.

INTRODUCTION

Dynamic loadings develop stress-strain relationships in soils which may be different from those developed under static loadings. The shape of the stress-strain curve provides information on the dynamic moduli and damping, which are the principal soil properties required for evaluation of wave propagation in soils and for dynamic soil-structure interaction studies. This paper considers field and laboratory methods for evaluating these dynamic properties of soils.

The shape of the stress-strain relationship for any particular soil depends upon the type of loading and boundary restraining conditions. Figure 1 shows typical behaviors of cylindrical samples of cohesionless soil when subjected to an increasing axial stress, σ_z, while a radial pressure, σ_r acts on its periphery. Curve A represents hydrostatic compression

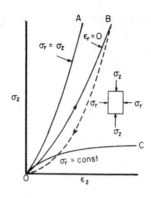

Fig.1 Stress-Strain Curves
      for Triaxial Tests of
      Sand with Various
      Lateral Restraints

($\sigma_r = \sigma_z$), a loading condition which develops a "strain-hardening" nearly elastic nonlinear type of stress-strain curve. Curve B represents the condition of constrained compression (no radial expansion permitted at the periphery of the sample) which is again a strain-hardening type of curve, but tests show an appreciable hysteresis loop developed upon unloading. Curve C illustrates the strain-softening type of stress-strain curve developed for a constant restraining boundary pressure as developed in a stress-controlled triaxial test.

At very low values of strain, the initial portion of curves A, B, or C may be approximated by a linear elastic stress-strain curve. One of the objectives of this discussion is to describe methods of evaluating the linear "elastic moduli" for soils and to note how these moduli are modified for dynamic loadings. Another objective is to present results of laboratory and field tests concerning dynamic stress-strain curves of type C.

EQUATIONS FOR STRESS-STRAIN RELATIONSHIPS

Because the different shapes of the stress-strain curves in Fig. 1 are produced by different proportions of volumetric compression and shearing strains, it is sometimes convenient to express the general stress-strain relations in the form of a constitutive equation. For example, Jackson (1969) has shown how laboratory test data can be interpreted to fit into an equation of the type,

$$\sigma_{ij} = K \bar{e} \delta_{ij} + 2G(\varepsilon_{ij} - \frac{1}{3} \bar{e} \delta_{ij}) \tag{1}$$

and a yield condition

$$\sqrt{J_2''} = f(\overline{\sigma}_o) \tag{2}$$

to describe dynamic behavior of soils under ground shock loadings. In Eq. (1), σ_{ij} = total stress tensor; ε_{ij} = total strain tensor; K = modulus of volume compressibility, or bulk modulus; $\overline{e}$ = $\varepsilon_x + \varepsilon_y + \varepsilon_z$ = cubic dilation, or volumetric strain; G = strain modulus; and δ_{ij} = Kronecker delta function (δ_{ij} = 1 when i = j, δ_{ij} = 0 when i $\neq$ j). In Eq. (2), J_2'' = second invariant of the stress deviation, and $\overline{\sigma}_o$ = average normal effective stress or octahedral normal stress.

Constitutive equations similar to Eq. (1) have often been incorporated into analytical studies of blast loadings on soils in which the soil was considered to deform primarily in one-dimensional compression (constrained compression). Both volumetric compression and change of shape are developed in constrained compression and the general stress-strain curve is similar to curve B in Fig. 1. However, for "low stress" levels (below about 3.5 kg/cm^2) Hadala (1973) noted that nonlinear curves of the strain-softening type were developed for many soils. This effect depended upon the type of soil, its initial relative density or degree of compaction, and the degree of saturation. Thus, the strain-softening type of stress-strain curve may be developed in most practical soil dynamics problems.

Theoretically, elastic shearing stress-strain relations developed by pure shear involve no change of volume, which causes $\overline{e}$ = 0 in Eq. (1). This leaves only the linear relation between shearing stress and shearing strain, which may be represented by

$$\tau_{xz} = G\,\gamma_{xz} \tag{3}$$

in the x-z plane (after converting the tensor symbols to ISSMFE symbols). Equation (3) relates shearing stress to shearing strain through the shear modulus, G. This expression may also be used to represent conditions at any point along a shearing stress-shearing strain curve, as shown in Fig. 2 for the strain-softening type curve. At shearing strains approaching zero, the value of G is a maximum and is designated as G_o. At a larger value of shearing strain, for example at a strain corresponding to point A in Fig. 2, the line connecting the origin to A represents the secant modulus, designated as G. The value of the secant modulus decreases as the shearing strain increases.

5

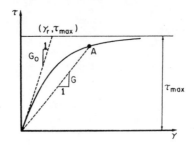

Fig.2 Basic Parameters for
Hyperbolic Stress-
Strain Curves.

The shape of the strain-softening shearing stress-strain curve for a particular soil can be adequately represented by a hyperbolic curve (Kondner, 1963, Hardin and Drnevich, 1972a, 1972b) which has a slope of G_o at $\gamma = 0$ and which is asymptotic to the horizontal line representing $\tau = \tau_{max}$ (Fig. 2). This curve is defined by the equation

$$\tau = \frac{\gamma}{1/G_o + \gamma/\tau_{max}} = \frac{\gamma\, G_o}{1 + \gamma/\gamma_r} \tag{4}$$

in which γ_r = "reference shearing strain" = τ_{max}/G_o. Hardin and Drnevich found that, by presenting test data in the normalized form of τ/τ_{max} vs. γ/γ_r, data from clay specimens fell nearly on a single curve while data from sand specimens fell along another curve. Neither curve fit the hyperbolic shape exactly but by devising a "hyperbolic strain" parameter, γ_h, they were able to adjust the curves for sand and for clay to fit on one curve on a τ/τ_{max} vs. γ_h plot.

The same hyperbolic type curves have been obtained from static and dynamic torsion tests. Test results from reversed torsion loading of a sand sample are shown in Fig. 3. Two complete hysteresis loops are shown, one with extremities at A-A' and a "hysteresis modulus" of G_1, and the second with extremities at B-B' and a hysteresis modulus G_2 which is smaller than G_1 because of the larger shearing strain reached at point B. The dots along the initial loading curve (OAB) represent extremities of hysteresis loops for other maximum strain values. Test data consistently show hysteresis loops which increase in width (increased loss of energy) and exhibit lower secant or hysteretic moduli as the maximum shearing stress in the loading cycle increases.

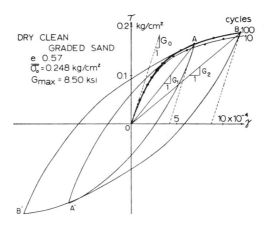

Fig.3 Stress-Strain Hysteresis Loops for Reversed
Loading (Hardin and Drnevich, 1972b)

Equation (4) is adequate to describe the loading curve
(OAB) in Fig. 3, but it is difficult to represent the hysteresis loops by hyperbolic segments. A convenient mathematical
relation to describe the loading curve OAB is given by

$$\frac{\gamma}{\gamma_r} = \frac{\tau}{\tau_{max}} \left[1 + \alpha \left| \frac{\tau}{C_1 \tau_{max}} \right|^{R-1} \right] \tag{5}$$

(Ramberg and Osgood, 1943) in which , R, and C_1, are constants
which permit adjustments of the shape and position of the curve.
For unloading from conditions (τ_1, γ_1) the Ramberg-Osgood (R-O)
curve follows the expression

$$\frac{\gamma - \gamma_1}{\gamma_r} = \frac{\tau - \tau_1}{\tau_{max}} \left[1 + \alpha \left| \frac{\tau - \tau_1}{2C_1 \tau_{max}} \right|^{R-1} \right] \tag{6}$$

Ramberg-Osgood curves have been used for more than a decade to
represent the stress strain behavior of structural elements,
and more recently (Constantopolous, et al, 1973; Streeter, et
al, 1974; Richart and Wylie, 1975; and Faccioli and Ramirez,
1976) the R-O curves have been adopted to represent dynamic
stress-strain behavior of soils.

7

SHEARING STRENGTH UNDER DYNAMIC LOADING CONDITIONS

Figure 2 and Eq. (4) illustrate that the two soil properties needed to evaluate the dynamic shearing stress-strain curve for a particular soil are the small strain shear modulus, G_o, and the ultimate shearing strength, τ_{max}. Both of these quantities must be evaluated for loading and restraining conditions anticipated for the soil in situ.

It is often convenient to establish the maximum shearing stress for a given soil under static loading conditions, then add corrections to account for dynamic effects. Under static conditions the maximum value of shearing strength available on a given plane through a soil mass depends (according to the Mohr-Coulomb failure theory) upon the normal pressure acting on that plane, the effective angle of internal friction, ϕ', and the cohesion intercept, c'. Of course, the shearing strength should be evaluated on the plane which will be subjected to the maximum static plus dynamic shearing stresses. For example, for vertical shear stress propagation through layers of soil, as is often considered in earthquake excitations, the maximum allowable shearing stress on the horizontal (and vertical) planes in a soil mass is given by

$$\tau_{max} = \left\{ \left[\left(\frac{1 + K_o}{2} \right) \bar{\sigma}_v \sin \phi' + c' \cos \phi' \right]^2 - \left[\frac{1 - K_o}{2} \bar{\sigma}_v \right]^2 \right\}^{1/2} \quad (7)$$

in which $\bar{\sigma}_v$ is the effective vertical stress, and K_o is the coefficient of earth pressure at rest ($\bar{\sigma}_h = K_o \bar{\sigma}_v$). The vertical stress on any plane is the total effective stress produced by the weight of the overburden plus pressures developed by construction of a building, etc.

Strain-Rate Effect

Tests to determine the increase in τ_{max} caused by an increase in rate of loading above normal static loading rates have been conducted during the past three decades by utilizing dynamic loading to failure in unconfined, triaxial, direct, or simple shear tests. The test results are described herein as $(\tau_{max})_{dynamic} = (\tau_{max})_{static} \times$ (strain rate factor). These results demonstrating the "strain rate effect" have been summarized by Whitman (1970) and Richart, et al (1970) who noted that: (a) for dry sands the strain rate factor was less than 1.10 to 1.15 for strain rates varying from about 0.02% per second to 1000% per second; (b) for saturated cohesive soils the strain rate factor was 1.5 to 3.0; and (c) for partially saturated soils the strain rate factor was 1.5 to 2.0. Dynamic triaxial tests on dry sand by Lee, et al (1969) showed a strain rate factor, for strain rates varying from 0.1% to 10000% per minute, of: (a) about 1.07 for loose sand at confining pressures up to 15 kg/cm^2 and for dense sand at low confining pressure (less than about 6 kg/cm^2), and (b) about 1.2 for dense sand at high confining pressures.

8

From this brief discussion it is evident that the strain rate effect is relatively unimportant for dry sands, but may be significant for clays at high rates of strain. For saturated sands and silty material the effect of strain rate includes the time-dependent build up of pore pressures and possible liquefaction.

Effect of Repeated Loadings

The strain rate effect described in the preceding section indicated that a maximum stress greater than the static maximum stress can be developed in the sample if an increasing load is applied rapidly. However, if stresses less than this dynamic maximum are repeated enough times, failure of the sample will occur. This constitutes a low-cycle fatigue type of failure which has been described by Seed and Chan (1966) and Murayama and co-workers (see Murayama 1970, for example) for clays. The number of repetitions of a particular stress level which may be applied before failure depends on the initial sustained stress level, the shape of the repeated stress pulse, the frequency of pulse application, the type of testing device, and of course, on the characteristics of the soil being tested.

Similar low-cycle fatigue behavior has been observed in repeated triaxial tests of saturated sands, but here the failure was usually developed by liquefaction. A special note should be made with respect to the reduction of the shearing strength of saturated sands by an increase in pore pressure, whether it is static or dynamic. An increase in pore pressure reduces the value of $\bar{\sigma}_o$ in Eq. (2) or $\bar{\sigma}_v$ in Eq. (7) and a lower value shearing strength is developed. Thus a given applied shearing stress will represent a greater proportion of the available shearing strength as pore pressures increase.

Repeated loadings of dry sand may lead to an actual strengthening of the material, because of reductions of void radio and more favorable grain orientations, but greater deformations may be developed than for the static case because of these progressive movements (see Toki and Kitago, 1974).

The shearing strengths of sands and clays are reduced if they are tested under conditions of increasing shear stresses plus superposed vibratory shearing stresses, or increasing shearing stresses while the normal stresses on the failure plane are subjected to vibratory pulsations. Tests of this type have been reported by Barkan (1962), Ermolaev and Senin (1968), and Satyavanija and Nelson (1971). The reduction in shearing stress is a function of the static confining pressures, magnitude of the vibratory stress, frequency and duration of the vibratory stresses, and the type of soil.

This brief discussion of the effects of strain rate and repeated loads on shearing strength of soils is presented to emphasize again that the numerical value of shearing strength to be included in analytical studies of dynamic soil motions must be determined for conditions likely to be encountered in the field.

9

LOW AMPLITUDE MODULI DETERMINED BY FIELD TESTS

Standard seismic tests permit evaluation of low strain-amplitude wave propagation velocities in soils and rock. Then using equations from elasticity theory, the appropriate modulus of elasticity is calculated. The compression wave velocity, v_P, is related to the constrained modulus, M, by

$$v_P = \sqrt{\frac{\lambda + 2G}{\rho}} = \sqrt{\frac{M}{\rho}} \qquad (8)$$

In Eq. (8), G is the shear modulus, ρ is the mass density (= unit weight, γ_s, divided by acceleration of gravity, g), and λ equals $2\nu G/1-2\nu$ where ν = Poisson's ratio. The compression wave velocity is governed by the volumetric compressibility of the medium. For saturated soils at shallow depths the volumetric compressibility of water in the pore spaces governs the propagation velocity and the compressibility of the soil skeleton has little effect (see Richart, Hall, and Woods, 1970, Chap. 5, and Ishihara, 1970). Consequently, compression wave velocities of 1500-1600 m/sec measured in the field usually represent the wave propagation velocity in water and give no effective information on soil properties. On the other hand, an abrupt increase to 1500 m/sec in compression wave velocities evaluated at increasing depths is a useful indication of the water table.

The shear wave velocity, v_s, is related to the shear modulus, G, by

$$v_s = \sqrt{\frac{G}{\rho}} \qquad (9)$$

and because water has no shearing strength, evaluation of v_s permits a direct evaluation of the shear modulus of the soil skeleton. The mass density, ρ, is influenced by water only because of the increase in unit weight, γ_s, as the water content increases. Ishihara (1970) has shown that the pore water moves in phase with the soil structure for saturated fine-grained soils and in coarse sands for frequencies of vibration up to about 50 Hz. Thus, the total unit weight is normally used in calculating ρ.

Because the shear wave velocity gives a reliable means to evaluate the dynamic soil property of shear modulus, and is not effected by the presence of the water table, only methods for field evaluation of shear wave velocity will be treated in the following sections. General seismic investigation procedures are described in textbooks and this information may be supplemented by technical publications directed toward shear wave evaluations (for example, Mooney, 1974 and Ballard and McLean, (1975), for soils in general, Roethlisberger(1972) for frozen soils, and Hamilton (1971) for submarine soils).

Evaluations of G_O by Explorations Conducted from the Surface

Seismic tests usually develop shearing strains in the field of 10^{-6} or less, therefore the shear modulus calculated from measured shear wave velocities obtained in these tests is designated as the "low amplitude shear modulus, G_O". This is one of the basic parameters noted in Fig. 2. Tests involving a seismic disturbance and response measurements at the ground surface are often used for evaluation of G_O throughout large zones of subsurface soil or rock.

Shear Wave Refraction

Shear waves polarized in the horizontal plane (SH-waves) may be developed at the surface of the ground by a horizontal impact on a loaded plank or footing in firm contact with competent soil. If the impact is directed North, for example, then a line of horizontal motion pick-ups are placed on an East-West line through the source and a standard refraction survey is run. The results may be enhanced by repeating each measurement with impacts directed North, then South, to reverse the direction of the shear wave arriving at each pick-up location. By superposing the records, the arrival of the shear wave is readily determined.   A general description of the method is given by Schwarz and Musser (1972) and by Mooney (1974), and the technique has been applied to underwater measurements of sea-bottom sediments by Schwarz and Conwell (1974). In the underwater application the excitation was developed by a 77 kg iron slug which impacts against the ends of a tube, actuated by right and left solenoids. The exciter and pick-ups are strung out along a line and laid down upon the sea bottom from a ship.

The shear wave refraction method has the same restrictions as the conventional refraction method in that the method is accurate only under conditions where seismic velocities increase with depth. A lower velocity stratum beneath a higher velocity stratum cannot be observed.

Steady-State Surface Vibrations (Rayleigh Wave Method)

Steady-state vertical vibration of a footing at the ground surface will set up a train of surface (Rayleigh) waves that propagate outward from the source. The velocity of the Rayleigh wave, v_R, is approximately equal to the shear wave velocity ($v_R/v_S = 0.933$ for $\nu = 0.33$, and $v_R/v_S = 0.95$ for $\nu = 0.45$) and can be used directly for most engineering applications.

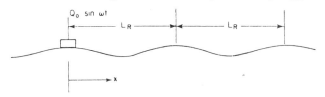

Fig.4 Deformed Shape of a Half-Space Surface

In a test, the exciter operates at a given frequency and the length, L_R, of the Rayleigh wave is measured along the ground surface by measuring the distance between ground motions at the same phase (i.e., distance from wave crest to crest as shown in Fig. 4). Then the distance vs. number of waves is plotted as shown in Fig. 5. For each frequency setting, a line is established on this diagram which corresponds to the wave length, L_R. Then from the relation, $v_R = f \cdot L_R$, the Rayleigh wave velocity is established. Finally, a diagram is made with depth as ordinate (positive downward) and wave velocity as abscissa and the wave velocity is plotted at a depth corresponding to $L_R/2$. By this procedure, Fig. 6 was developed from the information given on Fig. 5.

The steady state vibration method has been used successfully throughout the past two decades for evaluation of wave velocities in soils and rock (see Ballard, 1964, Ballard and McLean, 1975). Investigations have been made to depths greater than 61 m. by using large vibrators, but practically the test is limited by the availability of large vibrators and by site access. Often the limiting depth is about 30 m. The test results are not affected by the presence of the water table, and the method <u>does</u> identify softer layers beneath harder layers. An early application of this method was for evaluation of road construction in The Netherlands, carried out by van der Poel, Nijboer, Heukelom, and co-workers (see Heukelom and Foster, 1960, for a description of the method as applied to pavements and sub-bases).

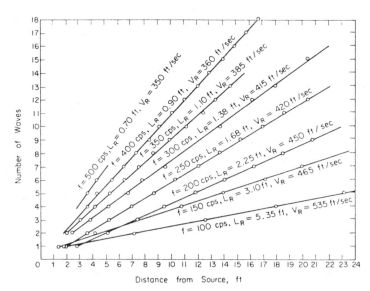

Fig.5 Determination of Average Wave Length of Rayleigh Wave.

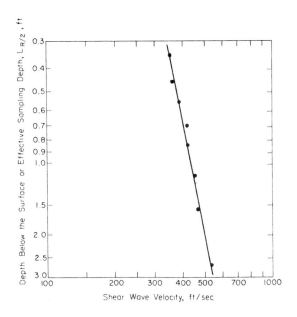

Fig.6 Shear Wave Velocity vs.Depth in Fine,Silty Sand

Evaluations of G_O by Measurements in Bore Holes

For detailed information on the variation of wave velocities with depth at a particular location, bore hole techniques can be used. Up-hole and down-hole tests can be performed
with one bore hole while cross-hole tests require two or more
bore holes. See Fig. 7 for arrangements of excitation and
sensors in these tests.

Up-Hole and Down-Hole Tests

In the up-hole method, the excitation is provided at various
depths with the bore hole and the sensor is placed at the surface.
Thus only the average value of wave propagation between the two
points is obtained.

In the down-hole method the excitation is provided at the
surface and one or more sensors are located down the bore hole.
Because the excitation is located at the surface, it is convenient to generate both compression and shear waves for determination of v_p and v_s. Compression waves can be generated by
vertical impact on a plate at the surface, and shear waves can
be generated by horizontal impact on a 15 cm x 15 cm plank about
2 m long which is pressed against the soil surface by driving
the two front wheels of a truck onto the plank. Then the plank
may be struck from the right and left ends to generate shear
waves of opposite sign. By spacing a series of sensors at relatively short intervals, the variations of shear wave velocities with depth can be readily identified. Details of the down-
hole method are given by Schwarz and Musser (1972), and Imai
(1977), for example.

13

The advantages of the down-hole method include, (a) only one bore hole is required and this can be prepared in advance of the seismic test, (b) average values of v_s along a vertical path are obtained, which may be significant measurement for vertical dynamic response of foundations on horizontally stratified media, and (c) the method can identify a lower velocity stratum between higher velocity strata. A disadvantage lies in the instrumentation required to measure travel times between sensors when attempting to identify relatively thin horizontal layers.

Cross-Hole Tests

In the cross-hole tests (White, 1965, Stokoe and Woods, 1972, Schwarz and Musser, 1972, for example) at least two boreholes are required, one for the impulse and one or more for sensors. As shown in Fig. 7(c) the impulse rod is struck at the top end and an impulse travels down the rod and is transmitted to the soil at the bottom. This shear impulse creates shear waves which travel horizontally through the soil to the vertical motion sensor located in the second hole, and the time required for the shear wave to traverse this known distance is measured. Because the distance between the impulse and pickup sensor is a critical quantity in evaluating v_s, it is necessary to determine the deviation from verticality of the bore holes by using a Slope Indicator or similar device.

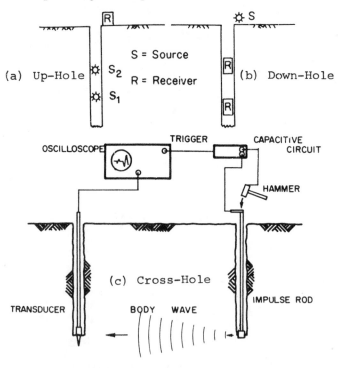

Fig. 7   Field Test Procedures to Evaluate v_s

In layered soils the spacing between boreholes is important, and refraction effects at interfaces (see Fig. 8) may introduce errors. For shallow depths in cohesionless soils, the wave velocity increases with depth and a curved-path travel distance, as shown in Fig. 9 (see Haupt, 1973), must be considered in order to obtain correct values for v_s.

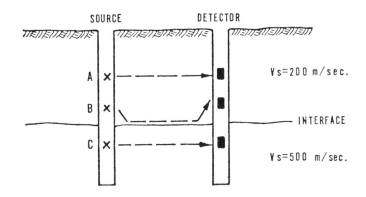

Fig.8 Seismic Wave Travel Paths for the Inter-Hole Method. Note Indirect Travel Path Caused by Refraction of Energy at Source Point B. (from Schwarz and Musser, 1972)

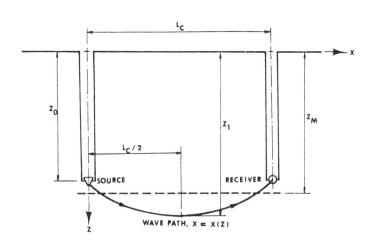

Fig.9 Diagram of Curved Wave Path Theory (Haupt, 1973)

The cross-hole method is particularly well suited for evaluating v_s at shallow depths at sites with limited access, such as beneath building or machine foundations. It is particularly useful at locations with a high level of ambient seismic background noise because of the relatively short distance between the source and pick-up. The cross-hole method is relatively cheap and easy to use, and by reversing the direction of impulse on successive tests (Schwarz and Musser, 1972) the shear wave signal may be enhanced for more precise identification of the shear wave. It is possible to run cross-hole tests in holes previously drilled and cased if a large shear impulse is developed by a drop hammer acting against an anvil jacked tightly against the casing, but packing between the casing and soil may introduce an error in the measurement of v_s (Stokoe and Abdel-razzak, 1975). Recent developments in the cross hole technique (Miller, et al, 1975) now permit evaluation of v_s at large strain amplitudes.

EVALUATION OF G_O FROM LABORATORY TESTS

The low amplitude shear modulus, G_O, may be evaluated from v_s by pulse tests or resonant column tests, or it may be determined from shearing stress-strain tests involving torsion of hollow cylindrical or ring-shaped samples or simple shear tests of cylindrical samples. In each test, it is necessary to establish G_O for shearing strains of 10^{-5} or less.

Pulse Tests

In the pulse method, an energy pulse usually of low magnitude is initiated at one point in a soil sample and the arrival of the wave transmitting the energy is recorded at another point. Then the velocity of wave propagation is calculated from the time required for this low energy pulse to travel the known distance. The pulse method using a compression wave has long been a standard laboratory method for measuring the mechanical properties of materials, especially for concrete (see Kessler and Chang, 1957). A summary of the various laboratory pulse techniques, types of equipment developed, and significant test results were presented by Whitman (1970).

Lawrence (1965) developed an apparatus which applied a torsional input to a sample. Twelve piezoelectric crystals were placed at each end of the sample in a radial pattern similar to spokes in a wheel. Each crystal was manufactured to produce a shear distortion when subjected to a sudden electric signal. Thus an electric signal developed the torsional input energy and the torsional wave at the other edge of the sample produced an electric signal by distorting the crystals. With this device, Lawrence studied the shear waves propagated in sands, Boston Blue Clay, and a pure kaolin clay. He also discussed at some length the major problems involved in obtaining reliable test data. The length of the specimen is one very important factor. If the sample is too long the shape of the wave front may be distorted to the point where it is undetectable either because of energy losses in the sample or because of extraneous signals traveling through the test equip-

16

ment. If the specimen is too short, the probable error in
measuring the travel time becomes significant with respect to
the total travel time. Furthermore, if the sample is too short
dispersion of the high frequency wave may occur in granular
soils. Lawrence has indicated that the thickness should be at
least 200 times the diameter of typical particles. Successful
tests have been run on samples 2 in. long which were confined
by a metal cylinder and 4 in. long for triaxial samples confined
by gas pressures.

The pulse test is not adaptable for studies of the effects
of stress (or strain) amplitudes, or frequency on the wave propagation velocities in soils. It provides no information on
internal damping within the soil sample. Furthermore, the
techniques involved in generating and interpreting the electrical signals are sufficiently complicated that the pulse
method as used in the laboratory is restricted primarily to
research investigations.

Resonant Column Tests

The resonant column method has been available for determining wave propagation velocities in soil samples for four
decades, but major developments have occurred during the past
18 years. More than 55 laboratories in the U.S.A., Canada,
Japan, Mexico, Germany, Venezuela, Taiwan, Yugoslavia, Norway,
and Turkey now use resonant column devices for research or soil
investigation. In the resonant column test, a cylindrical
(solid or hollow) column of soil is contained within a rubber
membrane, placed in a triaxial cell, and set into motion in
either the longitudinal or torsional mode of vibration. The
frequency of the electro-magnetic drive system is changed
until the first mode resonant condition is determined. This
resonant frequency, the geometry of the sample, and the conditions of end restraint provide the necessary information to
calculate the velocity of wave propagation in the soil under
the given testing conditions. During the past decade it has
often been found convenient to use hollow cylindrical samples,
particularly for cohesive soils, in tests including shearing
strain amplitude as one of the test variables.

In research investigations, it is always desirable to
change only one test variable during each series of tests.
Hardin and Black (1968) have noted the quantities which exert
an influence on the shear modulus of soils and have expressed
these as a functional relationship

$$G = f(\overline{\sigma}_o, e, A, t, H, f, C, \theta, \tau_o, S, T) \tag{10}$$

in which $\overline{\sigma}_o$ = average effective confining pressure; e = void
ratio; A = amplitude of shearing strain; t = secondary effects
that are functions of time and magnitude of time and magnitude
of stress increment; H = ambient stress history and vibration
history; f = frequency of vibration; C = grain characteristics;

17

θ = soil structure, τ_0 = octahedral shearing stress; S = degree of saturation, and T = temperature. Of course, several of these quantities may be related (for example e, C, and θ). When discussing the low amplitude shear modulus, G_0, the effect of A is eliminated and we have only the remaining parameters to consider.

Effect of Void Ratio and Confining Pressure

For clean sands, it has been found that G_0 is essentially independent of each of the variables except $\bar{\sigma}_0$ and e. This conclusion has been confirmed by Kuribayashi, et al (1975). Also the conclusion that particle size, shape, and distribution has negligible effect on the wave propagation velocity in sands was confirmed by Krizek, et al (1974). Analytical expressions have been presented for the shear modulus of clean sands as

$$G_0 = 700 \frac{(2.17-e)^2}{1+e} (\bar{\sigma}_0)^{0.5} \qquad (11)$$

for round-grained sands (e < 0.80) and as

$$G_0 = 326 \frac{(2.97-e)^2}{1+e} (\bar{\sigma}_0)^{0.5} \qquad (12)$$

for angular grained sands. In Eqs. (11) and (12), G_0 and $\bar{\sigma}_0$ have units of kg/cm^2. Both equations were originally established to correspond to shearing strains of 10^{-4} or less. Equation (11) was found to give values slightly lower than those obtained by pulse tests (Whitman and Lawrence, 1963), and recently Iwasaki and Tatsuoka (1977) have determined experimentally that

$$G_0 = 900 \frac{(2.17-e)^2}{1+e} (\bar{\sigma}_0)^{0.38} \qquad (13)$$

from tests on clean sands (0.61 < e < 0.86 and 0.2 < $\bar{\sigma}_0$ < 5 kg/cm^2) at shearing strain amplitudes of 10^{-6}. For shearing strains of 10^{-4} their results agreed with Eq. (11).

Effect of Time of Pressure Application

Equation (12) was modified slightly and proposed by Hardin and Black (1968) for use as a first estimate for the value of G_0 for cohesive soils, corresponding to 1-day's duration of the confining test pressure. However, the time effect (t) has been found to be important for resonant column tests of fine-grained soils (Afifi and Woods, 1971; Marcuson and Wahls, 1972; Afifi and Richart, 1973; and Anderson and Woods, 1976). Figure 10 illustrates a typical increase in v_s with time at constant confining pressure, $\bar{\sigma}_0$. Note that v_s was evaluated intermittently during the test by vibrating the sample less than 30 seconds to obtain each reading. The test curve in Fig. 10

18

shows two distinct zones, a primary behavior corresponding to that which might be anticipated during reconsolidation of the sample to its in-situ stress condition and the secondary behavior or "secondary time effect" which is perhaps analogous to secondary compression. It is important to note that it requires from 100 to 1000 minutes for the primary reconsolidation to be complete in cylindrical cohesive soil samples 3.57 cm in diameter. Larger samples would require greater times for the primary behavior to be complete.

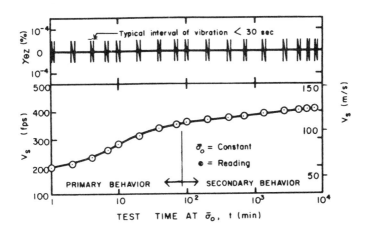

Fig.10 Time Effects in
Resonant Column
Tests (Anderson
and Woods,1976)

The secondary time effect has been described in terms of the increase in v_s (i.e., Δv_s) per log cycle of time expressed as a ratio of the value of v_s found at a cumulative testing time of 1000 minutes. Thus the ratio $\Delta v_s/v_{s1000}$ expresses the influence of the secondary time effect for different soils. When comparing laboratory test data to field conditions it is necessary to add to v_{s1000} the ratio $\Delta v_s/v_{s1000}$ multiplied by the number of log cycles corresponding to the time since the last major stress change occurred in the field conditions. For example, if a fill had been placed 20 years ago on a layer of soil where field v_s values were recently obtained, better agreement between laboratory and field values of v_s are produced if v_{s1000} is multiplied by $(1 + 4 \Delta v_s/v_{s1000})$. The secondary time effect was found to be unimportant for soils having $D_{50} > 0.04$ mm., but $\Delta v_s/v_{s1000}$ was found to be as large as 17% for some clays. In every laboratory test for v_s in cohesive soils, the secondary time effect must be evaluated (see Anderson and Woods, 1976, and Anderson and Stokoe, 1977, for data on secondary time effects).

The effect of temperature (T) of clay during resonant column testing was found to be unimportant by Anderson and Richart (1974). Tests of seven cohesive soils (total testing time of about 1 year) at 4°C and 22°C showed that v_s at 4°C was equal to or not more than 12% greater than v_s determined at 22°C. However, resonant column tests of frozen soils (Stevens, 1975) have shown a significant effect of changes in temperature near the freezing point.

Repeated Static Torsion Tests

Hardin (1965) devised a large amplitude-low frequency torsional device which accommodated a hollow cylindrical soil sample 23 cm long, 13 cm O.D., 10 cm I.D. in reversed torsion. This device permitted accurate measurements of moduli and damping for shearing strains as small as 10^{-5} up to 80% of the yield stress of the sample at frequencies of 1/12 Hz. The results of tests with this device were compared with those obtained from resonant column tests and it was found that both devices gave essentially the same results. Thus there appeared to be no frequency effect for tests on soils involving frequencies of 0 to about 100 Hz. An improved model of this low-frequency high-amplitude device using electromagnetic loading was developed by Hardin (1971). Iwasaki, et al, 1977, used a torsional shear device in which hollow cylindrical samples of sand 10 cm in height, 10 cm O.D., 6 cm I.D. were tested. At shearing strains of 10^{-4} the results were found to be the same as those obtained from resonant column tests. However, it shoul be noted that 10^{-4} is not a low enough strain to be considered "low amplitude", and the results from these torsional tests on hollow cylindrical samples essentially evaluate G at "large amplitude" shearing strains.

Yoshimi and Oh-Oka (1973) constructed a ring torsion device for testing samples 24 cm I.D., 2.4 cm wall thickness, and height varying from 2.0 cm at the inside diameter to 2.4 cm at the outside diameter. With this device data were obtained for the shear modulus of a sand sample for shearing strains varying from about 3×10^{-5} to 1×10^{-2}. Sherif and Ishibashi (1976) also developed a torsional simple shear device which can determine elastic moduli of samples in the range of 5×10^{-5} to 10^{-2}. Hollow samples of sand with 10.2 cm O.D., 5.1 cm I.D. and 2.5 cm high on the outside and 1.3 cm high on the inside were tested. The sample can be subjected simultaneously to a vertical stress, σ_v, inside and outside radial stresses σ_{ro} and σ_{ri}, and a cyclic shear stress, τ_{vr}.

The torsional shearing devices are particularly useful for evaluating the influence of the first few cycles of stress repetitions, and for determining the effects of higher amplitude shearing strains on reductions of G from G_o.

Repeated Simple Shear Tests

Simple shear testing devices were developed to evaluate shearing strength and volume changes during loading. The simple shear device designed and manufactured by the Norwegian Geotechnical Institute was modified for use in cyclic shear tests to give shearing stress-shearing strain data also. Descriptions of the equipment as well as significant test data were given by Silver and Seed (1971) and by Pyke (1973). In this device a cylindrical sample of soil 8 cm diameter and 2 cm high is constrained within a wire-reinforced rubber membrane. The sample base is mounted on a sliding table and the cap is fixed to a cross head which is restrained from horizontal motion but is permitted to move vertically. Vertical loading of the cross head provides vertical stress on the sample. A lever and cam arrangement drives the sliding table horizontally to produce cyclic shearing strain in the sample ranging from 10^{-4} to 5×10^{-3}.

COMPARISON OF G_O (or v_s) FROM FIELD AND LABORATORY TESTS

Pulse tests results from laboratory samples were compared with seismic field test results by Dobry and Poblete (1969) for a deposit primarily of basaltic sands and silts and a good agreement was found. Seismic and resonant column test results were obtained from clay soils in the Texcoco Basin (Martinez, et al, 1974) and higher values were found by the seismic method, with the difference increasing with depth of the sample. Resonant column and seismic results from limestone (Yang and Hatheway, 1976) showed appreciable differences. Possible sources of error were attributed to refraction through adjacent harder layers in seismic tests, possible nonhomogeneities in the laboratory samples, and time effects. Cunny and Fry (1973) reported on laboratory and field evaluations of G_O at 14 sites which included a variety of soils. The field method for evaluating G_O was the steady state surface vibration (Rayleigh wave method) and the resonant column test was used in the laboratory. From evaluation of their test data they found that the laboratory-determined shear and compression moduli ranged within $\pm 50\%$ of the in situ moduli. Discussions of this paper pointed out that the cross-hole method should give better values of v_s at the depth the undisturbed samples were taken, and that including the secondary time effect would bring the laboratory values for cohesive soils nearer to the field values. For tests of sands the secondary time effect is negligible, and Stokoe and Richart (1973) and Iwasaki and Tatsuoka (1977) have found good agreement between resonant column and cross-hole field test values. For cohesive soils, Trudeau et al (1973) and Anderson and Woods (1975) found agreement between resonant column and cross-hole test values of v_s after the secondary time effect correction was added to the laboratory values. Figure 11 shows the comparison of laboratory and field values of v_s, with the open symbols representing the 1000 minute value for laboratory data and the solid symbols showing the secondary time correction to the field history of 20 years. Thus, laboratory test values of v_s from the resonant column test involving shearing strains of 10^{-6} to 10^{-5} in undisturbed samples should be expected to agree within about 10% of the values of v_s obtained at the same location by cross-hole tests, if the secondary time correction is evaluated and applied to the laboratory test data.

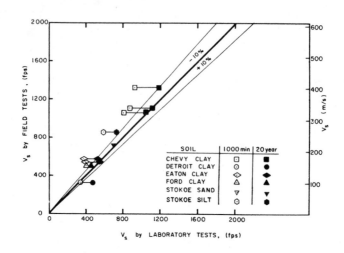

Fig.11 Comparison of Field and Laboratory Values of v_s (Anderson and Woods, 1975)

EVALUATION OF SHEAR MODULUS AT LARGE SHEARING STRAIN AMPLITUDES

It has been recognized for some time (see Hall and Richart, 1963) that an increase in shearing strain amplitude causes a decrease in v_s. Compilations of test data showing this trend were given by Seed and Idriss (1970) and Hardin and Drnevich (1972a). The Hardin-Drnevich data included resonant column tests and repeated torsion tests on hollow samples of particular materials. Recent tests by Iwasaki, Tatsuoka, and Takagi (1977) as shown in Fig. 12 illustrate the continuous curve obtained from resonant column tests $10^{-6} < \gamma < 3 \times 10^{-4}$ and from the repeated torsion shear tests $3 \times 10^{-5} < \gamma < 2 \times 10^{-2}$ on the same material. Good agreement is obtained in the region around $\gamma \approx 10^{-4}$ where the information overlaps.

The reduction of v_s with increasing shearing strain is evidence that the secant modulus G is lower than G_o as shown by Fig. 2. Hardin and Drnevich (1972a) found that curves of the type shown by Fig. 2 could adequately represent the torsional shearing stress-strain relations for sands and clays. If the test information was presented in dimensionless form with τ/τ_{max} as ordinate, and γ/γ_r as abscissa, then all data for sands fell on one curve and the data for clays fell on a

22

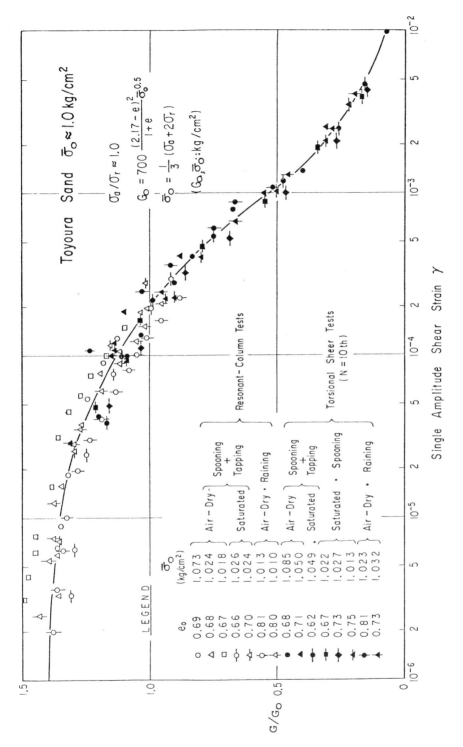

Fig.12 G/G₀ vs. Shearing Strain Relationship for Toyoura Sand (from Iwasaki, et al., 1977)

second curve, as shown in Fig. 13. The slope of the secant
to any point on one of these curves represents G/G_0, which
reduces as the shearing strain increases. For evaluating the
reduction in shear modulus because of earthquake induced
strains in soils, it is convenient to plot G/G_0 vs. γ/γ_r.
Figure 14 includes two solid curves which describe the reduction in shear modulus as the shearing strain increases, for
sands and for clays. Also shown in Fig. 14 is a dashed curve
representing the average of data presented by Seed and Idriss
(1970) for cohesive soils.

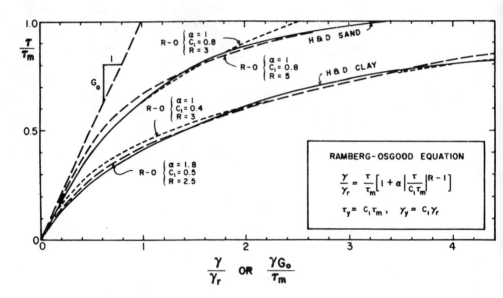

Fig.13 Fit of Ramberg-Osgood Curves to Soil Data

Often the numerical value of the shearing strain is used
as abscissa and G/G_0 as ordinate when presenting experimental
results. This may lead to errors in interpretation, particularl
for cohesionless soil. Consider a clean dry quartz sand for
which e = 0.65, ϕ = 30°, and K_0 = 0.6. A shearing strain of
γ = 0.001 is developed in the sand mass, and we need to estimate the G/G_0 values at a depth of 5 m and 25 m. From Eqs.
(7) and (11) values of τ_{max} and G_0 are calculated, then γ_r,
γ/γ_r and G/G_0 are as shown below:

Depth (m)	τ_{max} (kg/cm^2)	G_0 (kg/cm^2)	γ_r	γ/γ_r	G/G_0
5	0.28	750	3.7×10^{-4}	2.70	0.37
25	1.40	1690	8.2×10^{-4}	1.22	0.61

Because the ratio of τ_{max}/G_o increases as a function of $(\bar{\sigma}_o)^{0.5}$ the reference shearing strain increases with depth and the chosen strain of $\gamma = 0.001$ produces a smaller γ/γ_r at a deeper location. Therefore the value of G/G_o is larger at the greater depths. The influence of confining pressure on G/G_o was also noted by Shibata and Soelarno (1975) who developed an equation for the secant modulus reduction of sands at increasing shearing strains as

$$\frac{G}{G_o} = \frac{1}{1 + 1000 \dfrac{\gamma}{(\bar{\sigma}_o)^{0.5}}} \tag{14}$$

Note that Eq. (14) is quite similar to Eq. (4) since $\gamma_r = f(\bar{\sigma}_o)^{0.5}$. Thus Eq. (14) represents a modified hyperbola, comparable to the one developed by Hardin and Drnevich (1972b). The Shibata equation was found to fit recent test data (Yoshimi, 1976), and interpretation of this information led to development of the dotted curve shown in Fig. 14 which corresponds closely to the Hardin-Drnevich curve for sands.

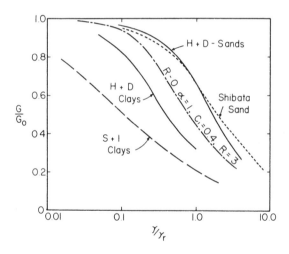

Fig.14 Dimensionless Curves for Shear Modulus Reduction with Increasing Shear Strain

In addition to resonant column tests, dynamic simple shear tests of sands (Silver and Seed, 1971, Park and Silver, 1975), free vibration and forced simple shear tests on clay (Kovacs et al, 1971), free torsional vibration tests of cohesive soils (Taylor and Parton, 1973; Zeevaert, 1973), ring torsion dynamic tests (Yoshimi and Oh-Oka, 1973), and high amplitude cross-hole field tests (Miller, et al, 1975) have given results showing the reduction of shear modulus with increasing shearing strain.

From resonant column test results, Anderson and Richart (1976) found G/G_o vs. γ/γ_r relations for six cohesive soils, which could be adequately represented throughout the range of $10^{-5} < \gamma < 10^{-3}$ by a Ramberg-Osgood curve with parameters $\alpha = 1$, $C = 0.4$, and $R = 3$. This R-O relation is shown in Fig. 14 as the dash-dot curve.

This discussion of reduction in G with increasing shearing strain amplitudes leads to the conclusion that for important installations which may be subjected to dynamic loadings, it is necessary to conduct laboratory tests to evaluate the G/G_o vs. γ/γ_r curve for each significant soil layer. However if previously published curves are to be adopted for preliminary studies, it should be noted that recent test data agree better with the Hardin-Drnevich curves than with the Seed-Idriss curves.

MATERIAL (HYSTERETIC) DAMPING IN SOILS

Material damping describes the energy losses within loaded soil masses caused by interparticle slip and friction of particle contacts. The energy losses in soils during cyclic or reversed loadings can be significant during vibratory or transient loadings which involve large strain amplitudes, as may be developed during earthquakes. Field methods for evaluating damping in soils have not yet been developed for practical use, consequently the following discussion will treat only laboratory methods.

Fundamental Relationships

Although interparticle friction develops hysteretic damping in sands, Hall and Richart (1963), and Hardin (1965) demonstrated that the results could be readily interpreted in terms of damping of a viscoelastic system.

Figure 15 shows a simple mass-spring-dashpot which is set into motion in the vertical direction, z, by a dynamic force $Q(t)$. The equation of motion for this one-degree-of-freedom system is

$$m\ddot{z} + c\dot{z} + kz = Q(t) \tag{15}$$

The damping constant, c, is often combined with the mass, m, and spring constant, k, to form a "damping ratio"

$$D = \frac{c}{2\sqrt{km}} = \frac{c}{c_{cr}} \tag{16}$$

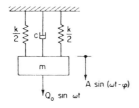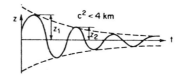

Fig.15 Mass-Spring-
Dashpot System

Fig.16 Free Vibration of a
Viscously Damped System

When the system shown in Fig. 15 is set into steady state vibration and the exciting force Q(t) is removed, the vibration amplitude will decay with time, as shown in Fig. 16, because of the damping. The amplitude ratio of decaying vibration is also a measure of damping and is designated as the "logarithmic decrement"

$$\delta = \ln \frac{z_1}{z_2} = \frac{2\pi D}{\sqrt{1-D^2}} \approx 2\pi D \qquad (17)$$

The loss of energy in viscoelastic systems may also be described by the strain energy lost during oscillations. The stress strain curves from reversed loading ($\tau - \gamma$ curves in Fig. 17) form an ellipse which has its major axis along a line at the slope of G. The slope of this major axis remains constant regardless of the magnitude of the shearing stress developed. The ratio of energy lost in one cycle of oscillation, ΔW, to the input energy, W, is often called the "specific damping capacity", and it is related to other damping terms by

$$\frac{\Delta W}{W} \approx 2\delta \approx 4\pi D \qquad (18)$$

for values of δ smaller than about 0.25. The relation between δ and $\Delta W/W$ is

$$\frac{W}{W} = 1 - e^{-2\delta} \qquad (19)$$

for a system having a modulus independent of strain amplitude.

27

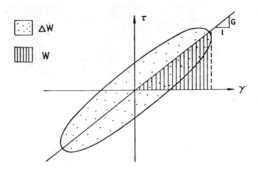

Fig.17 Shearing Stress-Strain
Ellipse for Viscoelastic
Material

Laboratory Measurements of Damping in Soils

An effective way of measuring damping in soils at small
strain amplitudes is by use of the resonant column test (see
Richart, Hall, and Woods, 1970). When the driving power is
turned off, internal damping within the soil sample causes
the vibration to die out after a series of oscillations, as
shown in Fig. 18a. Fig. 18b is a plot of relative amplitude
(log scale) vs. the number of cycles of oscillation from a
resonant column test of Ottawa sand. The test data plot along
a straight line on this diagram, as would be expected for a
viscoelastic material.

This apparent agreement with viscoelastic behavior has
been confirmed by many tests, and a comprehensive study by
Hardin (1965) confirmed that material damping in sands could
be represented by a Kelvin-Voigt model (viscous damping) if
the viscosity, μ, of the model was treated as varying with
frequency ω, to maintain the ratio $\mu\omega/G$ constant.

This ratio is related to the other terms for viscoelastic
damping by

$$\delta = \pi \left(\frac{\mu\omega}{G}\right) = 2\pi D \qquad\qquad (20)$$

Hardin also gave values of $\mu\omega/G$ which may be used in design,
which in terms of the logarithmic decrement are functions of
the shearing strain amplitude, γ, and average effective confining pressure, $\bar{\sigma}_o$, as

$$\delta = \pi 9 \ (\gamma)^{0.2} \ (\bar{\sigma}_o)^{-0.5} \qquad\qquad (21)$$

28

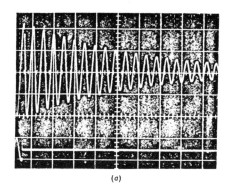

(a)

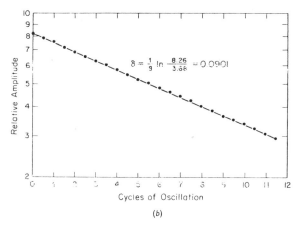

$$\delta = \frac{1}{9} \ln \frac{8.26}{3.66} = 0.0901$$

Cycles of Oscillation

(b)

Fig.18 Typical Free Vibration-Decay Curves Obtained
From Resonant Column Tests of Ottawa Sand.
(a) Amplitude-Time Decay Curves. (b) Amplitude
vs. Cycle Number Plot (from Richart, Hall,and
Woods, 1970)

Although Hardin and Drnevich (1972) and Drnevich (1972) have
improved Eq. (21) by including the influence of numbers of
repetitions (N) of the shearing strain, Kuribayashi, et al,
(1975) found that Eq. (21) provided a lower bound for damping
in several types of sands.

The effect of shearing strain amplitude on damping in
both sands and clays was illustrated on composite diagrams
by Seed and Idriss (1970), and Fig. 19 shows some data for
clays. This is useful general information, but for damping
of soils in a specific loading and strain environment, it is
preferable to obtain laboratory test results using the actual
values of $\bar{\sigma}_o$ and determinations of δ or D as a function of
γ or γ/γ_r.

29

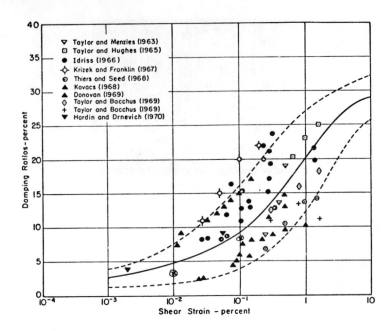

Fig.19 Damping Ratios for Saturated Clays
(from Seed and Idriss, 1970)

CONCLUSIONS

The shear modulus and damping are the two dynamic soil properties needed for evaluation of dynamic wave propagation and dynamic response of foundations.

Field methods are available for determining the low-amplitude values of shear modulus, G_O, (from v_s measurements) at a particular site. Laboratory methods, particularly the resonant column method, provide correlations for G_O when undisturbed samples are used, and when the secondary time effects are included. Laboratory methods also permit evaluation of the changes in shear modulus as the shearing strain amplitude is increased.

At the present time, hysteretic damping cannot be measured readily in the field, but can be evaluated from tests on laboratory samples by use of the resonant column method, or by the repeated torsion or direct shear methods.

ACKNOWLEDGEMENTS

Many of the studies described herein have been carried out at the University of Michigan during the past 15 years and have recently been described by Richart (1975) and Richart (1977). The author and his colleagues are particularly grateful to the National Science Foundation for support of their research during this period.

REFERENCES

Afifi, S. S., and Richart, F. E., Jr., (1973), "Stress-History Effects on Shear Modulus of Soils," Soils and Foundations (J. JSSMFE) Vol. 13, No. 1, March, pp. 77-95.

Afifi, S. S., and Woods, R. D., (1971), "Long-Term Pressure Effects on Shear Modulus of Soils," J. Soil Mech. and Found. Div. Proc. ASCE, Vol. 97, SM10, Oct., pp. 1445-1460.

Anderson, D. G., and Richart, F. E., Jr., (1974), "Temperature Effect on Shear Wave Velocity in Clays," J. GED, Proc. ASCE, Vol. 100, No. GT12, pp. 1316-1320.

Anderson, D. G., and Richart, F. E., Jr., (1976), "Effects of Straining on Shear Modulus of Clays," J. GED, Proc. ASCE, Vol. 102, No. GT9, pp. 975-987.

Anderson, D. G., and Woods, R. D., (1975), "Comparison of Field and Laboratory Shear Moduli," Proc. Conf. on In Situ Measurements of Soil Properties, N. Car. St. Univ., Raleigh, N.C., Vol. I, pp. 69-92.

Anderson, D. G., and Woods, R. D., (1976), "Time-Dependent Increase in Shear Modulus of Clay," J. GED, Proc. ASCE, Vol. 102, No. GT5, pp. 525-537.

Anderson, D. G., and Stokoe, K. H., II, (1977), "Shear Modulus: A Time-Dependent Material Property," Dynamic Soil and Rock Testing in the Field and Laboratory for Seismic Studies, ASTM STP.

Ballard, R. F., Jr., (1964), "Determination of Soil Shear Moduli at Depth by In-Situ Vibratory Techniques," U.S. Army Engineer Waterways Experiment Station, Vicksburg, Miss., Misc. Paper No. 4-691, Dec.

Ballard, R. F., and McLean, F. G., (1975), "Seismic Field Methods for In Situ Moduli," Proc. Conf. on In Situ Measurement of Soil Properties, N. Car. St. Univ., Raleigh, N.C., Vol. I, pp. 121-150.

Barkan, D. D., (1962), "Dynamics of Bases and Foundations," McGraw-Hill Book Co., New York, 434 pp.

Constantopoulos, I. V., Roesset, J. M., and Christian, J. T., (1973), "A Comparison of Linear and Exact Nonlinear Analyses of Soil Amplification," Proc. V WCEE (Rome), pp. 1806-1815.

Cunny, R. W., and Fry, Z. B., (1973), "Vibratory In Situ and Laboratory Soil Moduli Compared," J. SMFD, Proc. ASCE, Vol. 99, No. SM12, pp. 1055-1076.

Dobry, R., and Poblete, M., (1969), Discussion, J. SMFD, Proc. ASCE, Vol. 95, No. SM2, pp. 659-662.

Drnevich, V. P., (1972), "Undrained Cyclic Shear of Saturated Sand," J. SMFD, Proc. ASCE, Vol. 98, No. SM8, Aug., pp. 807-825.

Ermolaev, N. N., and Senin, N. V., (1968), "Shear Strength of Soil in the Presence of Vibrations," Soil Mech. and Found. Eng. (Transl. of Osnov., Fund. i Mekh. Grunt.) Consultants Bu., N.Y., No. 1, Jan.-Feb., pp. 15-18.

Faccioli, E., and Ramirez, J., (1976), "Earthquake Response of Non-Linear Hysteretic Soil Systems," Earthquake Eng. and Struc. Dynamics, Vol. 4, No. 3, pp. 261-276.

Hadala, P. F., (1973), "Effect of Constitutive Properties of Earth Media and Outrunning Ground Shock from Large Explosions," Tech. Rep. S-73-6, U.S. Army Waterways Exp. Sta., Vicksburg, Miss.

Hall, J. R., Jr., and Richart, F. E., Jr., (1963), "Dissipation of Elastic Wave Energy in Granular Soils," J. Soil Mech. and Found. Div. Proc. ASCE, Vol. 89, No. SM6, Nov., pp. 27-56.

Hamilton, E. L., (1971), "Elastic Properties of Marine Sediments," J. Geophysical Research, Vol. 76, No. 2, pp. 579-604.

Hardin, B. O., (1965), "The Nature of Damping in Sands," J. SMFD, Proc. ASCE, Vol. 91, No. SM1, Jan., pp. 63-97.

Hardin, B. O., (1965), "Dynamic Versus Static Shear Modulus for Dry Sand," Materials Research and Standards, ASTM, Vol. 5, No. 5, May, pp. 232-235.

Hardin, B. O., (1971), "Constitutive Relations for Airfield Subgrade and Base Course Materials. 1. Characterization and Use of Shear Stress-Strain Relation," University of Kentucky, Lexington, Ky., Technical Report UKY 32-71-CE5, Soil Mech. Series No. 4, Contract No. F29601-70-C-0040, with U. S. Air Force Weapons Lab., Albuquerque, N.M.

Hardin, B. O., and Black, W. L., (1968), "Vibration Modulus of Normally Consolidated Clay," J. Soil Mech. and Found. Div. Proc. ASCE, Vol. 94, No. SM2, Mar., pp. 353-369.

Hardin, B. O., and Drnevich, V. P., (1972a), "Shear Modulus and Damping in Soils: Measurement and Parameter Effects," J. Soil Mech. and Found. Div. Proc. ASCE, Vol. 98, No. SM6, June, pp. 603-624.

Hardin, B. O., and Drnevich, V. P., (1972b), "Shear Modulus and Damping in Soils: Design Equations and Curves," J. Soil Mech. and Found. Div. Proc. ASCE, Vol. 98, No. SM7, July, pp. 667-692.

Haupt, W. A., (1973), Discussion of, "In Situ Shear Wave Velocity by Cross-Hole Method," by K. H. Stokoe, II, and R. D. Woods, J. SMFD, Proc. ASCE, Vol. 99, No. SM2, Feb., pp. 224-228.

Heukelom, W., and Foster, C. R., (1960), "Dynamic Testing of Pavements," J. SMFD, Proc. ASCE, Vol. 86, No. SM1, Feb., pp. 1-28.

Imai, T., (1977), "P- and S-Wave Velocities of the Ground in Japan," Proc. 9ICSMFE (Tokyo), Vol. 2, pp. 257-260.

Ishihara, K., (1970), "Approximate Forms of Wave Equations for Water-Saturated Porous Materials and Related Dynamic Modulus," Soils and Foundations, Vol. X, No. 4, Dec., pp. 10-38.

Iwasaki, T., and Tatsuoka, F., (1977), "Dynamic Soil Properties with Emphasis on Comparison of Laboratory Tests and Field Measurements," Proc. VI WCEE (New Delhi).

Iwasaki, T., Tatsuoka, F., and Takagi, Y., (1977), "Shear Modulus of Sands under Cyclic Torsional Shear Loading," To appear in Soils and Foundations.

Jackson, J. G., Jr., (1969), "Analysis of Laboratory Test Data to Derive Soil Constitutive Properties," Misc. Paper S-69-16, U.S. Army Engineer Waterways Exp. Sta., Vicksburg, Miss.

Kessler, C. E., and Chang, T. S., (1957), "A Review of Sonic Methods for the Determination of Mechanical Properties of Solid Materials," Bull. ASTM, Oct., pp. 40-46.

Kondner, R. L., (1963), "Hyperbolic Stress-Strain Response: Cohesive Soils," J. Soil Mech. and Found. Div. Proc. ASCE, Vol. 89, No. SM1, Feb., pp. 115-143.

Kovacs, W. D., Seed, H. B., and Chan, C. C., (1971), "Dynamic Moduli and Damping Ratios for a Soft Clay," J. SMFD, Proc. ASCE, Vol. 97, No. SM1, pp. 59-75.

Krizek, R. J., McLean, F. G., and Giger, M. W., (1974), "Effect of Particle Characteristics on Wave Velocity," J. GED, Proc. ASCE, Vol. 100, No. GT1, pp. 89-94.

Kuribayashi, E., Iwasaki, T., and Tatsuoka, F., (1975), "Effects of Stress-Strain Conditions on Dynamic Properties of Sands," Proc. JSCE, No. 242, Oct., pp. 1-10.

Lee, K. L., Seed, H. B., and Dunlop, P., (1969), "Effect of Transient Loading on the Strength of Sand," Proc. 7th ICSMFE (Mexico City), Vol. 1, pp. 239-247.

Lawrence, F. V., Jr., (1965), "Ultrasonic Shear Wave Velocities in Sand and Clay," Rep. R64-05, Dept. Civil Eng., Mass. Inst. Tech., Cambridge, Mass., to WES, Vicksburg, Miss.

Marcuson, W. F., III, and Wahls, H. E., (1972), "Time Effects on Dynamic Shear Modulus of Clays," J. GED, Proc. ASCE, Vol. 98, No. SM12, pp. 1359-1373.

Martinez, B., Leon, J. L., Rascon, O., Villarreal, A., (1974), "Determination of the Dynamic Properties of Clays in the Texcoco Basin (Mexico)," (in Spanish), Ingenieria, Tacuba 5, Apartado M-6987, Mexico 1, D.F. No. 2, pp. 182-203.

Miller, R. P., Troncoso, J. H., and Brown, F. R., (1975), "In Situ Impulse Test for Dynamic Shear Modulus of Soils," Proc. Conf. In Situ Meas. Soil Prop., N. Car. St. Univ., Raleigh, N.C., Vol. I, pp. 319-335.

Mooney, H. M., (1974), "Seismic Shear Waves in Engineering," J. GED, Proc. ASCE, Vol. 100, No. GT8, pp. 905-923.

Murayama, S., (1970), "Dynamic Behavior of Clays," Proc. 5th Int. Conf. on Rheology, Vol. 2, U. Tokyo Press.

Park, T, K., and Silver, M. L., (1975), "Dynamic Triaxial and Simple Shear Behavior of Sand," J. GED, Proc. ASCE, Vol. 101, No. GT6, pp. 513-529.

Pyke, R. M., (1973), "Settlement and Liquefaction of Sands Under Multi-Directional Loading," Ph.D. Dissertation, University of California, Berkeley, California, Nov., 282 pp.

Ramberg, W., and Osgood, W. T., (1943), "Description of Stress-Strain Curves by Three Parameters," Tech. Note 902, NACA.

Richart, F. E., Jr., (1975), "Some Effects of Dynamic Soil Properties on Soil-Structure Interaction," J. GED, Proc. ASCE, Vol. 101, No. GT12, Dec., pp. 1193-1240.

Richart, F. E., Jr., (1977), "Dynamic Stress-Strain Relationships for Soils," Chap. 1 of "Soil Dynamics and Its Application to Foundation Engineering," by Y. Yoshimi, F. E. Richart, Jr., S. Prakash, D. D. Barkan and V. A. Ilyichev, Proc. 9th ICSMFE (Tokyo), Vol. 2, pp. 605-612.

Richart, F. E., Jr., Hall, J. R., Jr., and Woods, R. D., (1970), "Vibrations of Soils and Foundations," Prentice-Hall, Inc., Englewood Cliffs, N.J., 414 p. (Japanese transl. by T. Iwasaki and A. Shimazu, 1975, Kajima Inst. Publ. Co., Ltd. Tokyo).

Richart, F. E., Jr., and Wylie, E. B., (1975), "Influence of Dynamic Soil Properties on Response of Soil Masses," Symp. Struc. and Geotech. Mech., U. of Ill., Urbana, Ill., Oct. 2-5.

Roethlisberger, H., (1972), "Seismic Exploration in Cold Regions," Corps of Eng., U.S. Army, Cold Reg. Res. Eng. Lab., Hanover, N.H., CRREL Cold Reg. Sci. Eng., Monograph II-A 2a, 138 pp.

Satyavanija, P., and Nelson, J. D.,(1971), "Shear Strength of Clay Subjected to Vib. Loading," Proc. 4th Asian Reg. Conf. SMFE (Bangkok), pp. 215-220.

Schwarz, S. D., and Conwell, F. E., (1974), "A Technique for the In-Situ Measurements of Shear Wave Velocities (v_s) for Deep Marine Sediments," Offshore Technology Conference Preprints, Vol. I, Paper No. OTC2014.

Schwarz, S. D., and Musser, J. M., Jr., (1972), "Various Techniques for Making In-Situ Shear Wave Velocity Measurements," Proc. Int. Conf. Microzonation, Seattle, Wash., Vol. II, pp. 593-608.

Seed, H. B., and Chan, C. K., (1966), "Clay Strength Under Earthquake Loading Conditions," J. SMFD, Proc. ASCE, Vol. 92, No. SM2, pp. 53-78.

Seed, H. B., and Idriss, I. M., (1970), "Soil Moduli and Damping Factors for Dynamic Response Analyses," Earthquake Eng. Res. Cen., Univ. of Cal., Berkeley, Cal., Rep. No. EERC 70-10.

Sherif, M. A., and Ishibashi, I., (1976), "Dynamic Shear Moduli for Dry Sands," J. GED, Proc. ASCE, Vol. 102, No. GT11, Nov., pp. 1171-1184.

Shibata, T., and Soelarno, D. S., (1975), "Stress-Strain Characteristics of Sands under Cyclic Loading," Proc. Japanese Soc. Civil Engrs., No. 239, July, pp. 57-65 (in Japanese).

Silver, M. L., and Seed, H. B., (1971), "Deformation Characteristics of Sands Under Cyclic Loading," J. SMFD, Proc. ASCE, Vol. 97, No. SM8, pp. 1081-1098.

Stevens, H. W., (1975), "The Response of Frozen Soils to Vibratory Loads," Corps of Eng., U.S. Army Cold Regions Res. and Eng. Lab., Tech. Rep. 265, June, 98 pp.

Stokoe, K. H., II, and Abdel-razzak, K. G., (1975), "Shear Moduli of Two Compacted Fills," Proc. Conf. In Situ Meas. Soil Prop., N. Car. St. U., Raleigh, N.C., Vol. I, pp. 422-449.

Stokoe, K. H., II, and Richart, F. E., Jr., (1973), "In-Situ and Laboratory Shear Wave Velocities," Proc. VIII ICSMFE (Moscow), Vol. 1.2, pp. 403-409.

Stokoe, K. H., II, and Woods, R. D., (1972), "In Situ Shear Wave Velocity by Cross-Hole Method," J. SMFD, Proc. ASCE, Vol. 98, No. SM5, pp. 443-460.

Streeter, V. L., Wylie, E. B., and Richart, F. E., Jr., (1974), "Soil Motion Computations by Characteristics Method," J. GED, Proc. ASCE, Vol. 100, No. GT3, pp. 247-263.

Taylor, P. W., and Parton, I. M., (1973), "Dynamic Torsion Testing of Soils," Proc. VIII ICSMFE, (Moscow), Vol. 1.2, pp. 425-432.

Toki, S., and Kitago, S., (1974), "Strength Characteristics of Dry Sand Subjected to Repeated Loading," Soils and Found., Vol. 14, No. 3, pp. 25-39.

35

Trudeau, P. J., Whitman, R. V., and Christian, J. T., (1973), "The Shear Wave Velocity of Boston Blue Clay," Mass. Inst. of Tech., Cambridge, Mass., Soils Pub. No. 317, 62 pp.

White, J. E., (1965), "Seismic Waves: Radiation, Transmission, and Attenuation," McGraw-Hill Book Co., New York, 302 pp.

Whitman, R. V., (1970), "The Response of Soils to Dynamic Loads," Mass. Inst. Tech. Rep. 26 to U.S. Army Eng. Wat. Exp. Sta., Vicksburg, Miss., 200 pp.

Whitman, R. V., and Lawrence, F. V., (1963), Discussion, J. SMFD, Proc. ASCE, Vol. 89, No. SM5, pp. 112-118.

Yang, Z., and Hatheway, A. W., (1976), "Dynamic Response of Tropical Marine Limestone," J. GED, Proc. ASCE, Vol. 102, No. GT2, pp. 123-128.

Yoshimi, Y., (1976), Personal corresp. describing preliminary results obtained by Comm. on Dyn. Prop. Soils, Bldg. Res. Inst., Min. Constr., Japan.

Yoshimi, Y., and Oh-Oka, H., (1973), "A Ring Torsion Apparatus for Simple Shear Tests," Proc. VIII ICSMFE (Moscow), Vol. 1.2, pp. 501-506.

Zeevart, L., (1973), "Foundation Engineering for Difficult Subsoil Conditions," Van Nostrand Reinhold Co., 652 pp.

Note: J. SMFD, and J. GED, Proc. ASCE are, respectively, Journal of the Soil Mechanics and Foundation Division, and Journal of the Geotechnical Engineering Division, Proceedings of the American Society of Civil Engineers.

Parameters affecting elastic properties

RICHARD D. WOODS
University of Michigan, Ann Arbor, Mich., USA

INTRODUCTION

Many models of dynamic soil behavior make use of elastic properties of the soil. When a homogeneous, isotropic, linearly
elastic material is assumed to represent the soil, the elastic
properties required to fully account for the material behavior
are E, G and ν, or λ, G and ν where E is Young's Modulus, G is
shear modulus, ν is Poisson's ratio, and λ and G are Lame's constants. Table 1 lists mutual relationships among these elastic
constants and related quantities. In Table 1, E = Young's Modulus, k = bulk modulus, M = constrained modulus, λ and μ = Lame's
Constants, μ = G = shear modulus and ν = Poisson's Ratio.

IDEAL ELASTIC SPHERES

In mechanics we are usually considering material behavior in
terms of a homogeneous, isotropic, elastic continuum in which
self-weight or body forces are not considered. But our first
clue to understanding dynamic soil behavior comes from parti-
culate mechanics and specifically the behavior of perfect
spheres. Hertz (1881) described the distribution of stress at
the contact between two elastic spheres, Fig. 1, a and b. The
stress distribution on a circular section of radius "a" was
parabolic in shape, but more importantly the equation for the
tangent modulus for uniaxial loading had the form

$$E_T = K_1 \, \sigma_x^{1/3} \tag{1}$$

Similarly, the modulus of volume compression B, which relates
hydrostatic stress to volume compression had the form

$$B = K_2 \, \sigma_o^{1/3} \tag{2}$$

The clue to soil behavior provided by Eqs. 1 and 2 is that modulus is a function of the one-third power of axial or hydrostatic stress, and in soils these stresses are caused, at least
partially, by the soils own weight.

TABLE 1
Mutual Relations Among Elastic Constants

	E	k	M	λ	μ	ν
E,k			$\frac{3k(3k+E)}{(9k-E)}$	$\frac{3k(3k-E)}{(9k-E)}$	$\frac{3kE}{(9k-E)}$	$\frac{1}{2}\ \frac{E}{6k}$
E,M		$\frac{(3M-E+\omega_1)}{6}$		$\frac{(M-E+\omega_1)}{4}$	$\frac{(3M+E-\omega_1)}{8}$	$\frac{(E-M-\omega_1)}{4M}$
E,λ		$\frac{(E+3\lambda+\omega_2)}{6}$	$\frac{(E-\lambda+\omega_2)}{2}$		$\frac{(E-3\lambda+\omega_2)}{4}$	$\frac{(\omega_2-E-\lambda)}{4}$
E,μ		$\frac{E\mu}{3(3\mu-E)}$	$\frac{\mu(4\mu-E)}{(3\mu-E)}$	$\frac{\mu(E-2\mu)}{(3\mu-E)}$		$\left(\frac{E}{2\mu}-1\right)$
E,ν		$\frac{E}{3(1-2\nu)}$	$\frac{(1-\nu)E}{(1+\nu)(1-2\nu)}$	$\frac{\nu E}{(1+\nu)(1-2\nu)}$	$\frac{E}{2(1+\nu)}$	
k,M	$\frac{9k(M-k)}{(M+3k)}$			$\frac{(3k-M)}{2}$	$\frac{3}{4}(M-k)$	$\frac{(3k-M)}{(3k+M)}$
k,λ	$\frac{9k(k-\lambda)}{(3k-\lambda)}$		$(3k-2\lambda)$		$\frac{3}{2}(k-\lambda)$	$\frac{\lambda}{(3k-\lambda)}$
k,μ	$\frac{9k\mu}{(3k+\mu)}$		$\left(k+\frac{4}{3}\mu\right)$	$\left(k-\frac{2}{3}\mu\right)$		$\frac{(3k-2\mu)}{2(3k+\mu)}$
k,ν	$3k(1-2\nu)$		$\frac{3k(1-\nu)}{(1+\nu)}$	$\frac{3k\nu}{(1+\nu)}$	$\frac{3k(1-2\nu)}{2(1+\nu)}$	
M,λ	$\frac{(M+2\lambda)(M-\lambda)}{(M+\lambda)}$	$\frac{(M-2\lambda)}{2}$			$\frac{(M-\lambda)}{2}$	$\frac{\lambda}{(M+\lambda)}$
M,μ	$\frac{\mu(3M-4\mu)}{(M-\mu)}$	$\left(M-\frac{4}{3}\mu\right)$		$(M-2\mu)$		$\frac{(M-2\mu)}{2(M-\mu)}$
M,ν	$\frac{(1-2\nu)(1+\nu)M}{(1-\nu)}$	$\frac{(1+\nu)M}{3(1-\nu)}$		$\frac{\nu M}{(1-\nu)}$	$\frac{(1-2\nu)M}{2(1-\nu)}$	
λ,μ	$\frac{\mu(3\lambda+2\mu)}{(\lambda+\mu)}$	$\left(\lambda+\frac{2}{3}\mu\right)$	$(\lambda+2\mu)$			$\frac{\lambda}{2(\lambda+\mu)}$
λ,ν	$\frac{\lambda(1+\nu)(1-2\nu)}{\nu}$	$\frac{\lambda(1+\nu)}{3\nu}$	$\frac{\lambda(1-\nu)}{\nu}$		$\frac{\lambda(1-2\nu)}{2\nu}$	
μ,ν	$2\mu(1+\nu)$	$\frac{2\mu(1+\nu)}{3(1-2\nu)}$	$\frac{2\mu(1-\nu)}{(1-2\nu)}$	$\frac{2\mu\nu}{(1-2\nu)}$		

$$\omega_1 = \sqrt{(M-E)(9M-E)} \qquad \omega_2 = \sqrt{(E+\lambda)^2 + 8\lambda^2}$$

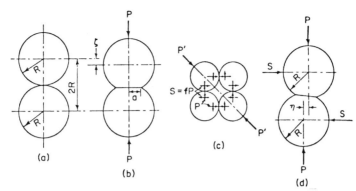

Fig.1 Behaviour of equal spheres in contact. (a) Spheres just
touching. (b) Deformation by normal force. (c) Shearing
forces between particles in cubic packing. (d) Lateral
deformation by shearing forces. (from Richart, Hall and
Woods, 1970).

Further studies by Cattaneo (1938), Mindlin (1949) and
Mindlin and Deresiewicz (1953) considered the influence of a
shearing force, S, as well as the axial force P on the behavior
of spheres in contact, Fig. 1, c and d. These investigators
found that shearing forces and displacements were a complicated
function of not only the initial state of loading but also the
entire past history of loading.
On a more practical vein, Duffy and Mindlin (1957) set out
to confirm the implication of the Hertz theory by determining
the wave speed "v_c" in rods consisting of assemblages of near
perfect spheres. They reasoned that if $E_T = K_1 \sigma_x^{1/3}$ and B =
$K_2 \sigma_0^{1/3}$, then from $V_c = \sqrt{E/\rho}$, $v_c \propto \sigma_0^{1/6}$. These investigators
used 1/8" diameter stainless steel spheres of two different
tolerences to find that experimental results did not perfectly
match theory, Fig. 2.

o Spheres $\frac{1}{8} \pm 10 \times 10^{-6}$ in. diameter

△ Spheres $\frac{1}{8} \pm 50 \times 10^{-6}$ in. diameter

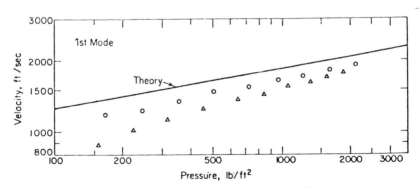

Fig.2 Variation of the velocity of the compression wave with
confining pressure for bars composed of stainless steel
spheres (from Duffy and Mindlin, 1957).

39

In view of the discrepancy caused by only slight variances from perfect sphericity, it was evident that wave propagation in real soils must be determined empirically. But at least one of the major parameters affecting elastic behavior, that is confining pressure, had been identified from theory.

VARIABLES AFFECTING ELASTIC MODULI

For most analytical methods which require elastic or dynamic soil moduli, it is shear modulus which is more frequently required, therefore most attention will be directed toward parameters affecting G. Young's modulus is similarly affected by most parameters but differences will be noted where they are known to exist.

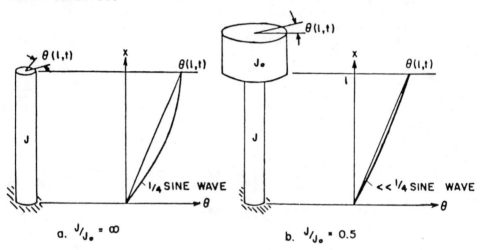

Fig.3 Schematic of "fixed-free" resonant column test (from Drnevich, 1967).

Most parametric studies of dynamic soil moduli have been performed using resonant column techniques. This technique is shown schematically in Fig. 3 and a photograph of a resonant column apparatus with all associated equipment is shown in Fig. 4.

Hardin and Black (1968) indicated in functional form the dependency of shear modulus of soils on the parameters known or suspected to be important as

$$G = f\ (\bar{\sigma}_0,\ e,\ C,\ A,\ H,\ t,\ S,\ \tau_0,\ f,\ \theta,\ T)$$

in which

$\bar{\sigma}_0$ = effective octahedral normal stress,
e = void ratio,
C = grain characteristics; shape, size, grading, mineralogy,

40

Fig.4 Resonant column apparatus

A = amplitude of strain,
H = ambient stress history and vibration history,
t = secondary time effects,
S = degree of saturation,
τ_0 = octahedral shear stress,
f = frequency of vibration,
θ = soil structure,
T = temperature.

PARAMETERS $\bar{\sigma}_0$, e, C, A

Hardin and Richart (1963) showed that for dry sand shear wave velocity and shear modulus could be expressed empirically as:

for round-grained sands

$$V_s = [170 - (78.2)e] \, (\bar{\sigma}_0)^{0.25} \tag{3}$$
$$\text{(ft/sec)} \qquad\qquad (\text{lb/ft}^2)$$

$$G = \frac{2630(2.17 - e)^2}{1 + e} \, (\bar{\sigma}_0)^{0.5} \tag{4}$$
$$\text{(lb/in}^2) \qquad\qquad\qquad (\text{lb/in}^2)$$

for angular-grained materials

$$V_s = [159 - (53.5)e] \, (\bar{\sigma}_0)^{0.25} \tag{5}$$
$$\text{(ft/sec)} \qquad\qquad (\text{lb/ft}^2)$$

$$G = \frac{1230(2.97 - e)^2}{1 + e} \, (\bar{\sigma}_0)^{0.5} \tag{6}$$
$$\text{(lb/in}^2) \qquad\qquad\qquad (\text{lb/in}^2)$$

Equations 3 and 5 show that shear wave velocity varies as the one-quarter power of confining pressure rather than the one-sixth power predicted by Hertz theory. Similarly, Eqs. 4 and 6 show shear modulus varies as the one-half power of confining pressure rather than the one-third power. For other dimensional units, equations 3-6 can be rewritten,

(Metric)

Eq. (3) becomes $V_s = [348-(160)e] \, (\bar{\sigma}_0)^{0.25}$ (3a)
$\text{(m/sec)} \qquad\qquad (\text{kg/cm}^2)^+$

(4) becomes $G = \frac{697(2.17)-e)^2}{1 + e}(\bar{\sigma}_0)^{0.5}$ (4a)
$\text{(kg/cm}^2) \qquad\qquad (\text{kg/cm}^2)$

(5) becomes $V_s = [326-(110)e] \, (\bar{\sigma}_0)^{0.25}$ (5a)
$\text{(m/sec)} \qquad\qquad (\text{kg/cm}^2)$

(6) becomes $G = \frac{326(2.97-e)^2}{1 + e} \, (\bar{\sigma}_0)^{0.50}$ (6a)
$\text{(kg/cm}^2) \qquad\qquad (\text{kg/cm}^2)$

+ Note: kp/cm^2 may be replaced by kg/cm^2, bars, or ton/ft^2, (ton = 2000 pounds).

or (SI)

Eq. (3) becomes v_s = $[19.7-(9.06)e]$ $(\bar{\sigma}_0)^{0.25}$ (3b)
    (m/s) $(N/m^2$ or Pa)

(4) becomes G = $\dfrac{6906(2.17-e)^2}{1+e}$ $(\bar{\sigma}_0)^{0.5}$ (4b)
    (kN/m^2) $(kN/m^2$ or kPa)

(5) becomes v_s = $[18.4-(6.20)e]$ $(\bar{\sigma}_0)^{0.25}$ (5b)
    (m/s) (N/m^2)

(6) becomes G = $\dfrac{3230(2.97)-e)^2}{1+e}$ $(\bar{\sigma}_0)$ (6b)
    (kN/m^2) (kN/m^2)

Initially Eqs. 5 and 6 were established for angular-grain sands but these equations have been confirmed for clays with low surface activity, Hardin and Black (1968).

Shearing strain amplitude "A" was included in the heading of this paragraph and the discussion up to now has been appropriate for low shearing strain amplitudes. Low amplitudes are those shearing strain amplitudes below which shearing strain has no influence on modulus. Low amplitude shearing strains are generally considered to be less than 0.00001 (10^{-5}) in/in or 0.001 (10^{-3}) %. As shearing strain amplitude increases beyond this threshold, shear modulus decreases. For very low strain amplitude, 10^{-6} or 10^{-4} %, Kuribayashi, Iwasaki and Tatsuoka (1975) found a small increase in shear modulus and propose the following equation:

$$G = \frac{900\,(2.17-e)^2}{1+e}\,(\bar{\sigma}_0)^{0.38} \qquad (7)$$
$$(kg/cm^2) \qquad\qquad\qquad (kg/cm^2)$$

Compare Eq. 7 with Eq. 4a.

The decrease in shear modulus as a function of shearing strain amplitude is shown in Fig. 5.

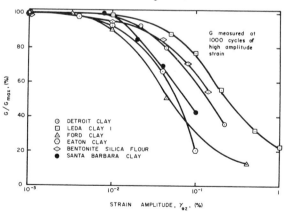

Fig.5 Effect of shearing strain amplitude on shear modulus (from Anderson, 1974).

This behavior is particularly important for analysis of earthquake excitation of structures on soil sites. This non-linear behavior of the soil must be taken into account to correctly predict the ground surface motion. There are at least two analytical methods of expressing the strain amplitude behavior soils. The first from Hardin and Drnevich (1972) is based on the maximum shearing strength of the material and a reference strain, Fig. 6.

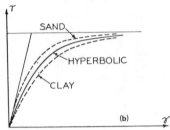

Fig.6 Hyperbolic shearing stress-strain curves for sand and clay (from Hardin and Drnevich, 1970).

The shearing stress, τ, in Fig. 6 can be expressed as

$$\tau = \frac{\gamma}{\frac{1}{G_{max}} + \frac{\gamma}{\tau_{max}}} \tag{8}$$

Then by introducing

$$\gamma_r = \frac{\tau_{max}}{G_{max}} \tag{9}$$

and rearrangig

$$\frac{G}{G_{max}} = \frac{1}{1 + \frac{\gamma}{\gamma_r}} \tag{10}$$

Differences in material behavior were included by defining hyperbolic shearing strain as

$$\gamma_h = \frac{\gamma}{\gamma_r} [1 + a\ e^{b(\frac{\gamma}{\gamma_r})}] \tag{11}$$

in which a and b are soil dependent parameters. Introducing Eq. 11 into Eq. 10 we get

$$\frac{G}{G_{max}} = \frac{1}{1 + \gamma_h} \tag{12}$$

Equation 12 and a similar equation for damping can be shown on a normalized diagram, Fig. 7. (Damping will be covered in separate paper).

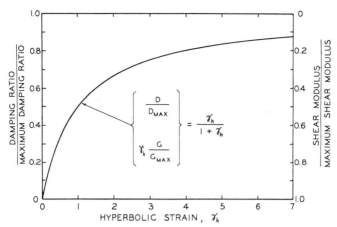

Fig.7 Normalized shear modulus and normalized damping ratio
for all soils vs. hyperbolic strain (from Hardin and
Drnevich, 1970).

The second method of expressing strain dependent modulus is
with the Ramberg-Osgood equation, Ramberg and Osgood (1943).
This equation is written

$$\frac{G}{G_{max}} = \frac{1}{1 + \alpha \left(\frac{\tau}{\tau_y}\right)^{R-1}}$$

(13)

in which α is a shape factor, R is a correlation number and τ_y
is shearing stress at yield. With the proper selection of a
and b in the Hardin-Drnevich equations (Eqs. 11 and 12) or
and R in the Ramberg-Osgood equation (Eq. 13) the modulus vs.
strain amplitude behavior of most soils can be modeled, Figs.
8 and 9. The Seed and Idriss (1970) curves fall consistently
below the measured data in Fig. 8.
    Stress history plays an important role in conditioning the
soil and influencing the dynamic modulus. Hardin and Black
(1969) introduced preconsolidation effects by modifying Eq. 6
to

$$G = \frac{1230(2.97-e)^2}{1 + e} (OCR)^K (\bar{\sigma}_0)^{0.5}$$

(14)

in which OCR is the overconsolidation ratio and K is a function
of plasticity index as shown in Fig. 10. Afifi also demonstrated the influence of overconsolidation ratio on dry fine sand,
Fig. 11.
    Anderson and Richart (1976) showed that shearing stiffness
that was destroyed by high amplitude straining will be re-
gained after sufficient time at rest or low amplitude straining,
Fig. 12. They also showed that the time required to regain
G_{max} after prestraining depended upon the shearing strain amplitude and the soil type, Fig. 13. The number of cycles of high
strain amplitude also influenced the time to regain maximum
shear modulus, Fig. 14.

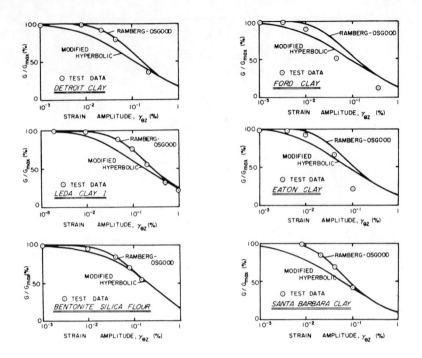

Fig.8 Comparison of high amplitude test results for several clays (after Anderson and Richart, 1976).

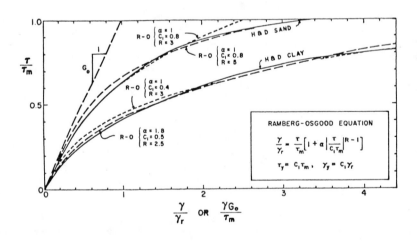

Fig.9 Fit of Ramberg-Osgood curves to soil data (from Richart and Wylie, 1977).

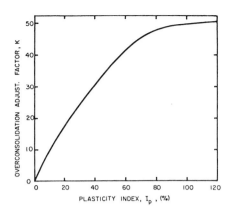

Fig.10   Overconsolidation adjustment factor, K, vs. plasticity
index, Ip (after Hardin and Black, 1969).

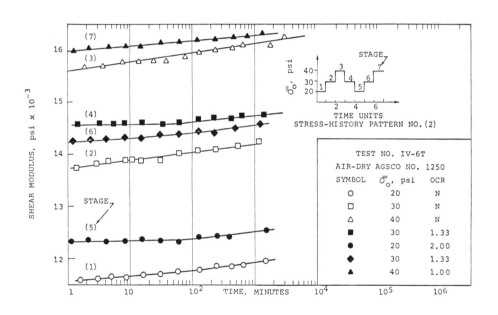

Fig.11 Variation of shear modulus with time and pressure for
air-dry, crushed quartz sand (after Afifi, 1971).

47

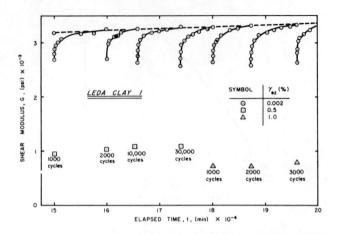

Fig.12 Regain in G_{max} after high amplitude cycling of Leda Clay I (after Anderson and Richart, 1976).

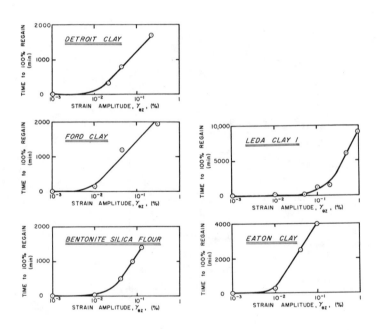

Fig.13 Time to 100% regain in G_{max} for 5 days (after Anderson and Richart, 1976).

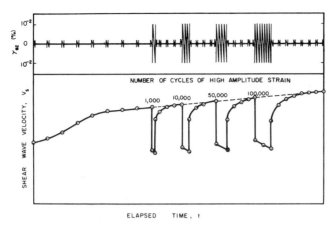

Fig.14 Typical high amplitude test sequence (after Anderson, 1974).

PARAMETER t

Secondary time effects represent rearrangements in particle to particle contact as time increases. These contacts become stronger with a resulting increase in shear modulus. Afifi and Woods have shown the influence of time on air dry sand, silt, and clay, Fig. 15. Anderson and Woods have shown the same for natural clays, Fig. 16. The amount of increase in shear modulus per logarithmic cycle of time depends on the type of soil and the reference time for normalizing the modulus. For most soils 1000 minutes is a convenient time to normalize on and increases in modulus over the 1000 minute value vary from 0 % for coarse sands to 15 % for dry clays. Figures 17 and 18 show the relationship between normalized modulus increase and D_{50} particle size.

In making comparisons between laboratory and field measured moduli, it is necessary to account properly for the age of the soil by correcting the laboratory values for secondary time effects. Anderson and Woods (1975) have shown good correlation between field and lab when secondary time effects have been included, Fig. 19.

PARAMETER S

The effects of degree of saturation can be accounted for by using the insitu mass density "ρ" used in computing shear modulus from $G = \rho v_s^2$, Fig. 20. However, the degree of saturation has an important effect on the compression wave velocity. The theory of wave propagation in saturated or nearly saturated poro-elastic solids is fairly complex and the reader is referred to Richart, Hall and Woods (1970) paragraph 5.3, and Ishihara (1968 and 1970).

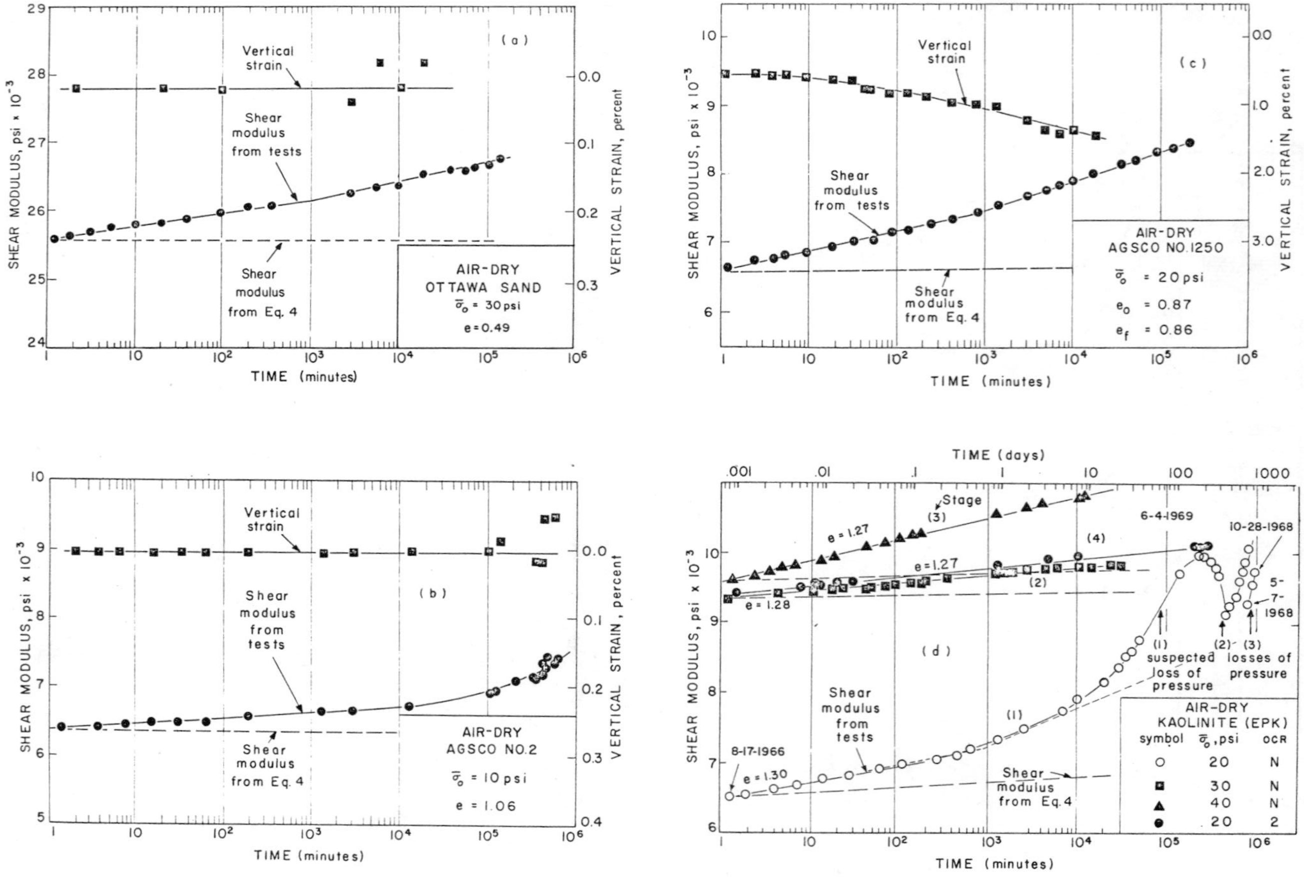

Fig.15 Shear modulus vs. time for four soils (after Afifi, 1970).

50

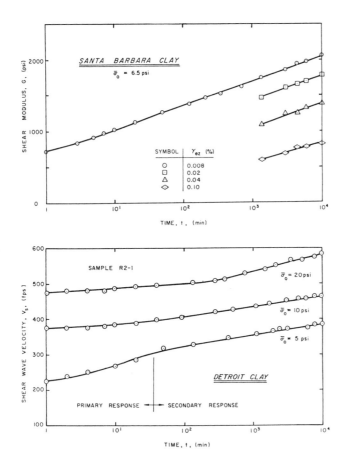

Fig.16 Shear modulus vs. time for natural clays (after
       Anderson, 1974).

The compression wave velocity in the fluid of a saturated
media is also dependent upon the percent of dissolved gas in
the water.

PARAMETERS τ_0, f

The theory for waves in elastic spheres and the early experimental work on elastic wave propagation in soils pointed to a
dependence of G on the octahedral normal stress. Hardin and
Black (1966) demonstrated that the shear modulus is independent
of octahedral shearing stress, τ_0,

51

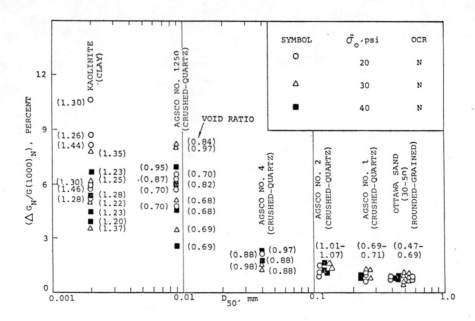

Fig.17 Effects of particle size and shape, void ratio and confining pressure on shear modulus of air-dry soils (after Afifi, 1970).

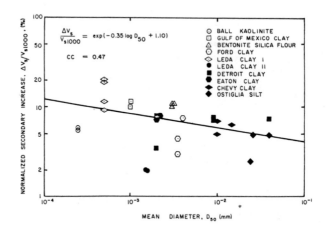

Fig.18 Relationship between logarithm of $\Delta V_s/V_{s1000}$ and logarithm of mean particle diameter (from Anderson and Woods, 1976).

$$\tau_0 = \frac{1}{3} \sqrt{(\sigma_1 - \sigma_2)^2 + (\sigma_2 - \sigma_3)^2 + (\sigma_3 - \sigma_1)^2}$$

Recent work by the writer indicates that Young's modulus may be dependent upon the state of shearing stress, or more discriptively, the directions of the major and minor principal stresses.

Hardin and Richart (1963) showed that the frequency of vibration had no measurable effect on G for frequencies less than 2500 Hz.

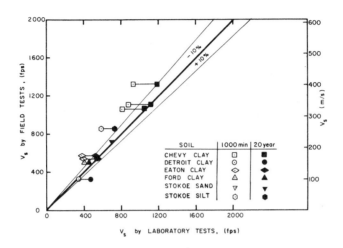

Fig.19 Comparison between V_S laboratory and V_S field (from Anderson and Woods, 1975).

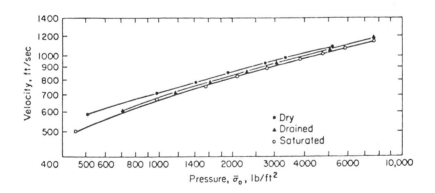

Fig.20 Variation of shear wave velocity with confining pressure for a specimen of Ottawa Sand, (dry, saturated and drained) and e = 0.55. (after Hardin and Richart, 1963).

53

Soil structure or fabric is just recently (1970's) receiving attention in terms of trying to include it in describing elasti‹ properties of soils. Evidence of the influence of structure or fabric in dynamic soil testing was reported by Skoglund, Marcuson and Cunny (1976). These authors reported on resonant column tests performed by 6 agencies or individuals. Tables 2 and 3 show variations in moduli measured by the 6 reporters. Part of the difference in reported moduli can be attributed to different specimen preparation techniques which resulted in specimens of differing structure or fabric.

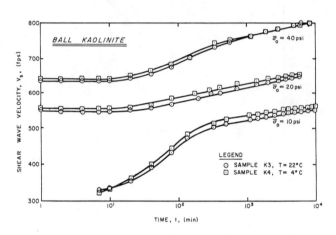

Fig.21 Comparison of V_S with time at 4^OC and 22^OC for Detroit Clay (from Anderson and Richart, 1974).

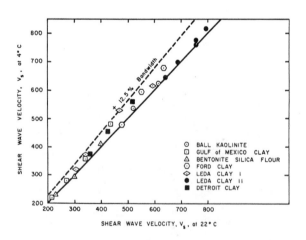

Fig.22 Relationship between V_{s1000} at 4^OC and V_{s1000} at 22^OC (from Anderson and Richart, 1974).

54

Table 2
Shear Moduli For Sand

Participant	Shear Modulus, ksi				Difference From Avg, %			Strain	Adjusted Strain For Edge Peak-To-Peak
	0.7 kg/cm²	1.4 kg/cm²	2.1 kg/cm²	2.8 kg/cm²	0.7 kg/cm²	1.4 kg/cm²	2.8 kg/cm²		
Drnevich	10.3	16.2	-	23.5	-19	-11	- 7	0.99 to 1.06 × 10^{-4} cm/cm Avg zero-to-peak	3.0 to 3.2 × 10^{-4} cm/cm
Hardin	11.3	17.6	-	24.7	-11	- 4	- 3	0.99 to 1.04 × 10^{-4} cm/cm Avg peak-to-peak	1.5 × 10^{-4} cm/cm
Woods	14.4	19.7	23.8	26.8	+13	+ 8	+ 5	4.00 × 10^{-5} cm/cm Edge peak-to-peak	0.4 × 10^{-4} cm/cm
Shannon & Wilson	13.3	17.1	-	21.0	+ 5	- 7	-17	4.30 to 7.10 × 10^{-7} percent Edge zero-to-peak	0.9 to 1.4 × 10^{-8} cm/cm
CRREL	15.8	21.9	-	32.2	+24	+20	+27	2.54 × 10^{-4} rad/cm	
WES	11.3	17.2	-	23.9	-11	- 6	- 6	2.60 to 4.80 × 10^{-5} cm/cm Edge peak-to-peak	0.3 to 0.5 × 10^{-4} cm/cm
Avg	12.7	18.3	-	25.4					

Table 3

Shear Moduli For Clayey Silt

Participant	Shear Modulus, ksi				Difference From Avg, %			Strain	Adjusted Strain For Edge Peak-To-Peak
	0.7 kg/cm²	1.4 kg/cm²	2.1 kg/cm²	2.8 kg/cm²	0.7 kg/cm²	1.4 kg/cm²	2.8 kg/cm²		
Ornevich	6.9	10.9	-	18.0	- 9	- 9	+ 0.7	1.00 to 1.07 x 10^{-4} cm/cm Avg zero-to-peak	3.0 to 3.2 x 10^{-4} cm/cm
Hardin	-	10.3	13.7	-	- 9	-13	- 6	1.02 to 1.03 x 10^{-4} cm/cm Avg peak-to-peak	1.5 x 10^{-4} cm/cm
	6.9	-	-	16.8	-	-	-	--	--
Woods	10.1	14.9	18.7	22.1	+32	+25	+24	2.00 x 10^{-5} cm/cm Edge peak-to-peak	0.2 x 10^{-4} cm/cm
Shannon & Wilson	8.1	11.2	-	15.9	+ 7	- 6	-11	3.90 to 10.60 x 10^{-7} percent Edge zero-to-peak	0.8 to 2.1 x 10^{-8} cm/cm
CRREL	6.6	13.4	-	18.0	-13	+12	+0.7	2.54 x 10^{-4} rad/cm	--
WES	6.9	10.8	-	16.4	- 9	- 9	- 8	3.70 to 7.10 x 10^{-5} cm/cm Edge peak-topeak	0.4 to 0.7 x 10^{-4} cm/cm
Avg	7.6	11.9	-	17.9					

The influence of specimen temperature on shear modulus was reported by Anderson and Richart (1974). These investigators studied the effects of temperature from 4°C to 22°C on saturated coheseve soils. The main purpose of these tests was to examine the potential difference between the laboratory temperature environment and the insitu temperature environment, but not to examine the influence of freezing on modulus. In general the influence of temperature was found to be small, Figs. 21 and 22.

They also showed that the effect of a rapid temperature change was only temporary, Fig. 23.

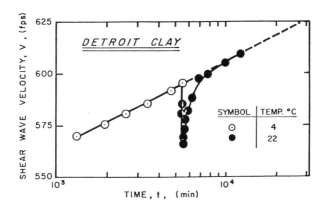

Fig.23 Effect of rapid temperature change on V_s (from Anderson and Richart, 1974).

REFERENCES

Afifi, S.S. (1970), "Effects of Stress-History on the Shear
    Modulus of Soils", Ph.D. dissertation, University of
    Michigan.
Afifi, S.S., and Woods, R.D., (1971), "Long-term Pressure
    Effects on Shear Modulus of Soils", Journal of the Soil
    Mechanics and Foundations Division, ASCE, Vol. 97, No.
    SM10, Proc. Paper 8475, Oct., pp. 1445-1460.
Anderson, D.G. (1974), "Dynamic Modulus of Cohesive Soils",
    Ph.D. dissertation, University of Michigan, 311 pp.
Anderson, D.G., and Richart, F.E., Jr., (1974), "Temperature
    Effect on Shear Wave Velocity in Clays", Technical Note,
    Journal of the Geotechnical Engineering Division, ASCE,
    Vol. 100, No. GT12, Dec., pp. 1316-1320.
Anderson, D.G., and Richart, F.E., Jr., (1976), "Effects of
    Straining on Shear Modulus of Clays", Journal of the Geo-
    technical Engineering Division, ASCE, Vol. 102, No. GT9,
    Sept., pp. 975-987.
Anderson, D.G., and Woods, R.D. (1975), "Comparison of Field
    and Laboratory Shear Moduli", Proceedings of the Conference
    on In situ Measurement of Soil Properties, North Carolina
    State University, Raleigh, N.C., Specialty Conference of
    the Geotechnical Engineering Division, ASCE, June, pp. 69-
    92.
Anderson, D.G., and Woods, R.D. (1976), "Time-Dependent In-
    crease in Shear Modulus of Clay", Journal of the Geotechni-
    cal Engineering Division, ASCE, Vol. 102, No. GT5, May,
    pp. 525-537.
Cattaneo, C. (1938), "Sul contatto di due corpi elastici",
    Accademia dei Lincei Rendiconti, Ser. 6, Vol. 27, pp. 342-
    348, 434-436, 474-478.
Drnevich, V.P. (1967), "Effect of Strain History on the Dy-
    namic Properties of Sand", Ph.D. dissertation, University
    of Michigan, 151 pp.
Duffy,J., and Mindlin, R.D.(1957), "Stress-Strain Relations
    of a Granular Medium", J. Appl. Mech., Trans. ASCE, Dec.,
    pp. 585-593.
Hertz, H. (1881), "Über die Berührung fester elastischer Kör-
    per", J. reine u. angew. Math., Vol. 92, pp. 156-171.
Hardin, B.O., and Black, W.L. (1966), "Sand Stiffness Under
    Various Triaxial Stresses", J. Soil Mech. and Found. Div.,
    Proc. ASCE, Vol. 92, No. SM2, March, pp. 27-42.
Hardin, B.O., and Black, W.L. (1968), "Vibration Modulus of
    Normally Consolidated Clay", J. Soil Mech. and Found. Div.,
    Proc. ASCE, Vol. 94, No. SM2, March, pp. 353-369.
Hardin, B.O., and Black, W.L. (1969), Closure to "Vibration
    Modulus of Normally Consolidated Clay", Journal of the Soil
    Mechanics and Foundations Division, ASCE, Vol. 95, No. SM6,
    Nov., pp. 1531-1539.
Hardin, B.O., and Drnevich, V.P. (1970), "Shear Modulus and
    Damping in Soils: Design Equations and Curves", Technical
    Report UKY 27-70-CE3, Soil Mechanics Series No. 2, Univer-
    sity of Kentucky.

Hardin, B.O., and Richart, F.E., Jr. (1963), "Elastic Wave Velocities in Granular Soils", J. Soil Mech. and Found.Div., Proc. ASCE, Vol. 89, No. SM1, Feb., pp. 33-65.

Ishihara, K. (1968), "Propagation of Compressional Waves in a Saturated Soil", Proceedings of the International Symposium on Wave Propagation and Dynamic Properties of Earth Materials, University of New Mexico Press, pp. 195-206.

Ishihara, K. (1970), "Approximate Forms of Wave Equations for Water-Saturated Porous Materials and Related Dynamic Modulus", Soils and Foundations (Japan), Vol. 10, No. 4, pp. 10-38.

Kuribayashi, E., Iwasaki, T., and Tatsuoka, F., (1975), "Effects of Stress-Strain Conditions on Dynamic Properties of Sands", Submitted to Japanese Society of Civil Engineers for publication.

Mindlin, R.D. (1949), "Compliance of Elastic Bodies in Contact", JAM, Sept., pp. 259-268.

Mindlin, R.D., and Deresiewicz, H. (1953), "Elastic Spheres in Contact under Varying Oblique Forces", J. Appl. Mech., Trans., ASME, Sept., pp. 327-344.

Ramberg, W., and Osgood, W.T., (1943), "Description of Stress-Strain Curves by Three Parameters", Tech. Note 902, National Advisory Committee for Aeronautics.

Richart, Frank E., Jr., (1975), "Some Effects of Dynamic Soil Properties on Soil-Structure Interaction", Journal of the Geotechnical Engineering Division, ASCE, Vol. 101, No.GT12, Dec., pp. 1193-1240.

Richart, F.E., Jr., Hall, J.R., Jr., and Woods, R.D. (1970) Vibrations of Soils and Foundations, Prentice-Hall, Inc., Englewood Cliffs, N.J.

Richart, F.E., Jr., and Wylie, E.B. (1977), "Influence of Dynamic Soil Properties on Response of Soil Masses", Structural and Geotechnical Mechanics, Prentice-Hall, Inc., Englewood Cliffs, New Jersey, USA, edited by W.J. Hall, pp. 141-162.

Seed, H.B., and Idriss, I.M., "Soil Moduli and Damping Factors for Dynamic Response Analysis", Earthquake Engineering Research Center, College of Engineering, University of California, Berkeley, California, Report No. EERC 70-10, Dec., 1970.

Parameters affecting damping properties

B. PRANGE
University of Karlsruhe, Karlsruhe, Germany

SYNOPSIS

The physical reason for material damping in granular media is investigated. From different experimental data the damping is found to be hysteretic. A simple two-sphere model explains most of the phenomena connected with damping. The influence of material damping on SSI and wave propagation is studied.

INTRODUCTION

The reaction of a system to a dynamic load consists of an in-phase response and an off-phase response which can be identified as the spring-reaction and the damping-reaction, respectively. The system may be a structural system, a dynamically loaded foundation or an element of the subsoil. Since we deal with soil-dynamical problems, it will mainly be one of the latter ones. Consequently, we must investigate the parameters which affect damping properties involved in soil-structure-interaction problems and parameters which affect the material damping of the soil itself. We will confine ourselves to the problems of material damping and the related influence upon SSI-problems. Geometric damping involved in wave propagation problems will not be considered here, neither will the problem of damping constants of soil-structure-interaction from the mere point of view of lumped parameters be dealt with. Structural systems will only be referred to in so far as dynamic model testing is concerned.
    Hence, the investigation will concern the following points:
- Damping phenomena in soils and their physical background
- Implication of material damping with respect to wavefields
- Implication of material damping with respect to dynamic model testing and soil-structure-interaction problems.

VISCO-ELASTIC MATERIAL

To examine as a first step the damping properties of soils, we look at a soil element in pure shear and establish a relation between dynamic shear stress and shear strain, i.e. the soil element subjected to shear-wave propagation. We could, nevertheless, have considered a soil element in compression beeing subject to compressive-wave propagation. However, the theoretical results below will be checked against dynamic experiments in the resonant column test. In this test, cylindrical soil samples in the torsional (shearing) mode only do not exhibit dispersion, i.e. the shear-wave velocity v_s will not depend on the ratio wavelength to sample radius λ/r_0 and consequently on frequency. The simpliest relation between stress and strain for a visco-elastic material is the following

$$\tau = G \; \gamma + \; \mu \; \dot{\gamma} \qquad (1)$$

where τ = dynamic shear stress
γ = dynamic shear strain
$\dot{\gamma}$ = shear strain velocity
G = dynamic shear modulus
μ = coefficient of shear viscosity

Experiments have shown, that this linear equation is valid for shear strains below 10^{-5}.
We assume the shear wavefield to which the soil element is subjected beeing periodic and stationary, in which case the time dependency of the shear strain can be written as

$$\gamma = \gamma_0 \; e^{i\omega t} \qquad (2)$$

where γ_0 = shear strain amplitude
$e^{i\omega t}$ = complex unit vector
ω = angular frequency

Putting equ.2 into equ.1 we find

$$\tau = G \; \gamma_0 \; (1 + i \; \omega \; \mu) e^{i\omega t} \qquad (3)$$

The complex term in the parenthesis of equ.3 can be presented in the complex plane:

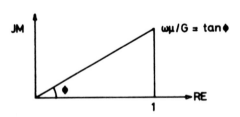

Fig.1   Vector presentation of τ in complex plane

From fig.1 we immediately find, that the dynamic stress leads the dynamic strain by a phase angle ϕ. Similarly, the amplitude

62

of the stress vector can be deduced from fig.1 to be

$$\tau_0 = \sqrt{1 + \tan^2\phi} \ G \ \gamma_0 \approx G \ \gamma_0 \qquad (4)$$

if tanϕ is small (the error beeing 0.5 % for tanϕ = 0.1).

In the next step we will follow the lines of Dobry (1970) to establish the damping capacity of a soil element.
The elastic strain energy W compiled by the elastic shear stress along the resp. strain equals

$$W = \frac{1}{2} \tau_0 \ \gamma_0 \approx \frac{1}{2} G \ \gamma_0^2 \qquad (5)$$

if we insert the stress amplitude from equ.4.
The internal work W performed during one cycle equals the product of shear stress and shear strain velocity, integrated over the period time:

$$\Delta W = \int_0^{2\pi/\omega} \tau(t) \ \dot{\gamma}(t) dt \qquad (6a)$$

Putting the values of equ.2 and equ.4 into equ.6 we have

$$\Delta W = G \ \gamma_0^2 \ \sqrt{1 + \tan^2\phi} \int_0^{2\pi/\omega} e^{i(\omega t+\phi)} \ \frac{\partial}{\partial t} \ e^{i\omega t} \ dt \qquad (6b)$$

where the integral yields the term $\pi\sin\phi$. We could have gained the same result had we plotted stress versus strain:

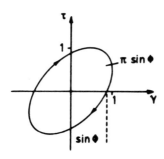

Fig.2   Stress-strain hysteresis loop

The internal work of the unit stress vector along the unit strain vector equals the area of the resulting ellipse, $\pi\sin\phi$, in fig.2. We find the internal work than:

$$\Delta W = G \ \gamma_0^2 \ \sqrt{1 + \tan^2\phi} \ \pi\sin\phi \qquad (6c)$$

Defining the damping capacity ψ as the ratio of the internal work ΔW and the elastic strain energy W we arrive at

$$\psi = 2\pi \sqrt{1 + \tan^2 \phi} \, \sin\phi = 2\pi \tan\phi \tag{7}$$

(because $\sqrt{1 + \tan^2} = 1/\cos^2$).

Due to a shear wave propagating through the soil element, the element will be subjected to an energy flow input and output.

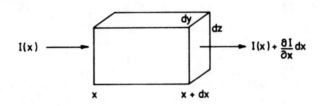

Fig.3   Energy flow through soil element

Because of the material damping, the strain amplitude will be attenuated along the direction of wave propagation

$$\gamma(x) = \gamma_0 \, e^{-\delta x} \tag{8a}$$

where γ_0 = strain amplitude at $x = 0$
δ = damping exponent

Applying equ. 4 we similarly find for the shear stress

$$\tau(x) = G \gamma_0 \, e^{-\delta x} \tag{8b}$$

The elastic work W performed by the stress along the resp. strain is

$$W = \frac{1}{2} \tau \gamma \, dV = \frac{1}{2} G \gamma_0^2 \, e^{-2\delta x} \, dx \, dy \, dz \tag{9}$$

which must be equal to the energy flow input at x during the time increment dt:

$$W = I(x) \, dy \, dz \, dt \tag{10a}$$

Putting equ.9 into equ.10 we find

$$I(x) = \frac{1}{2} G \gamma_0^2 \, e^{-2\delta x} \, \frac{dx}{dt} \tag{10b}$$

and knowing that $\frac{dx}{dt} = v_s$ we arrive at

$$I(x) = \frac{1}{2} G_o{}^2 e^{-2\delta x} v_s \qquad (10c)$$

describing the specific energy flow.

Let us again consider the energy dissipation W during one cycle, equalling the difference between the energy flow input and output:

$$W = \frac{\partial I}{\partial x} \; dV \quad 2\pi/\omega \qquad (11a)$$

Putting equ.10c into equ.11a we find

$$W = - \; G_o{}^2 e^{-2\delta x} v_s \; dV \; 2\pi/\omega \qquad (11b)$$

The damping capacity ψ, by the same definition as in equ.7 equals the ratio of equ.11b and equ.9:

$$\psi = \; W/W = 4 \; \pi/\omega \; v_s \; \delta \qquad (12)$$

Equalling the values for the damping capacity ψ in equ.12 and equ.7 we can now relate the damping exponent to the coefficient of shear viscosity as follows:

$$\delta = \frac{\omega}{2v_s} \; \tan\phi = \frac{\omega}{2v_s} \; \omega\mu/G \qquad (13)$$

WAVE ATTENUATION

We now have to find out wether the coefficient of shear vis-cosity is a constant or depends on frequency. Let us first inspect experimental wavefield data. It we plot the surface vibration amplitudes on a semilogarithmic scale versus distance from the wave source, we can distinguish between the attenuation of the propagating wave due to geometric damping and material damping. In fig.4 the solid line indicates e.g. the geometric

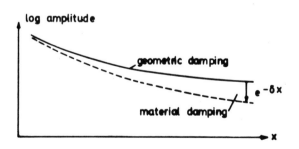

Fig.4   Attenuation due to geometric and material damping

damping of the surface wave, $x^{-0.5}$, whilst the dashed line includes both geometric and material damping. In a semilogarithmic plot, the difference must equal the material damping effect
alone, depending on distance x. If equ.8 holds, the difference
must be represented by a straight line with an inclination
tanα = -δ. Only one example of the experimental field data shall
be given here as reference, (Prange 1978).

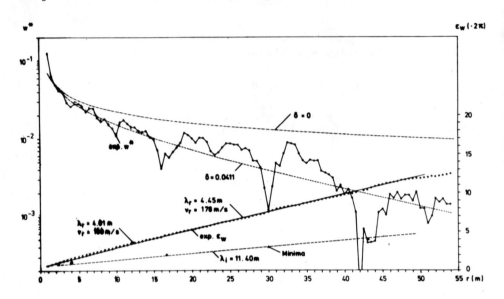

Fig.5 Experimental wavefield, vertical component axis 5; 40 Hz

If we plot the experimental results of measuring δ from a number
of tests with different frequencies in a double-logarithmic
plot, fig.6, we find some very interesting results.
    Not only lie all the values of $\delta(f)$ for the respective axes
on a straight line, but this line is inclined at 45°.
This means, that δ is directly proportional to f or ω, respectively.
    From this fact we immediately deduce that tanϕ is a constant,
see equ.13, or, similarly, that $\omega\mu$ is a constant. This can only
be true if the coefficient of shear viscosity is inversely proportional to ω. Our basic finding is, therefore, that the granular soils investigated do not behave viscous in the strict
sense.
    We will have to prove next what kind of damping behaviour is
represented by tanϕ = constant.

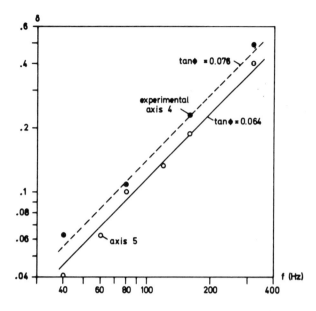

Fig.6 Variation of damping exponent δ with frequency

RESONANT COLUMN TESTS

A powerful tool to investigate the dynamic properties of soils is the resonant column test. Fig.7 shows the details with I_t representing the top inertia mass with the drive mechanism applying a torsional moment M(t) to the sample. G and μ are the visco-elastic constants already introduced, ρ the specific gravity of the soil material.

We may remind ourselves of the governing equations of motion and wave propagation, respectively:

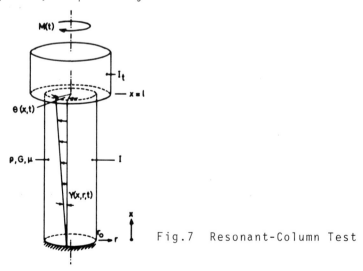

Fig.7 Resonant-Column Test

67

$$\frac{\partial^2 \Theta}{\partial t^2} = G \frac{\partial^2 \Theta}{\partial x^2} + \mu \frac{\partial}{\partial t} \frac{\partial^2 \Theta}{\partial x^2} \qquad (14)$$

with the mass-term on the left hand side and the spring-and damping-terms on the right hand side of the equation. The wave-propagation along the sample is controlled by

$$\Theta(x,t) = \Theta_0 \, e^{i(\omega t - kx)} \, e^{-\delta x} \qquad (15)$$

where we find the time dependency and the phase shift in the first exponent and the material damping in the second one (no geometrical damping due to one-dimensional wave propagation along the sample).

Hardin (1965) calculated the relevant damping parameters as well as the resulting wave velocity by putting equ.15 into equ.14 and solving for δ. The results are the following equations 16 - 20:

$$\delta_x = \omega \sqrt{\frac{\rho}{G}} \, \sqrt{\frac{\sqrt{1 + \tan^2\phi} - 1}{2(1+\tan^2\phi)}} \qquad \approx \frac{\omega}{2v_s} \tan\phi \qquad (16)$$

$$\theta_x = \delta_x \lambda = 2\pi \sqrt{\frac{\sqrt{1 + \tan^2\phi} - 1}{1 + \tan^2\phi + 1}} \qquad \approx \pi\tan\phi \qquad (17)$$

$$\delta_t = \frac{G}{\mu}(1 - \sqrt{1 - \tan^2\phi}) \qquad \approx \frac{\omega}{2} \tan\phi \qquad (18)$$

$$\theta_t = \delta_t \frac{2\pi}{\omega} = 2\pi \frac{1 - \sqrt{1-\tan^2\phi}}{\tan\phi} \qquad \approx \pi\tan\phi \qquad (19)$$

$$v_s = \sqrt{\frac{G}{\rho}} \, \sqrt{\frac{2(1+\tan^2\phi)}{1+\tan^2\phi + 1}} \qquad \approx \sqrt{\frac{G}{\rho}} \qquad (20)$$

where δ_x = damping exponent with respect to coordinate
δ_t = damping exponent with respect to time
θ_x = logarithmic decrement with respect to coordinate
θ_t = logarithmic decrement with respect to time
v_s = shear wave velocity

The sign $\approx$ holds for small values of $\tan\phi$.

The results for θ and v_s are plotted in fig.8.

We first find the coincidence between equ.16 and equ.19 for small values of $\tan\phi$. Another finding is that for small values of $\tan\phi$ the logarithmic decrements with respect to coordinate and with respect to time are equal. This means that the decay of free vibrations per period and per wavelength is equal for the same material.

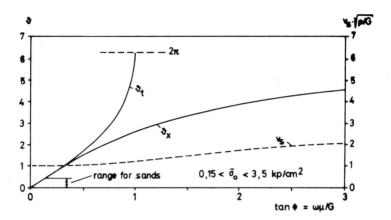

Fig.8   Variation of log. decrement with tan φ
        (from Hardin, 1965)

Hardin (1965) also calculated the resonance curves in
the first three eigenmodes for the cases of μ beeing constant
and μ beeing inversely proportional to ω. The results are shown
in figure 9a and 9b.

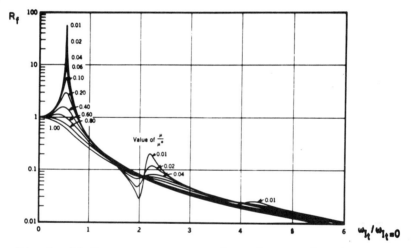

Fig.9a RC-Resonance curves for viscous damping
        (from Hardin, 1965)

In fig.9a, μ/μ^* is a dimensionless viscosity factor, μ^* the
critical viscosity for which no resonance maximum in the amplitude curve occurs (corresponding to the value $D = 1/\sqrt{2}$ in lumped
parameter systems). The resonance curves are plotted versus a
dimensionless frequency, $\omega(I_t=0)$ beeing the resonance frequency
of the cylindrical sample without top-mass.

The experimental results of the resonant column test of
Hardin (1965) are given in fig.10 for two stress levels, together with the theoretical response.

69

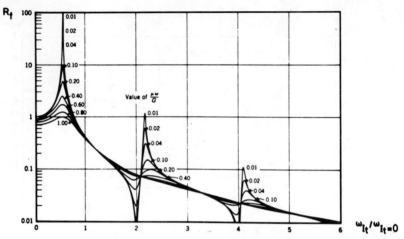

Fig.9b RC-Resonance curves for hysteretic damping
(from Hardin, 1965)

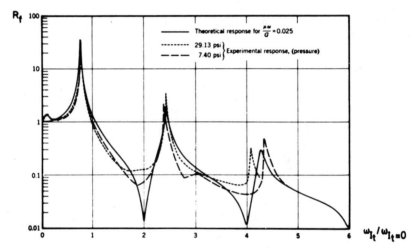

Fig.10 Experimental results for hysteretic damping
(from Hardin, 1965)

The results coincide well with a theoretical response for $\mu\omega/G = 0.025$, indicating that again μ must be inversely proportional to frequency.

PHYSICAL MODEL FOR MATERIAL DAMPING

For the further investigation of the nature of material damping we will now take an insight into the frictional behavior of two grains in contact with each other. Mindlin (1954) established a model which may directly be applied to soils as grain structures.

As the simpliest model, Mindlin considered two spheres of grain material in contact, subjected to a normal force N and a tangential force T. The normal force causes a parabolic normal stress distribution σ in the circular contact area, the tangential force a respective shear stress distribution τ. The friction in the contact is assumed to be of the Coulomb type, the maximum possible shear stress beeing f-times the normal stress. As seen in fig.11, the elastic shear stress is smaller than f σ for r ≤ c. However, for a ≥ r > c, the possible shear stress is smaller than the elastic shear stress and therefore slip will occur in this region.

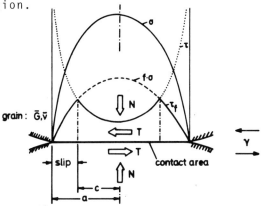

Fig.11 Contact stress-distribution; 2-sphere model
(after Mindlin, 1954)

The resulting shear-stress distribution τ_f consists of an elastic region and a plastic region, the transient zone r = c depending on the intensity of the tangential force T. The resulting relative displacement, for the sake of simplicity, was called γ due to the immediate relation to the shear strain in the granular soil structure (see below). Mindlin (1954) calculated γ as a function of the tangential force T and the normal force N:

$$\gamma = \frac{3(2-\bar{\nu})}{8\bar{G}a} \, f \, N \, (1-(1-T^*)^{2/3})$$ (21)

where a = radius of contact area
$\bar{G}$ = shear modulus of grain material
$\bar{\nu}$ = Poisson's ratio of grain material
T^* = T/f N
f = coefficient of friction

The effective shear modulus of the two spheres in contact is than

$$G = \frac{dT}{d\gamma} = \frac{4\,\bar{G}\,a}{(2-\bar{\nu})} \, (1 - T^*)^{1/3}$$ (22)

Mindlin also calculated the dissipated energy per cycle due to the plastic components of the shear stress distribution:

$$\Delta W = \frac{9(2-\bar{\nu})}{5\,\bar{G}\,a} \, f^2 N^2 [1-(1-T^*)^{5/3}- \frac{5}{G} \, T^*(1+(1-T^*)^{2/3})]$$ (23)

71

It we calculate the maximum possible values of γ and G and define normalized parameters by dividing the resp. equations by the respective maximum values we find:

$$\gamma^* = \gamma/\gamma(T^*=1) = 1 - (1 - T^*)^{2/3} \tag{24}$$

and

$$G^* = G/G_{max} = (1 - T^*)^{1/3} \tag{25}$$

To attain the relation between damping and strain amplitude we formally calculate the dissipated energy ΔW from eq.6c and equ.23 assuming that G varies with γ as indicated in equ.21 and equ.22. From equ.6c we have

$$\Delta W = G \gamma^2 \pi \tan\phi \tag{26}$$

The well known relation between the logarithmic decrement θ and the damping ratio (damping coefficient over critical damping) D of lumped parameter systems together with equ.19 yields

$$\theta_t = \frac{2 \pi D}{\sqrt{1-D^2}} = 2\pi \frac{1 - \sqrt{1-\tan^2\phi}}{\tan\phi} \tag{27}$$

Putting the value of $\tan\phi$ from equ.26 into equ.27 we find

$$\frac{D}{\sqrt{1-D^2}} = \frac{1 - \sqrt{1-[\Delta W/(\pi G \gamma^2)]^2}}{\Delta W/(\pi G \gamma^2)} \tag{28}$$

with the term $\Delta W/(\pi G \gamma^2)$ taken from equ.21-23:

$$\tan\phi = \Delta W/(\pi G \gamma^2) = \frac{64}{20\pi} \frac{1-(1-T^*)^{5/3}- \frac{5}{6} T^*\left[1+(1-T^*)^{2/3}\right]}{(1-T^*)^{1/3} \left[1-(1-T^*)^{2/3}\right]^2} \tag{29}$$

We see that for $T^* = 1$ the left-hand term of equ.29 becomes infinite. This is due to the fact that we have taken the value of G at γ, whilst in equ.6a γ (and consequently G) varies periodically between 0 and $\pm \gamma$.

The experiments of Hardin and Drnevich (1972) suggest, that in the hysteresis loop immediately after reversal we have $G = G_{max}$, independent of the strain level (fig.12).

The same holds true for the two-sphere model (Mindlin et al. 1951).

If we put G_{max} instead of G into equ.29, the term $(1-T^*)^{1/3}$ reduces to 1 and equ.29 remains finite even for $T^* = 1$.

We do of course introduce an error into equ.28 by assuming $G = G_{max}$. If we assume, however, that the loop area in fig.12 is a constant percentage of the area a-b-c-d, controlled by γ, τ and G_{max}, (see Hardin and Drnevich, (1972), this error will cancel if we normalize D to D^*, as seen below.

Since D remains finite at $T^* = 1$, we define in analogy to equ.24 and equ.25:

$$D^* = D/D(T^*=1) = D/D_{max} \tag{30}$$

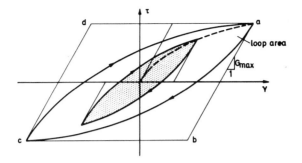

Fig.12 Loop-area of stress-strain hysteresis loop

The normalized values T^* (from equ.24), G^* (from equ.25) and D^* (from equ.30) are plotted versus γ^* in fig.13.

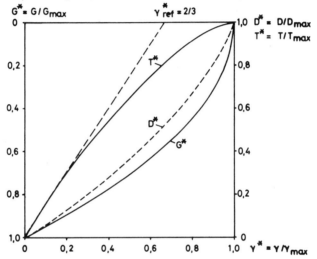

Fig.13 Normalized values T^*, G^* and D^* versus γ^*

Before we proceed to the discussion of fig.13 we will investigate the effects of a transition from forces T, N and displacements γ in Mindlins model fig.11 to stresses τ, σ and shear-strains $\bar{\gamma}$ in real soil structures.
    It the radius of the sphere is denoted R, the forces can be described by stresses as follows (spheres occupying cubes):

$$T = \tau(2R)^2$$
$$N = \sigma(2R)^2$$

(31a)

Similarly the shear strain $\bar{\gamma}$ becomes

$$\bar{\gamma} = \gamma/2R$$

(31b)

If we write equ.21 in terms of $\bar{\gamma}$, τ and σ we have to multiply equ.21 by 2R, leaving equ.24 unchanged.
    Similarly, equ.22 has to be divided by 2R, leaving equ.25 unchanged.
    Finally, equ.23 must be multiplied by $16R^4$, leaving equ.30 unchanged (reminding ourselves that ΔW of equ.26 is the work per unit-volume).
    Therefore, fig.13 is still valid if we replace T by τ and define γ^* as the normalized shear strain.
    From fig.13 we see immediately, that within the limitations of the simple two-sphere model, the following well known equation (Hardin and Drnevich, 1972) holds

$$D^* = 1 - G^* \tag{32}$$

If a reference strain γ_r is defined by $\gamma_r = \tau_{max}/G_{max}$ (Hardin and Drnevich, 1972) the normalized reference strain γ_r in fig.13 is found from equ.21 and equ.22 to be 2/3.
    In fig.14, experimental results of resonant column tests by Hardin and Drnevich (1972) are plotted together with the curves for G^* and D^* of fig.13 versus normalized shear strain γ/γ_r.

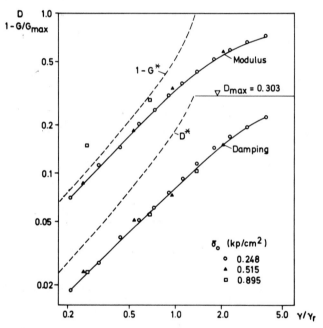

Fig.14 Experimental and 2-sphere model data for G^* and D^*
(after Hardin and Drnevich, 1972)

The accuracy of the two-sphere model is quite good for small strains ($\gamma/\gamma_r < 0.5$). The respective curves deviate for larger strains which of course reflects the fact that in real sands, with increasing shear strain, more contacts between grains are established resulting in a reduction of the gradient of G^* and D^*.

74

If we plot G^* and D^* versus hyperbolic strain γ_h

$$\gamma_h = \frac{\gamma}{\gamma_v} (1 + a\ e^{-b(\gamma/\gamma_r)}) \qquad (33)$$

(Hardin and Drnevich 1972) and take a = -0.5; b = 0.16 for sand, we arrive at fig.15, where the solid line indicates the well known relationship für $D^* = \gamma_h/(1+\gamma_h)$. We see, that the two-sphere model is capable of describing the dynamic behavior of sand to hyperbolic shear-strains of up to 0.4 corresponding to shear stresses $\tau = 0.6f\sigma$ in the sphere model.

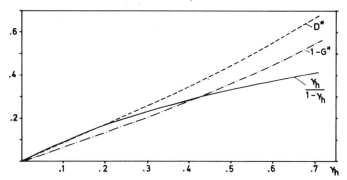

Fig.15 Theoretical and 2-sphere model data for G^* and D^* versus hyperbolic strain

If σ is increased, T^* in the above equations decreases, resulting in decreasing material damping. The influence of normal stress on the logarithmic decrement θ_t was shown by Hardin 1965 (fig.16).

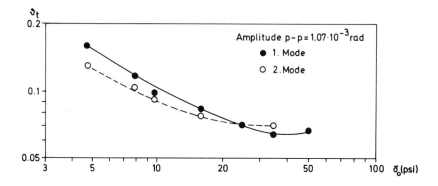

Fig.16 Variation of log.decrement δ_+ with octahedral stress $\bar{\sigma}_o$ (experimental data from Hardín, 1965)

The effect of stress and strain on material damping is described by the well known emperical formula (Richart, Hall and Woods, 1970) resulting from a great number of tests.

$$\tan\phi = \frac{\omega\mu}{G} = 0.2 \ \gamma^{0.2} \ \bar{\sigma}_0^{-0.5} \quad (kp/cm^2) \tag{33}$$

$$\text{with } \bar{\sigma}_0 = \frac{1}{3} (\sigma_x + \sigma_y + \sigma_z)$$

For $\tan\phi$ small, $\tan\phi$ may be expressed by D following equ.19 and equ.27:

$$\tan\phi = 2D \tag{34}$$

and with equ.7

$$D = \psi/4\pi \tag{35}$$

INFLUENCE OF MATERIAL DAMPING IN LUMPED PARAMETERS

The influence of the relatively small amount of material damping in granular soils upon the lumped parameters of soil-structure-interaction is similarly small. In the case of a rigid circular foundation vibrating vertically e.g., only the values of the damping constant f_r calculated after Bycroft (1956) with D = 0.03 differ slightly from the case D = 0 as seen in fig.17.

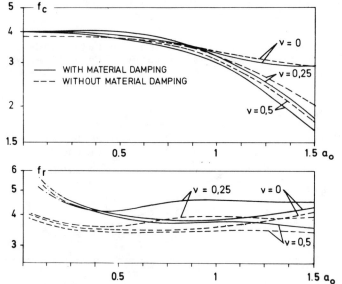

Fig.17 Spring and damping coefficients of circular foundations with and without damping (after Bycroft, 1956)

where f_c = spring constant vertical translation
         f_r = damping constant vertical translation
         r_0 = radius of circular foundation
         a_0 = normalized frequency = $\omega r_0/v_s$

76

COEFFICIENT OF RAYLEIGH WAVE VISCOSITY

So far we have considered the coefficient of shear wave viscosity only. To determine material damping from experiments of surface wavefields we need the respective values for the Rayleigh wave. The damping exponent for the Rayleigh wave may be calculated after Dobry (1970), following Hardins (1965) result that for $\nu = 0.25$ the ratio of the coefficient of compressive wave viscosity η to shear wave viscosity μ_S is $\eta/\mu_S = 1.88$. The respective calculation yields $\delta_r = 1.051 \; \delta_S$.

An other way is to lock at the ratio of strain velocity and shear-strain velocity of a Rayleigh wave versus depths. If we relate the coefficient of compressive wave viscosity η to $\dot{\varepsilon}$ and the coefficient of shear wave viscosity μ_S to $\dot{\gamma}$, assuming again that $\eta/\mu_S = 1.88$, we arrive at fig.18.

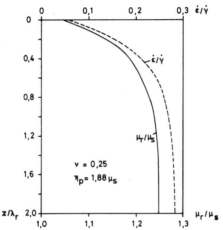

Fig.18 Coefficient of Rayleigh-wave viscosity vs. depth

Fig. 18 shows the variation of the coefficient of Rayleigh wave viscosity μ_r with depth z/λ_r. If we assume the main portion of the Rayleigh wave energy travelling at a depth $z/\lambda_r = 0.5$ (Richart, Hall and Woods, 1970), we find $\mu_r/\mu_S = 1.2$.

In any case, the material damping of the Rayleigh-wave is slightly larger than that of the shear wave.

CONCLUSIONS

The dynamic stress-strain behavior of granular media can best be represented by a visco-elastic relation, in which the material damping is described by the coefficient of shear viscosity. Experimental data of resonant column tests and wavefields prove that the coefficient of shear viscosity is inversely proportional to frequency resulting in the material damping beeing of hysteretic nature. A physical model of two spheres is capable of explaining qualitatively the behavior of granular media and the stress- and strain dependency of the respective elastic and

plastic moduli (real and imaginary part of the modulus).
     For granular media, the material damping is generally small.
Lumped parameters of SSI-problems will therefore not be re-
markably changed considering material damping, if the main
source of energy dissipation is wave propagation.
     If wavefields are concerned, the material damping plays an
important role in attenuation, especially in dependency on frequency. Since the imaginary part of the shear modulus is small,
the effective modulus and hence the wave-velocity will only
slightly be increased.

REFERENCES

Bycroft, G.N.   1965, Forced Vibrations of a Rigid Circular Plate
     on a Semiinfinite Elastic Space and on an Elastic Stratum.
     Phil.Trans.Royal Soc. London Ser.A Vol.248
Dobry, R.   1970, Damping in Soils: Its Hysteretic Nature and
     the Linear Approximation. MIT Soils Publ. No.253
Hardin, B.O.   1965, The Nature of Damping in Sands.
     J.Soil Mech.and Found.Div., Proc.ASCE, SM1
Hardin, B.O. and Drnevich, V.P.   1972, Shear Modulus and Damping
     in Soils: Measurement and Parameter Effects.
     J.Soil Mech.and Found.Div., Proc. ASCE SM6, June 1972
Hardin, B.O. and Drnevich, V.P.   1972, Shear Modulus and Damping
     in Soils: Design Equations and Curves
     J.Soil Mech.and Found.Div., Proc. ASCE SM6, July 1972
Mindlin, R.D.   1954, Mechanics of Granular Media.
     Proc. 2nd US Nat.Congr. of Appl.Mech., Ann Arbor
Mindlin, R.D., Mason, W.P., Osmer, T.F. and Deresiewicz, H.
     1951, Effects of an Oszillating Tangential Force on the Con-
     tact Surfaces of Elastic Spheres. Proc.1st.US Nat.Congr.
     Appl. Mech., ASME, June 1951
Prange, B.   1978 , Primary and Secondary Interferences in Wave-
     fields. Proceedings Dynamical Methods in Soil and Rock
     Mechanics DMSR 1977, Vol.I, Balkema, Rotterdam
Richart, F.E., Hall, J.R. and Woods, R.D.   1970, Vibrations of
     Soils and Foundations. Englewood Cliffs, N.J., Prentice Hall

Lumped parameter models

RICHARD D. WOODS
University of Michigan, Ann Arbor, Mich., USA

INTRODUCTION

The lumped parameter models currently available for analysis of dynamic foundation response are based on the behavior of a rigid block on an elastic half-space. The rigid block, as shown in Fig. 1, has six degrees of freedom, three translational and three rotational.

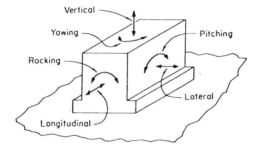

Translational Modes	Rotational Modes
Vertical	Rocking
Longitudinal	Pitching
Lateral	Yawing

Fig.1   Six modes of vibration for a foundation (from Richart, 1960).

Ideally, each degree of freedom has associated with it both stiffness and damping terms. Figure 2 shows schematically the stiffness and damping constants for two translational and one rotational degrees of freedom. It is the purpose of this paper to develop the stiffness and damping coefficients required for lumped parameter models for dynamic foundation response.

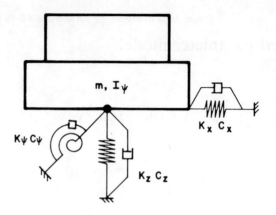

Fig.2 Lumped parameter model.

BASIC DEVELOPMENT (Vertical Mode)

E. Reissner while working for Deutsche Forschungsgesellschaft
für Bodenmechanik (DEGEBO) developed expressions for the vertical response of a flexible, circular, uniform load of intensity q_0 on the surface of an elastic half-space. The equation for vertical displacement at the center of the footing,
according to Reissner (1936), was

$$z_0 = \frac{P_0 \exp(i\omega t)}{G r_0} (f_1 + i f_2) \qquad (1)$$

in which

$$
\begin{array}{ll}
P_0 & = \text{amplitude of applied force,} \\
\omega & = \text{circular frequency of applied force,} \\
G & = \text{shear modulus of the half-space,} \\
r_0 & = \text{radius of circular contact area,} \\
f_1, f_2 & = \text{Reissner's "displacement functions".}
\end{array}
$$

The functions f_1 and f_2 are complex functions of Poisson's
ratio, ν, and a dimensionless "frequency ratio", a_0,

$$a_0 = \omega r_0 \sqrt{\rho/G} = \frac{\omega r_0}{v_s} \qquad (2)$$

in which

$$
\begin{array}{ll}
\rho & = \text{mass density of half-space material,} \\
v_s & = \text{shear wave velocity in half-space.}
\end{array}
$$

Reissner also established a second dimensionless term designate
as the "mass ratio", b,

$$b = \frac{m}{\rho r_0^3} \qquad (3)$$

in which m is the total mass of the footing and the exciting mechanism. The mass ratio will be important in subsequent development of half-space theory.

While the predictions of motion by Reissner's theory did not agree well with measured motion in experimental programs due to several reasons, his work remains as the basis for nearly all further development in this area.

Two papers were published about the same time (Quinlan (1953) and Sung (1953)) which extended and corrected Reissner's earlier work. These investigators considered alternate pressure distributions and developed expressions for the response of a footing subject to an eccentric mass excitation, Fig. 3.

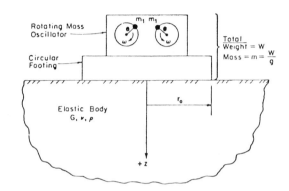

Fig.3   Rotating mass oscillator on elastic half-space (from Richart, Hall & Woods, 1970).

Figure 4a shows the response curves for three assumed load distributions over a circular area. The shapes of the three curves in Fig. 4a indicate different values of damping, but how can this be in an "elastic material".

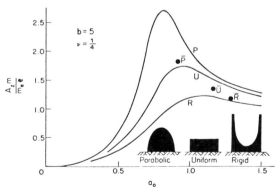

Fig.4a Effect of pressure distribution on theoretical response curves for vertical footing motion (after Richart and Whitman, 1967).

81

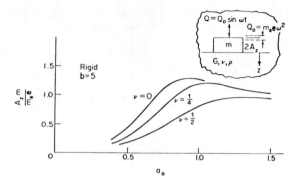

Fig.4b Effect of Poisson's ratio on theoretical response curves
for vertical footing motion (after Richart and Whitman,
1967).

The damping comes about through radiation of energy into the
half-space. This type of damping is also called "geometric
damping". As the mass ratio, b, decreases the geometric damping
of the system increases, Fig. 5, and for b less than about 10,
vertical oscillations are so highly damped that dynamic magnification is negligible. The influence of Poisson's ratio on the
response for the rigid footing assumption is shown in Fig. 4b.

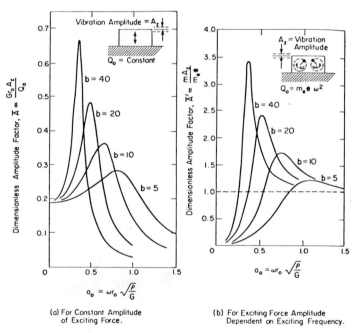

(a) For Constant Amplitude
of Exciting Force.

(b) For Exciting Force Amplitude
Dependent on Exciting Frequency.

Fig.5 Amplitude vs. frequency for vertical motion of rigid
circular footing on elastic half-space (ν = 1/4)
(after Richart, 1962).

82

Hsieh (1962) reorganized Reissner's basic equations into the form of the equilibrium equation for a damped single-degree-of-freedom system.

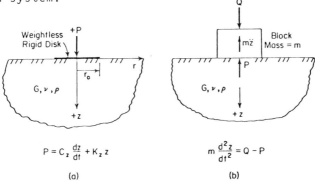

$$P = C_z \frac{dz}{dt} + K_z z$$

(a)

$$m \frac{d^2z}{dt^2} = Q - P$$

(b)

Fig.6   Notation for Hsieh's equations (from Richart, Hall & Woods, 1970).

He first considered a weightless rigid disk at the surface of the half-space, Fig. 6, excited by a periodic loading

$$P = P_0 \exp(i\omega t) \tag{4}$$

From Eq. 1

$$z = \frac{P_0 \exp(i\omega t)}{G r_0} (f_1 + i f_2) \tag{5}$$

then, differentiating Eq.(5) with respect to time,

$$\frac{dz}{dt} = -\frac{P_0 \omega \exp(i\omega t)}{G r_0} (i f_1 - f_2) \tag{6}$$

and using Eqs. (5) and (6) in the expression

$$f_1 \omega z - f_2 \frac{dz}{dt} \tag{7}$$

Hsieh found that

$$P = -\frac{G r_0}{\omega} \frac{f_2}{(f_1^2 + f_2^2)} \frac{dz}{dt} + G r_0 \frac{f_1}{(f_1^2 + f_2^2)} z \tag{8}$$

In Eq. (8) the coefficients of dz/dt and z are damping and stiffness terms so Eq.(8) can be rewritten

$$P = c_z \frac{dz}{dt} + k_z z \tag{9}$$

in which

$$c_z = \frac{r_0^2}{a_0} \sqrt{G\rho} \left(\frac{-f_2}{f_1^2 + f_2^2}\right) \tag{10}$$

and

83

$$k_z = G\, r_0 \frac{f_1}{f_1^2 + f_2^2} \tag{11}$$

Now, if a rigid cylindrical footing of radius r_0 and total weight W is placed on the half-space and excited by a harmonic force Q, we may write an equilibrium equation as

$$\frac{W}{g} \frac{d^2z}{dt^2} = Q - P \tag{12}$$

and then by combining with Eq. (9), obtain

$$m \frac{d^2z}{dt^2} + c_z \frac{dz}{dt} + k_z\, z = Q_0 \exp(i\omega t) \tag{13}$$

Equation (13) is the key to the lumped parameter model.

Before leaving this section, however, the implications of Eqs. (10) and (11) should be noted. In particular that the Reissner function f_1 is primarily associated with stiffness and f_2 is primarily associated with damping.

Up to this point in the development, the functions f_1 and f_2 have veen dependent on a_0 and ν, Fig. 7.

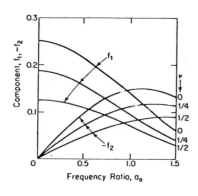

Fig.7 Displacement functions for vertical motion of rigid circular footing on elastic halfspace (after Bycroft, 1956).

Lysmer (1965) found that the dependency of f_1 and f_2 on ν could be essentially eliminated by multiplying Reissner's functions by $4/1-\nu$. The modified displacement functions are shown in Figs. 8 and 9. Similarly Lysmer eliminated the dependency of the mass ratio b on ν by multiplying by $(1-\nu/4)$,

$$B_z = \frac{1-\nu}{4}\, b = \frac{1-\nu}{4} \frac{m}{\rho\, r_0^3} \tag{14}$$

Lysmer then developed response curves based on the modified dis placement functions and mass ratio, and a static spring con-stant

$$k_z = \frac{4\, G_0\, r_0}{1 - \nu} \tag{15}$$

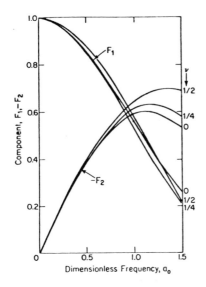

Fig.8 Variation of components of F with Poisson's ratio (after Lysmer and Richart, 1966).

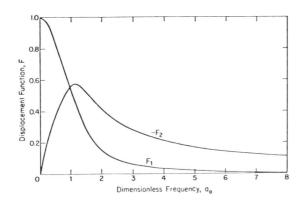

Fig.9 Displacement function F for vertical vibration of a weightless rigid circular disk (after Lysmer and Richart, 1966).

See the solid lines on Fig. 10. Lysmer also noted that for a_0 in the practical range, a damping term could be found which when used in Eq. (13) gave close agreement, dashed lines in Fig. 10. The damping term was expressed as

$$c_z = \frac{3.4 \ r_0^2}{(1-\nu)} \sqrt{\rho G} \tag{16}$$

Now Lysmer presented an analog for vertical motion of a circular footing on an elastic half-space

85

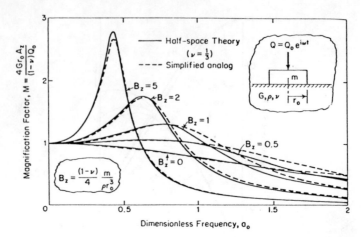

Fig.10 Response of rigid circular footing to vertical force
developed by constant force excitation (from Lysmer
and Richart, 1966).

$$m \ddot{z} + \frac{3.4 \, r_o^2}{(1-\nu)} \sqrt{\rho G} \, \dot{z} + \frac{4 \, G \, r_o}{1-\nu} \, z = Q \qquad (17)$$

Extending the analog, the damping ratio D can be obtained
from

$$c_c = 2 \sqrt{k_z m} = 2 \sqrt{\frac{4 \, G \, r_o \, m}{1-\nu}} \qquad (18)$$

and

$$D = \frac{c_z}{c_c} = \frac{0.425}{B_z} \qquad (19)$$

Furthermore, the frequency of maximum response and maximum
amplitude of vertical displacement can also be computed from
single-degree-of-freedom equations

Constant force amplitude

$$f_m = \frac{1}{2\pi} \frac{v_s}{r_o} \sqrt{\frac{B_z - 0.36}{B_z}} \qquad (20)$$

$$A_{zm} = \frac{Q_o (1-\nu)}{4 \, G_o \, r_o} \frac{B_z}{0.85 \sqrt{B_z - 0.18}} \qquad (21)$$

Rotating mass excitation

$$f_{mr} = \frac{v_s}{2\pi \, r_o} \sqrt{\frac{0.9}{B_z - 0.45}} \qquad (22)$$

$$A_{zmr} = \frac{m_e \ e}{m} \quad \frac{B_z}{0.85 \ B_z - 0.18} \tag{23}$$

The results obtained from Eqs. 20 - 23 can also be obtained from Figs. 11 and 12.

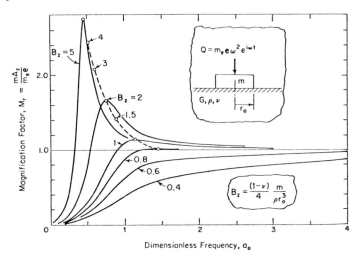

Fig.11 Response of rigid circular footing to vertical force developed by rotating mass exciter (from Richart, Hall and Woods, 1970).

The major contribution by Lysmer was providing the link between the elastic half-space theory and the mass-spring-dashpot system. By obtaining representative values of damping and stiffness, Lysmer removed the frequency dependency from the problem, which led to the simplified analog.

OTHER UNCOUPLED MODES

Torsion

For torsion the spring constant and damping coefficients are respectively

$$k_{\theta S} = \frac{16}{3} \ G \ r_o^{\ 3} \tag{24}$$

$$c_{\theta} = \frac{4 \sqrt{B_{\theta} \ \rho \ G}}{1 + 2 \ B_{\theta}} \tag{25}$$

$$B_{\theta} = \frac{I_{\theta}}{\rho \ r_o^{\ 5}} \tag{26}$$

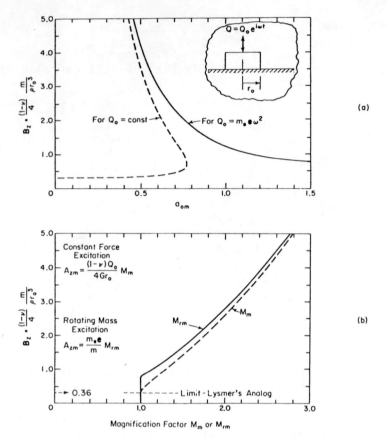

Fig.12 Vertical oscillation of rigid circular footing on
elastic half-space. (a) Mass ratio vs. dimensionless
frequency at resonance. (b) Mass ratio vs. magnification factor at resonance. (from Richart, Hall and
Woods, 1970).

and

$$D = \frac{0.50}{1 + 2 B_\theta} \tag{27}$$

in which I_θ is the mass moment of inertia about the axis of rotation. Figure 13 provides graphical solutions for both constant force and rotating mass excitation.

Rocking

Based on analytical solutions for the rocking of a rigid disk
on an elastic half-space by Arnold; Bycroft and Warburton (1955
and Bycroft (1956), Hall (1967) formulated the lumped parameter

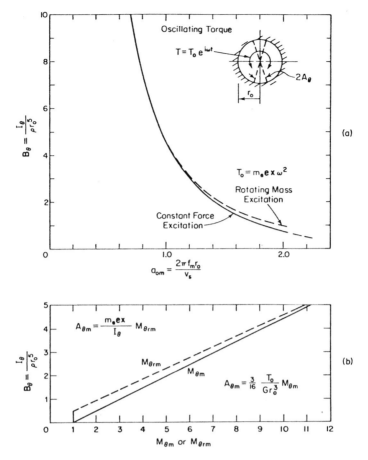

Fig.13 Torsional oscillation of rigid circular footing on elastic half-space. (a) Mass ratio vs. dimensionless frequency at resonance. (b) Mass ratio vs. magnification factor at resonance. (from Richart, Hall and Woods, 1970).

coefficients for rocking motion,

$$k_{\psi S} = \frac{8G\, T_0^{\,3}}{3(1-\nu)} \tag{28}$$

$$c_\psi = \frac{0.80\, r_0^{\,4}\, \sqrt{G\rho}}{(1-\nu)(1+B_\psi)} \tag{29}$$

$$B_\psi = \frac{3(1-\nu)}{8} \frac{I_\psi}{\rho\, r_0^{\,5}} \tag{30}$$

and

$$D_\psi = \frac{0.15}{(1+B_\psi)\sqrt{B_\psi}} \tag{31}$$

89

in which I_ψ is the mass moment of inertia of the footing about the axis of rotation. Figure 14 shows the agreement between the elastic half-space solution to the rocking mode and the Hall analog with a modified B_ψ where

$$B_{\psi eff} = n_\psi B_\psi \qquad (32)$$

and

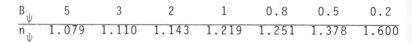

B_ψ	5	3	2	1	0.8	0.5	0.2
n_ψ	1.079	1.110	1.143	1.219	1.251	1.378	1.600

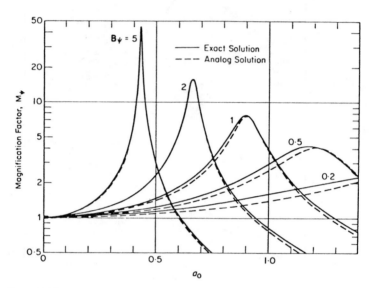

Fig.14 Magnification factor vs. dimensionless frequency relations for pure rocking of rigid circular footing on elastic half-space (from Hall, 1967).

Figure 15 provides a graphical solution for rocking.

Sliding

Hall (1967) also provided the coefficients for sliding of a rigid disk on an elastic half-space,

$$k_{xs} = \frac{32(1-\nu)}{7-8\nu} G r_0 \qquad (33)$$

$$c_x = \frac{18.4(1-\nu)}{7-8\nu} r_0^2 \sqrt{\rho G} \qquad (34)$$

$$B_x = \frac{7-8\nu}{32(1-\nu)} \frac{m}{\rho r_0^3} \qquad (35)$$

and

$$D_x = \frac{0.288}{\sqrt{B_x}} \qquad (36)$$

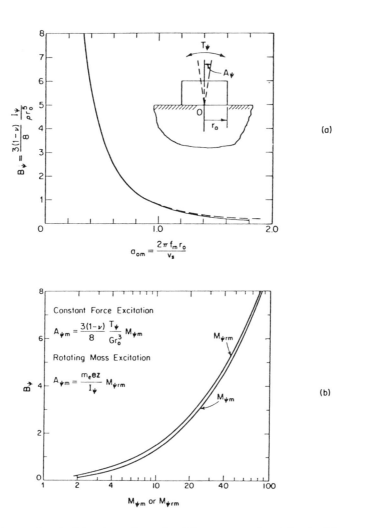

Fig.15 Rocking of rigid circular footings on elastic half-space. (a) Mass ratio vs. dimensionless frequency at resonance. (b) Mass ratio vs. magnification factor at resonance. (from Richart, Hall and Woods, 1970).

Figure 16 shows the fit of Hall's analog with half-space theory and Fig.17 provides a graphical solution.

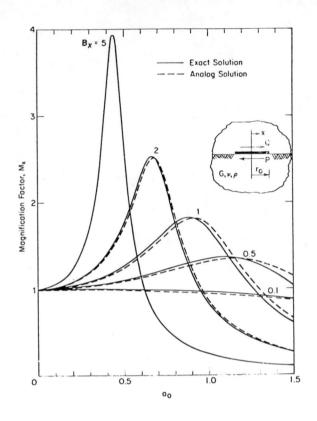

Fig.16 Magnification factor vs. dimensionless frequency relations for pure sliding of rigid circular footing on elastic half-space (from Hall, 1967).

GEOMETRIC DAMPING

It is very important to reflect for a moment on the values for damping ratio proposed for each of the single-degree-of-freedom systems described so far. By plotting the damping ratio for all modes on a common graph, Fig. 18, the relative damping for each mode is clear.

Vertical and sliding translational modes are highly damped while torsion is moderately damped and rocking is lightly damped.

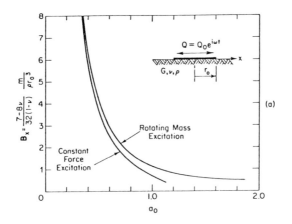

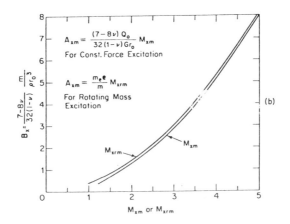

Fig.17 Sliding oscillation of rigid circular disk on elastic
half-space. (a) Mass ratio vs. dimensionless frequency
at resonance. (b) Mass ratio vs. magnification factor
at resonance. (from Richart, Hall and Woods, 1970).

NONCIRCULAR FOOTINGS

The development of analogs so far has been concerned only with
circular footings. Elorduy, Nieto and Szekely (1967) demonstrated that for rectangular footing subject to vertical oscillation and with width to length ratios up to 2, an equivalent
circular footing gave satisfactory results. Chae (1969) reported that the equivalent circular footing was adequate for pre-
dicting frequency but not for amplitude of motion. For most

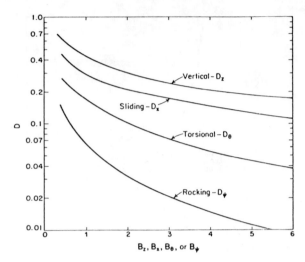

Fig.18 Equivalent damping ratio for oscillation of rigid
circular footing on elastic half-space (from Richart,
Hall and Woods, 1970).

problems Richart, Hall and Woods (1970) recommend the expressions in Table 1 for computing equivalent r_0 for use in footing
response analogs.

COUPLED ROCKING AND SLIDING

The most common mode of footing vibration is coupled rocking
and sliding. This coupled mode of vibration occurs because the
center of sliding resistance does not coincide with the unbalanced force of the exciting system resulting in a couple
which causes rocking. Figure 19 shows the notation for rocking
and sliding. In the development of the equations of motion for
rocking and sliding, the subscript "g" refers to the center of
gravity and "b" refers to the center of the base.
   The translation of the base can be expressed as

$$x_b = x_g - h_o \psi \tag{37}$$

The horizontal force P_x acting on the base of the footing is
given by

$$P_x = - c_x \dot{x}_b - k_x x_b \tag{38}$$

The resistance of the half-space to rocking is given by

$$R_\psi = -c_\psi \dot{\psi} - k_\psi \psi \tag{39}$$

The equation of motion for translation of the c·g is

$$m\ddot{x} = P_x = -c_x \dot{x}_b - k_x x_b \tag{40}$$

94

Table 1. Rigid Circular Disk on an Elastic Half-Space

Mode	Spring Constant	Damping Constant	Mass Ratio	Damping Ratio
Vertical	$k_z = \dfrac{4 G r_0}{1-\nu}$	$C_z = \dfrac{3.4 r_0^2}{1-\nu}\sqrt{\rho G}$	$B_z = \dfrac{1-\nu}{4}\dfrac{m}{\rho r_0^3}$	$D_z = \dfrac{0.425}{\sqrt{B_z}}$
Horizontal	$k_x = \dfrac{32(1-\nu)G r_0}{7-8\nu}$	$C_x = \dfrac{18.4(1-\nu)r_0^2}{7-8\nu}\sqrt{\rho G}$	$B_x = \dfrac{7-8\nu}{32(1-\nu)}\dfrac{m}{\rho r_0^3}$	$D_x = \dfrac{0.288}{\sqrt{B_x}}$
Rocking	$k_\psi = \dfrac{8 G r_0^3}{3(1-\nu)}$	$C_\psi = \dfrac{0.8 r_0^4 \sqrt{\rho G}}{(1-\nu)(1+B_\psi)}$	$B_\psi = \dfrac{3(1-\nu)}{8}\dfrac{I_\psi}{\rho r_0^5}$	$D_\psi = \dfrac{0.15}{(1+B_\psi)\sqrt{B_\psi}}$
Torsion	$k_\Theta = \dfrac{16}{3} G r_0^3$	$C = \dfrac{4\sqrt{B_\Theta \, \rho G}}{1+2 B_\Theta}$	$B_\Theta = \dfrac{I_\Theta}{\rho r_0^5}$	$D_\Theta = \dfrac{0.50}{1+2 B_\Theta}$

M = W/g = mass of footing

I_ψ = mass moment of inertia in rocking about y-axis

I_Θ = mass moment of inertia in torsion about z-axis

Equivalent Radius for rectangular footing length 2c, width 2d

For Translation $\qquad r_0 = \sqrt{\dfrac{4\,cd}{\pi}}$

For Rocking $\qquad r_0 = \sqrt[4]{\dfrac{16\,cd^3}{3\pi}}$

For Torsion $\qquad r_0 = \sqrt[4]{\dfrac{16\,cd(c^2+d^2)}{6\pi}}$

Equations for mass moments of inertia:

For cylinder of radius r_0 and height h about horizontal axis ψ through base of cylinder

$$I_\psi = \frac{\pi r_0^2\, h\gamma}{g}\left(\frac{r_0^2}{4} + \frac{h^2}{3}\right)$$

For a rectangular prism length b, width c, height a about axis x through the center (parallel to b)

$$I_{xx} = \frac{1}{12}\,\frac{W}{g}\,(a^2 + c^2)$$

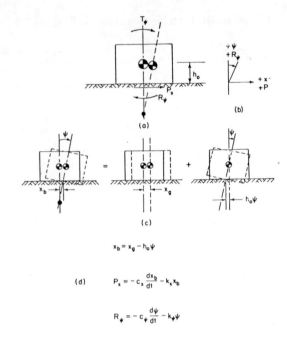

$$x_b = x_g - h_o \psi$$

$$(d) \qquad P_x = -c_x \frac{dx_b}{dt} - k_x x_b$$

$$R_\psi = -c_\psi \frac{d\psi}{dt} - k_\psi \psi$$

Fig.19 Notation for rocking and sliding mode of vibration (from Richart, Hall and Woods, 1970).

and after substituting from Eq. (37) we get

$$m \ddot{x}_g + c_g \dot{x}_g + k_x x_g - h_o c_x \dot{\psi} - h_o k_x \psi = 0 \qquad (41)$$

The equation of motion for rotation about the c·g is

$$I_g \ddot{\psi} = T_\psi - R_\psi - h_o P_\psi \qquad (42)$$

in which I_g is the moment of inertia of the footing about the c·g. Substituting from Eqs. (37), (38) and (39) we get

$$I_g \ddot{\psi} + (c_\psi + h_o^2 c_x)\dot{\psi} + (k_\psi + h_o^2 k_x)\psi - h_o c_x \dot{x}_g - h_o k_x x_g = T_\psi \qquad (43)$$

Then for harmonic excitation let

$$x_g = A_{x1} \sin \omega t + A_{x2} \cos \omega t \qquad (44)$$

$$\psi = A_{\psi 1} \sin \omega t + A_{\psi 2} \cos \omega t \qquad (45)$$

$$T_\psi = T_{\psi o} \sin \omega t \qquad (46)$$

Introducing Eqs. (44), (45) and (46) into (41) and (43) yields four equations from which the coefficients A_{x1}, A_{x2}, $A_{\psi 1}$ and

and $A_{\psi 2}$ can be obtained for each chosen value of ω. Then A_x and A_ψ given by

$$A_x = \sqrt{A_{x1}^2 + A_{x2}^2} \tag{47}$$

$$A_\psi = \sqrt{A_{\psi 1}^2 + A_{\psi 2}^2} \tag{48}$$

can be plotted as functions of ω to obtain the maximum amplitudes and frequencies of maximum response.

The coefficient for damping and stiffness in Eqs. (41) and (43) can be taken from the previously developed analogs.

EMBEDDED FOUNDATIONS

The elastic half-space analogs considered up to now have dealt only with footings at the surface. Often footings are partially or fully embedded into the supporting soil. Several investigators, Stokoe (1972), Erden and Stokoe (1974) and Gupta (1972) for example, investigated the influence of embedment by experimental methods. From these studies the general trends associated with footing embedment were established. These included (a) amplitude of vibration decreased with depth of embedment, (b) frequency of maximum response increased with embedment, and (c) damping increased with embedment. Furthermore, the condition of the side contact between footing block and surrounding soil were shown to be very important and the influence of embedment varied from full effect for footings cast against the raw earth to no effect for footings separated from the soil by a small crack.

Novak and his co-workers, Novak (1973), Beredugo and Novak (1972), and Novak and Sachs (1973), adopted an analytical approach suggested by Baranov (1967) to provide closed form solutions for coefficients required in the half-space analogs for embedded foundations. The approach assumed a footing on an elastic half-space with multiple independent elastic layers on top of the half-space providing the embedment, see Fig. 20.

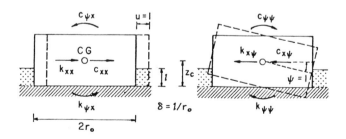

Fig.20 Translation and rocking components of coupled motion or related stiffness and damping coefficient of embedded footings (from Novak, 1973).

For vertical vibration the coefficients for the single-degree-of-freedom analog are

$$k_z = G\ r_0\ (c_{\omega 1} + \frac{G_s}{G}\ \frac{H}{r_0}\ S_{\omega 1}) \tag{49}$$

$$c_z = r_0^2\ \sqrt{\rho G}\ (\bar{c}_{\omega 2} + \bar{S}_{\omega 2}\ \frac{H}{r_0}\ \sqrt{\frac{\rho_s G_s}{\rho G}}) \tag{50}$$

in which

G = shear modulus of the half-space,
G_s = shear modulus of the side layers,
H = depth of embedment
ρ = mass density of half-space
ρ_s = mass density of side layers

and $c_{\omega 1}$, $S_{\omega 1}$, $\bar{c}_{\omega 2}$ and $\bar{S}_{\omega 2}$ can be found in Table 2.
For torsional vibration the coefficients are

$$k_\zeta = G\ r_0^3\ (c_{\zeta 1} + \frac{G_s}{G}\ \frac{H}{r_0}\ S_{\zeta 1}) \tag{51}$$

$$c_\zeta = r_0^4\ \sqrt{\rho G}\ (\bar{c}_{\zeta 2} + \bar{S}_{\zeta 2}\ \frac{H}{r_0}\ \sqrt{\frac{\rho_s G_s}{\rho G}}) \tag{52}$$

for which $c_{\zeta 1}$, $\bar{c}_{\zeta 2}$, $S_{\zeta 1}$ and $\bar{S}_{\zeta 2}$ cand be found in Table 2.
For coupled rocking and sliding Beredugo and Novak (1973) provide the following equations for complex amplitude of coupled motion

$$[(k_{xx} - m\ \omega^2) + i\omega c_{xx}]\ u_c + (i\omega c_{x\psi} + k_{x\psi})\psi_c = Q_0 \tag{53}$$

$$[(k_{\psi\psi} - I\ \omega^2) + i\omega c_{\psi\psi}]\ \psi_c + (i\omega c_{x\psi} + k_{x\psi})u_c = T_0 \tag{54}$$

in which u_c and ψ_c are complex displacement amplitudes at the c.g. of the foundation and the coefficients are as follows

$$k_{xx} = G\ r_0\ (c_{u1} + \frac{G_s}{G}\ \frac{H}{r_0}\ S_{u1}) \tag{55}$$

$$k_{\psi\psi} = G\ r_0^3 [c_{\psi 1} + (\frac{z_c}{r_0})^2\ c_{u1} + \frac{G_s}{G}\ \frac{H}{r_0}\ S_{\psi 1} +$$

$$+ \frac{G_s}{G}\ \frac{H}{r_0}\ (\frac{1}{3}\ \frac{H^2}{r_0^2} + \frac{z_c^2}{r_0^2} - \frac{H\ z_c}{r_0^2})\ S_{u1}] \tag{56}$$

$$k_{x\psi} = -G\ r_0^2\ [\frac{z_c}{r_0}\ c_{u1} + \frac{G_s}{G}\ \frac{H}{r_0}\ (\frac{z_c}{r_0} - \frac{H}{2r_0})\ S_{u1}] \tag{57}$$

$$c_{xx} = \sqrt{\rho G}\ r_0^2\ (\bar{c}_{u2} + \frac{H}{r_0}\ \sqrt{\frac{\rho_s G_s}{\rho G}}\ \bar{S}_{u2}) \tag{58}$$

98

Table 2. Stiffness and Damping Parameters

Motion	Soil	Side Layer		Half Space	
Sliding	coh.	$S_{u1} = 4.1$	$\bar{S}_{u2} = 10.6$	$C_{u1} = 5.1$	$\bar{C}_{u2} = 3.2$
	gran.	$S_{u1} = 4.0$	$\bar{S}_{u2} = 9.1$	$C_{u1} = 4.7$	$\bar{C}_{u2} = 2.8$
Rocking	coh.	$S_{\psi1} = 2.5$	$\bar{S}_{\psi2} = 1.8$	$C_{\psi1} = 4.3$	$\bar{C}_{\psi2} = 0.7$
	gran.			$C_{\psi1} = 3.3$	$\bar{C}_{\psi2} = 0.5$
Torsion	coh.	$S_{\zeta1} = 10.2$	$\bar{S}_{\zeta2} = 5.4$	$C_{\zeta1} = 4.3$	$\bar{C}_{\zeta2} = 0.7$
	gran.				
Vertical	coh.	$S_{w1} = 2.7$	$\bar{S}_{w2} = 6.7$	$C_{w1} = 7.5$	$\bar{C}_{w2} = 6.8$
	gran.			$C_{w1} = 5.2$	$\bar{C}_{w2} = 5.0$

with

$$C_{u1} = \frac{8}{2-\nu} k_1 \qquad \bar{C}_{u2} = \frac{8}{2-\nu} c_1$$

$$C_{\psi1} = \frac{8}{3(1-\nu)} k_2 \qquad \bar{C}_{\psi2} = \frac{8}{3(1-\nu)} c_2$$

and

ν	0	0.25	0.333	0.40	0.50
$\frac{32(1-\nu)}{1-8\nu}$	4,57	4.80	4.92	5.05	5.33
$\frac{8}{2-\nu}$	4.00	4.57	4.80	5.00	5.33

$$c_{\psi\psi} = \sqrt{\rho G}\, r_o^4 \left\{ \bar{c}_{\psi 2} + \left(\frac{z_c}{r_o}\right)^2 \bar{c}_{u2} + \frac{H}{r_o} \sqrt{\frac{\rho_s G_s}{\rho G}} \left[\bar{S}_{\psi 2} + \right.\right.$$

$$\left.\left. \left(\frac{1}{3}\frac{H^2}{r_o^2} + \frac{z_c^2}{r_o^2} - \frac{H\, z_c}{r_o^2}\right) \bar{S}_{u2} \right] \right\} \tag{59}$$

$$c_{x\psi} = -\sqrt{\rho G}\, r_o^3 \left[\frac{z_c}{r_o}\bar{c}_{uz} + \frac{H}{r_o}\sqrt{\frac{\rho_s G_s}{\rho G}} \left(\frac{z_c}{r_o} - \frac{1}{2}\frac{H}{r_o}\right) \bar{S}_{u2} \right] \tag{60}$$

The c and S coefficients are found in Table 2 and have slightly different values for sands and clays.

Equations (53) and (54) are formally equal to the equations of motion for a two-degree-of-freedom system, therefore the solution proceeds as before finding amplitude of rocking and sliding as functions of frequency.

REFERENCES

Anandakrishnan, M., and Krishnaswamy, M.R. (1973), "Response of Embedded Footings to Vertical Vibrations", Journal of the Soil Mechanics and Foundations Division, ASCE, Vol. 99, No. SM10, Oct., pp. 863-883.

Arnold, R.N., Bycroft, G.N., and Warburton, G.B. (1955), "Forced Vibrations of a Body on an Infinite Elastic Solid", J.Appl.Mech., Trans. ASME, Vol.77, pp. 391-401.

Beredugo, Y.O., and Novak, M. (1972), "Coupled Horizontal and Rocking Vibration of Embedded Footings", Canadian Geotechnical Journal, Vol. 9, pp. 477-497.

Bycroft, G.N. (1956), "Forced Vibrations of a Rigid Circular Plate on a Semi-Infinite Elastic Space and on an Elastic Stratum", Philosophical Trans., Royal Society, London, Ser. A, Vol. 248, pp. 327-368.

Chae, Y.S. (1969), "Vibration of Noncircular Foundations", Journal of the Soil Mechanics and Foundations Division, ASCE, Vol. 95, No. SM6, Nov., pp. 1411-1430.

Elorduy, J., Nieto, J.A., and Szekely, E.M. (1967), "Dynamic Response of Bases of Arbitrary Shape Subjected to Periodic Vertical Loading", Proc. International Symposium on Wave Propagation and Dynamic Properties of Earth Materials, Albuquerque, N.M., Aug.

Erden, S.M., and Stokoe, K.H., II (1975), "Effects of Embedment on Foundation Response", Meeting Preprint 2536, ASCE National Convention, Denver, Colorado, Nov.

Gupta, B.N. (1972), "Effect of Foundation Embedment on the Dynamic Behavior of the Foundation-Soil System", Geotechnique, Vol. 22, No. 1, March, pp. 129-137.

Hall, J.R., Jr., (1967), "Coupled Rocking and Sliding Oscillations of Rigid Circular Footings", Proc. International Symposium on Wave Propagation and Dynamic Properties of Earth Materials, Albuquerque, N.M., Aug.

Lysmer, J. (1965), Vertical Motion of Rigid Footings, Dept. of Civil Eng., Univ. of Michigan Report to WES Contract Report No. 3-115, under Contract No. DA-22-079-eng-340; also a Ph.D. dissertation, Univ. of Michigan, Aug.

Lysmer, J., and Richart, F.E., Jr. (1966), "Dynamic Response of Footings to Vertical Loading", J. Soil Mech. and Found. Div., Proc. ASCE, Vol. 92, No. SM1, Jan., pp. 65-91.

Novak, M. (1973), "Vibrations of Embedded Footings and Structures", Meeting Preprint 2029, ASCE National Convention, San Francisco, California, April.

Novak, M., and Sachs, K. (1973), "Torsional and Coupled Vibrations of Embedded Footings", Journal of Earthquake Engineering and Structural Dynamics, Vol. 2, pp. 11-33.

Quinlan, P.M. (1953), "The Elastic Theory of Soil Dynamics", Symposium on Dynamic Testing of Soils, ASTM STP No. 156, pp. 3-34.

Reissner, E. (1936), "Stationäre, axialsymmetrische, durch eine schüttelnde Masse erregte Schwingungen eines homogenen elastischen Halbraumes", Ingenieur-Archiv, Vol. 7, Part 6, Dec., pp. 381-396.

Richart, F.E., Jr., (1960), "Foundation Vibrations", Journal of the Soil Mechanics and Foundations Division, ASCE, Vol. 86, No. SM4, Aug., pp. 1-34.

Richart, F.E., Jr., (1962), "Foundation Vibrations", Trans. ASCE, Vol. 127, Part 1, pp. 863-898.

Richart, F.E., Jr., and Whitman, R.V. (1967), "Comparison of Footing Vibration Tests with Theory", J. Soil Mech. and Found. Div., Proc. ASCE, Vol. 93, No. SM6, Nov., pp. 143-168.

Richart, F.E., Jr., Hall, J.R., and Woods, R.D. (1970), Vibrations of Soils and Foundations, Prentice-Hall, Englewood Cliffs, N.J.

Stokoe, K.H., II, (1972), "Dynamic Response of Embedded Foundations", Ph.D. Dissertation, University of Michigan, Jan., 251 pp.

Sung, T.Y. (1953), "Vibrations in Semi-Infinite Solids due to Periodic Surface Loadings", Symposium on Dynamic Testing of Soils, ASTM-STP No. 156, pp. 35-64.

Dynamic soil-structure interaction

ULRICH HOLZLÖHNER
Bundesanstalt für Materialprüfung, Berlin, Germany

SYNOPSIS

This report gives the state-of-the-art in the field of dynamic
soil-structure interaction. The problem ist analyzed and different methods of solution are described. The interaction effects
on typical structures are studied using the results obtained by
different authors. To conclude, the advantages and disadvantages
of the different methods of analysis are discussed.

1. INTRODUCTION

If a soil-structure system is loaded, displacements are produced in the structure and in the soil below. The mutual de-
pendancy of the displacements is called soil-structure interaction.

Soil-structure interaction is largely dependent on the deformability of the soil. Here, only dynamic interaction is
considered. The structure may either be loaded directly by
forces or may be excited through the ground. In both cases,
the vibrations and the residual deformations of the soil and
the structure are mutually dependent.

The various problems can be classified by the non-linear
effects involved as follows:
The soil and the structure are assumed to behave purely elastic
with constant material parameters. Here, one single linear analysis is sufficient.

The weight of the structure influences the elastic modulus
which in turn influences the spatial variation of the stiffness. Also in this case, one single linear analysis is sufficient.

The values of the strain amplitudes influence the modulus
of the soil. An iterative linear analysis must be performed.
This is normally done in seismic analysis of soil-structure
systems.

Residual settlements are induced by the vibrations. Residual settlements and liquefaction are not considered in this
report.

2. STRUCTURES EXCITED THROUGH THE GROUND

In the following, an elastic problem which may be non-linear is considered. It is also assumed that the structure is excited through the ground. The problem can be formulated more exactly as follows, see Fig. 1.

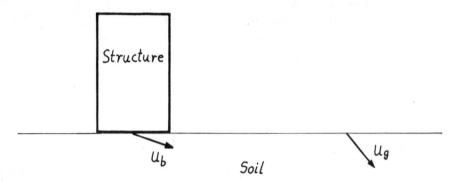

Fig. 1. Structure excited through the ground.

A soil and a structure may be given. The distribution of the mass and the stiffness may be given for both soil and structure. The soil may be subjected to a certain free-field displacement which shall be known as a function of time and space. The free-field displacement is the displacement of the soil without any structure. The problem is, to calculate the motion of the soil-structure system.

It is easy to see that the motion of the base of the structure will be different from the free-field displacement. The difference between the two displacements will be the larger the softer the soil is. Parmelee (1967) has illustrated this classical problem by a simple interaction model, Fig. 2.

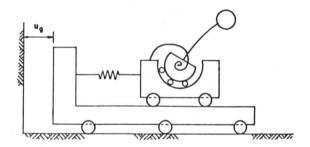

Fig. 2. Simple interaction model (from Parmelee 1967)

The upper car is the idealization of the flexible building. The free-field displacement is applied not directly at its base but through a spring representing the deformability of

the soil. The softer the spring, the larger is the difference
between the free-field displacement U_g and the displacement of
the base of the structure U_b. If the spring stiffens more and
more the difference between the two displacements will vanish.
Then, there is no interaction anymore.

3. METHODS OF ANALYSIS

The simple interaction model of Parmelee already suggests a
method of analysis. The soil is represented by a spring which
is coupled to the base of the structure. At the other end of
the soil spring, the free-field displacement is applied. This
method has been called "coupling method" by Holzlöhner (1972)
because the soil springs or impedance functions are coupled to
the structure. Perhaps the expression "substructure method",
Gutierrez and Chopra (1975), describes the essentials of the
method even better. This name expresses the fact that the soil-
structure system is first divided into two substructures, the
soil and the structure, which are calculated separately. Here,
both names "substructure method" and "coupling method" are
used. In the references, different other names can be found:
"lumped parameter method", "half space analysis" or "continuum
approach". As these names refer only to the method which has
been applied to evaluate the response of the soil, they do not
reflect the particular problem of soil-structure interaction.
    The interaction problem can also be dealt with in a funda-
mentally other way: Soil and structure can be treated together
in a combined analysis. This method is called here the "direct
method". In the references this method is mostly called "finite
element analysis". But the finite element method must not
necessarily be used with the direct method and, on the other
hand, the FEM has actually been used in order to evaluate
lumped springs (Johnson et al. 1975).
    The main difference between the two methods seems to be
that, by using the direct method, soil and structure are treated
in only one analysis whereas by the substructure method a special analysis of the soil alone yields results, which determine
the properties of the "springs" to be coupled to the structure
in the second analysis.

3.1 The direct method

Here, both, the structure and the soil are modelled by finite
elements, Fig. 3. The boundary below is identified with the
bedrock. The method has mainly been developed in California e.g.
by Seed et al. (1975), where bedrock usually is to be found in
small depths. If there is no bedrock, the layer must be thick
enough so that there is no effect of the vibrations at the
lower boundary on the vibrations of the structure. In case of
local vibration sources, the source must be included in the
model. The direct method is applied to the evaluation of seismic interaction effects. Normally, the free-field displacement
is given at the surface of the soil. Then, the analysis is performed in two steps.

105

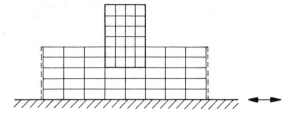

Fig. 3. Direct method. Finite element model.

First step: Evaluation of the base rock motion in such a way that the surface of the soil without structure displays the free-field motion of the surface.

Second step: The finite element model shown in the figure is subjected to the previously computed base rock motion. The combined soil-structure FE-analysis provides all displacements and forces in the structure and in the soil.

3.2 The substructure method

The substructure or coupling method is especially suited for the understanding of the phenomenon of interaction. By utilizing this method, also some basic effects of interaction can easily be understood.

3.2.1 Theory

The method of analysis suggested by the simple model of Parmelee, Fig. 2, needs some theoretical justification. Though there is not much theory about the coupling method it seems to be necessary for the scope of this paper to present the basic equations. This will be done in four steps (Holzlöhner 1972):

1. Let us first consider a structure cut free from the ground. If the base of the structure is subjected to deformations {U}, forces {P} are necessary, see Fig. 4 (a):

$$[B] \{U\} = \{P\} \tag{1}$$

The matrix [B] includes stiffness, damping and mass of the structure. The elements of {P} and {U} are complex amplitudes of harmonic varying functions. All quantities may be dependent on frequency. As the Fourier analysis and synthesis can be applied to transient processes, Eq. (1) is sufficient for all kinds of vibrations.

2. At a part of the surface of the soil which is equal in size and shape to the base area of the structure the deformations {U} are applied. The forces {Q} are produced, see Fig. 4 (b):

106

$$[A] \{U\} = \{Q\} \tag{2}$$

$[A]$ denotes the stiffness matrix of the ground, the elements of $[A]$ depending on frequency.

In both steps, $\{U\}$ is a boundary displacement which is imposed at the soil surface or at the structure.

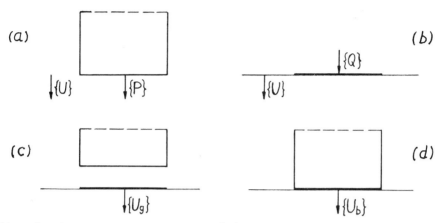

(a) *(b)* *(c)* *(d)*

Fig. 4. Substructure method: (a) structure cut free from the ground, (b) soil loaded at the surface, (c) soil subjected to free-field displacement, (d) coupled motion of soil and structure.

3. Now, the soil shall be subjected to some free-field displacement $\{U_g\}$ which is due to an event at some distance within the body of soil. Then, of course, no forces are generated at the surface of the soil, see Fig. 4 (c).

4. A structure is coupled with the soil which moved with $\{U_g\}$ before. After coupling, the common contact area moves with $\{U_b\} \neq \{U_g\}$, see Fig. 4 (d). Then, the forces produced at the structure

$$[B] \{U_b\} = \{P_b\} \tag{3}$$

and the forces produced at the soil surface

$$[A] (\{U_b\} - \{U_g\}) = \{Q_b\} \tag{4}$$

must be in equilibrium:

$$[B] \{U_b\} + [A] (\{U_b\} - \{U_g\}) = 0 \tag{5}$$

The free-field displacement $\{U_g\}$ is known, the matrices $[A]$ and $[B]$ can be evaluated in two separate analyses. Then, equation (5) is solved with respect to $\{U_b\}$.

3.2.2 Interaction stiffness matrix for use in a structural stiffness analysis

With Eq. (5), the interaction problem is solved in principle. It is convenient, however, to express Eq. (5) in terms suitable for a stiffness analysis. In order to account for the interaction effect, the stiffness matrix

$$\begin{bmatrix} A & -A \\ -A & A \end{bmatrix} \begin{Bmatrix} U_b \\ U_g \end{Bmatrix} = \begin{Bmatrix} P_b \\ P_g \end{Bmatrix} \tag{6}$$

where the submatrix $[A]$ is defined by Eq. (2), has to be assembled to the stiffness matrix of the structure. The upper equation of Eqs. (6) is identical to Eq. (4), as can easily be seen. After assembling, the forces at the points b are set equal to zero, of course. As $\{U_g\}$ is prescribed when solving the assembled system of equations, $\{P_g\}$ can be calculated. $\{P_g\}$ are the forces in the soil-structure interface.

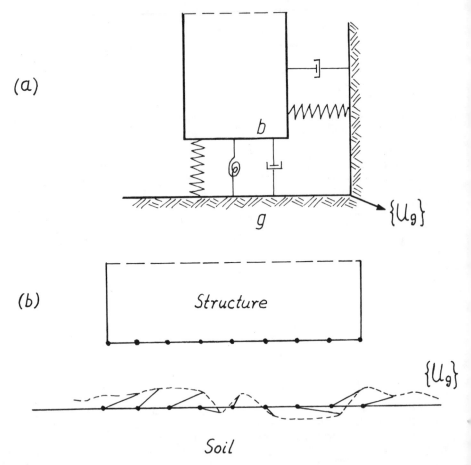

Fig. 5.   Soil structure systems: (a) rigid base (b) flexible b

The geometric model underlying Eq. (6) is shown in Fig. 5a. The springs and dashpots represent the stiffness matrix of the ground [A]. The springs and dashpots are extended between nodal points with index b at the soil-structure interface and the boundary g, which is deformed by {Ug}. Though the distance between the interface b and the boundary g corresponds to no real distance, the model represents the soil-structure interaction, as has been shown above.

The vectors {Ub} and {Ug} have as many components as are necessary for the description of the deformation of the base slab. If the base of the structure is rigid, there generally are six independent components of deformation and [A] is a six by six matrix. The coupling method is not restricted to rigid bases, (Holzlöhner 1972). Fig. 5b shows how to proceed if the base slab is flexible. The points marked at the contact area of the base slab are nodal points of the discrete system of the structure. Each node generally has three, in plane problems two degrees of freedom. The size of the stiffness matrix is determined by the total of the number of degrees of freedom. Gutierrez and Chopra (1975) have analyzed a structure applying the substructure method where the structure foundation interface possessed 12 degrees of freedom.

3.2.3 Determination of the stiffness matrix of the ground

The force-displacement relationship of a mass-less rigid plate fixed at the surface of the ground is often expressed in the form, (Richart et al. 1970)

$$W = \frac{P\ r_o}{G}\ (f_1 + if_2). \qquad (7)$$

This equation relates, e.g., the vertical amplitude W to the vertical force P, the radius r_o of the circular contact area and the constants G and ν of the elastic half-space. The complex flexibility $f_1 + if_2$, or compliance function, is a function of the frequency parameter

$$a_o = \frac{\omega\ r_o}{v_s} \qquad ; \qquad v_s = \sqrt{\frac{G}{\rho}} \qquad (8)$$

where ω = angular frequency, v_s = shear wave velocity and ρ = density. Lysmer and Kuhlemeyer (1969) use the functions F_1 and F_2 defined by

$$F = \frac{4}{1 - \nu}\ f,$$

where ν = Poisson's ratio. In Fig. 6, F_1 and F_2 are plotted versus the frequency ratio a_o · F_2, being the imaginary part of F can be taken as the damping. Fig. 6 shows that the damping is dependent on frequency and that it is high compared to damping values in structures which normally amount only to a few percent of the critical. As Fig. 6 shows the results of an analysis where an elastic half space has been assumed, the high damping of the soil is due only to radiation.

109

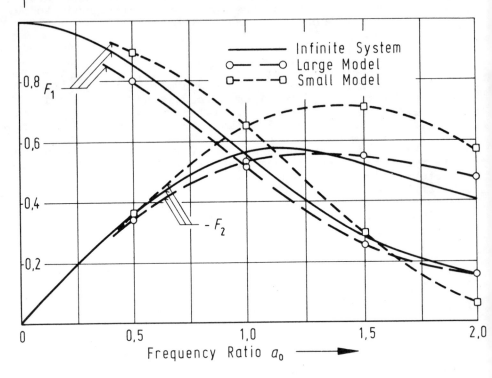

Fig. 6. Compliance function for vertical vibrating mass-less rigid disc on half space. Analytic solution and finite element approximations (From Lysmer and Kuhlemeyer 1969).

Some authors, e.g. Luco and Hadjian (1974), prefer impedance functions which are the reciprocal of the compliance functions

$$k + ia_0c \sim \frac{1}{F_1 + iF_2} \qquad (9)$$

The impedance functions form the elements of the stiffness matrix [A]. Closed form solutions for impedance or compliance functions exist for all degrees of freedom of a rigid plate on the elastic half space, on visco-elastic half space and on some special layered systems. In some cases, the functions can be "lumped" to constant stiffness and damping values, as has been shown by Whitman and Richart (1967) and Woods (1977). For the evaluation of the impedance functions, analytical or numerical methods, e.g. the FEM, or experimental methods can be used.

4. INTERACTION EFFECTS

4.1 Qualitative considerations

The main effects of interaction can be easily understood by studying the behaviour of simple systems as those shown in Fig. 7.

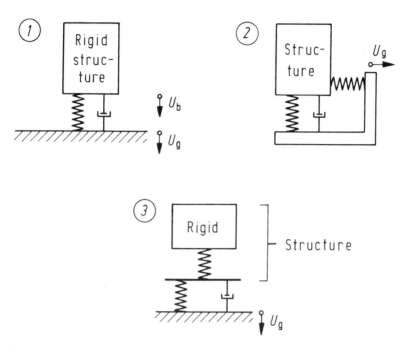

Fig. 7. Simple structures on deformable ground.

The rigid structure ① is able to vibrational motion only if the interaction springs are introduced to the system. The horizontal free-field displacement induces rocking to the soil-structure model ②. Without the interaction springs no rocking motion would be possible. Similarly, asymmetric structures can be excited to torsional vibrations. Without interaction damping, model ③ would display infinite amplitudes at resonance. If there is only little structural damping, the damping of the soil which is always present is important.

4.2 Behaviour of typical structures

Almost all the results, presented here, refer to the seismic response of structures. Seismic excitation consists of components of different frequency. Soil-structure interaction affects both the resonant frequencies and the maximum amplitudes

of the systems. The presentation will have to distinguish between the interaction effects on these two quantities.

4.2.1 Rigid blocks

The simplest structure is the rigid block indicated by (1) in
Fig. 7. If the spring and the dashpot are included in the complex spring $\bar{c}$, the interaction effect can be expressed by the
ratio

$$\frac{U_b}{U_g} = \frac{\bar{c}}{\bar{c} - m\omega^2} \tag{10}$$

where m = mass of the rigid block. In the case of the elastic
half space, Eq. (10) specializes to, (Bycroft 1977),

$$\left|\frac{U_b}{U_g}\right| = \frac{1}{\sqrt{(1 - f_1 ba_o^2)^2 + (ba_o^2 f_2)^2}} \tag{11}$$

where the mass ratio b is defined by

$$b = \frac{m}{\rho r_o^3} \tag{12}$$

For consistency, the sign of f_1 had to be changed in Eq. (11)
because, here, f_1 is defined positive when $a_o = o$ and $c > o$.
Eqs. (10) and (11) are only valid for stationary harmonic
vertical or horizontal translation.
    According to Eq. (11), the motion of the mass can be great
or smaller than the motion of the ground. Bycroft (1977) show
that the response of a rigid mass to an impulse always decrea
with increasing b and a_o, where the angular frequency corresponding to the half sine impulse is inserted in a_o.

4.2.2 Single-story buildings

Parmelee and Wronkiewicz (1971, 1973) consider the simple mod
Fig. 8, which represents a typical single story building wher
subjected to earthquakes. Fig. 9 shows the effect of the inte
action on the vibrational behaviour of the building. The uppe
curves show the increase of the fundamental period T_v of the
system with decreasing shear wave velocity v_s of the soil. T_r
is the fundamental period of the structure without any interaction. Note that for infinite shear wave velocity, i.e. rigi
soil, there is no interaction. Therefore, all curves are
asymptotic to unity when $v_s \rightarrow \infty$. The parameter essentially is
the height of the structure. The higher the building, the mor
severe are the interaction effects. The lower curves show the
decrease of the maximum displacement with decreasing shear wa
velocity. $\theta*$ is the ratio of the maximum horizontal deflectic
of the top mass to the corresponding quantity of a single mas
oscillator with rigid base having the same fundamental period
T_v.

112

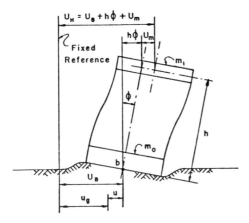

Fig. 8. Single-story building in deflected position (From Parmelee 1967).

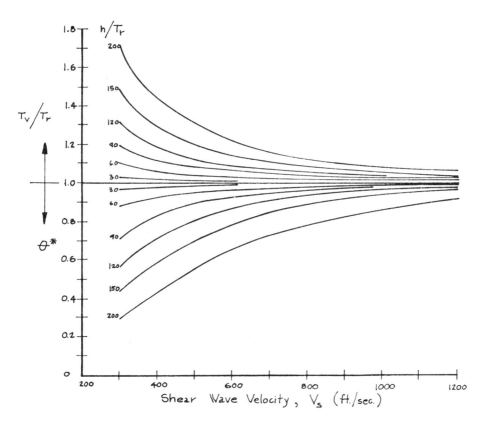

Fig. 9. Interaction effect on response of single-story buildings (From Parmelee and Wronkiewicz 1973).

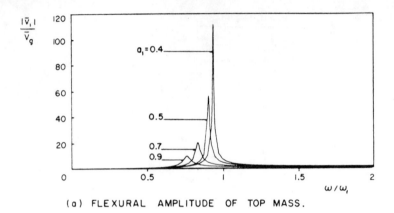

(a) FLEXURAL AMPLITUDE OF TOP MASS.

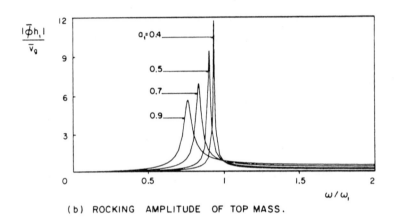

(b) ROCKING AMPLITUDE OF TOP MASS.

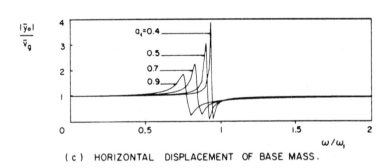

(c) HORIZONTAL DISPLACEMENT OF BASE MASS.

Fig. 10. Interaction behaviour of undamped building (From Jennings and Bielak 1972).

The results compiled in this diagram have been obtained by numerically analyzing single story structures. Some parameters have been varied within the following limits: height of the structure h = 12÷240 ft, radius of the base slab r = 14÷56 ft. The fundamental period T_v displayed values between 0.2 and 4 seconds. The structure was assumed to have viscous damping equal to one percent of the critical (η = 0.01).

The fundamental period always increases when the soil becomes softer; the tendency of the modification of the maximum amplitude, however, is dependent on the size of the damping.

Fig. 10 shows the response of a single-story building investigated by Jennings and Bielak (1972). The three degrees of freedom of the system have been considered. The dimensionless parameter a_1 is reciprocal to shear wave velocity, a_1 = 0 means rigid soil. In this case, infinite response to unit harmonic excitation results. This is due to the fact that in Fig. 10 no structural damping has been assumed, see also Sec. 4.1. The same tendency is shown by Fig. 9, where small damping has been assumed.

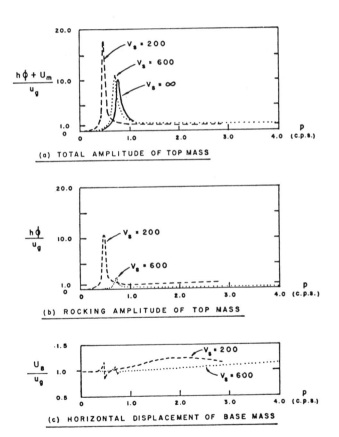

Fig. 11. Interaction behaviour of damped building, η = 0.05 (From Parmelee 1967).

115

However, higher damped structures exhibit another behaviou
Fig. 11 shows the results obtained by Parmelee (1967) where
5 percent of the critical damping has been assumed. Here, the
maximum displacements increase with decreasing shear wave
velocity. This has been confirmed by Lee and Wesley (1971).
Till now, all results refered to conventional buildings. Lee
and Wesley (1971) have also calculated the unit harmonic response of a prestressed concrete reactor vessel using the same
structural model, Fig. 8. Though the same structural damping
of $\eta = 0.05$ has been considered, Fig. 12 shows that, the maxi
amplitude of the top mass decreased with decreasing shear wav
velocity. Lee and Wesley (1971) explain this result by the
presence of a large amount of radiation damping of the reacto
vessel which is a low, heavy structure.

The results can be summarized as follows: If the ratio of
structural to radiation damping is high, the maximum displace
ment amplitudes increase with decreasing shear wave velocity,
otherwise they decrease.

4.2.3 Multi-story buildings

Jennings and Bielak (1972) have also investigated multi-story
buildings. They established the fact that only the fundamenta
frequency decreases significantly as the soil becomes softer.
For all values of the shear wave velocity the second and the
higher resonant frequency did not differ by more than one per
cent from the corresponding frequency of the rigidly founded
buildings. Chu, Agrawal and Singh (1973) have investigated a
multistory nuclear power plant. Fig. 13 shows the response
spectra of the basement and of the third floor to a special
earthquake. The solid line refers to the interaction model an
the dashed line to the rigidly founded structure. The differe
between the responses is more severe for low frequencies. The
amplitudes may be reduced or increased by considering the int
action.

4.3 Traffic induced vibrations

Whereas the interaction effects on seismic responses have oft
been investigated, there is not much information about vibrations induced by other sources, e.g. by traffic. Compared to
seismic excitations, traffic induced vibrations contain highe
frequencies which excite single structural members in their
resonant frequencies. The vertical component of the free-fiel
displacement now is very important. A typical problem is that
vibrations of the floor slabs of residential buildings exceed
admissible values so that the convenience of the inhabitants
is affected. Preliminary investigations showed, (Holzlöhner a
Rücker 1978), that the modification of the maximum displaceme
amplitudes follow the same trends as under seismic excitation

For large values of the frequency factor a_o, approximate
solutions for the compliance functions of the elastic half-sp
have been given by Bycroft (1977).

This paper is restricted to structures with shallow founda

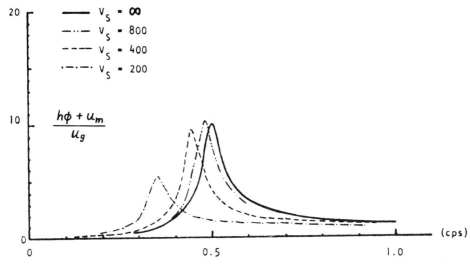

Fig. 12. Amplitude of top mass of single-story prestressed concrete reactor vessel, $\eta = 0.05$ (From Lee and Wesley 1971).

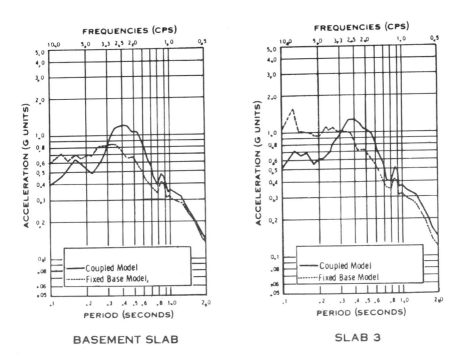

BASEMENT SLAB　　　　　　SLAB 3

Fig. 13. Response of multi-story nuclear power plant to earthquake (From Chu, Agrawal and Singh 1973).

Other important fields of interaction as the behaviour of structures on piles, dams and earth retaining structures are not studied here.

5. COMPARISON OF METHODS OF ANALYSIS

5.1 Summary

In section 3, two methods of analysis suited for soil-structure interaction have been introduced. Here, one method has been called "direct method", the other "coupling method" or "substructure method".

In the last years, there has been a controversy about the advantages and the limitations of the two methods. The direct method has been propagated especially by Seed, Lysmer and Hwang (1975, 1977) whereas the coupling method has been recommended by Hadjian, Luco and Tsai (1974) and Hadjian (1976). The result of this controversy was that both methods have been developed to a degree that today they both will yield good solutions for most practical problems. In the following, it is summarized how the two methods can be applied to typical problems today (1977).

Direct method   Substructure or coupling method

Dimensionality:

Two-dimensional analysis;   Three-dimensional
three-dimensional effects
are approximately covered
by additional damping

Inhomogeneity, dependency of the modulus on strain:

Arbitrary layered soils   Half space solutions are
and continuous variation   available only for some one-
of material properties   or two -layer systems. How-
can be considered   ever, the impedance functions
                                  can be calculated by FEM

Embedded structures, flexible base:

No problems   Embedment can be approximately
                                  considered

High frequencies:

The higher the frequencies   No problems
the smaller the required
element size

Damping:

Radiation damping and material damping of the soil can be considered in both analysis methods.

5.2 Comparison of impedance functions

It seems to be necessary to give some comments on certain
assertations which are found in the references and which are
not correct any longer. Luco et al. (1974) compare impedance
functions calculated by different methods. In Fig. 14 their
results are given for the vertical vibration of a rigid strip.

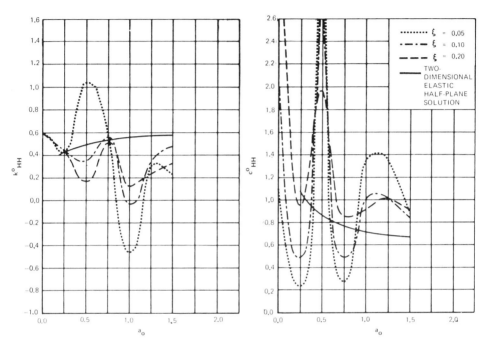

Fig. 14. Impedance functions for rigid strip footing. Analytic
and various finite element solutions (From Luco et al. 1974).

The solid lines give the results obtained with half space
theory. The dotted and dashed lines correspond to finite element solutions. From the poor agreement of the curves the
authors conclude that because of the reflections from the
boundaries of the model the FEM is not suited to the calculation of impedance functions. However, Lysmer and Kuhlemeyer
(1969) and Waas (1972) have developed special boundary elements which largely suppress reflections. Lysmer and Kuhlemeyer
(1969) show by comparison of F_1, F_2 functions calculated by FE
and analytical methods that their finite element model sufficiently represents the half space, see Fig. 6.

5.3 Limitations of the direct method

The most important limitations of the direct method are listed
by Chopra and Gutierrez (1977) as follows: The lower horizontal

119

boundary of the finite element model is usually assumed to be rigid. This is only reasonable if the boundary corresponds to a real soil rock interface. As only vertically propagating shear waves are assumed no wave propagation effects, see Section 6.2, can be taken into account. The computational effort cannot be significantly reduced by using model analysis because the modes of the soil-structure system are inefficient in representing deformations of the structure.

5.4 Embedment

On the other hand, the direct method can easily be applied to nonhomogeneous soil and embedded structures though the critici of Seed et al. (1975) is not to the point. When applying the coupling method to an embedded structure the problem arises that the free-field displacements at the soil-structure inter-face cannot be used as input but have to be calculated by a preceding analysis, taking the soil to be excavated as a struc-ture, (Holzlöhner 1972). Kausel et al. (1977) have presented the conditions for consistency of the results obtained by the direct method and the substructure method and they have deve-loped an approximative method for the evaluation of the sub-grade stiffness in case of embedment. They have got good re-sults with the approximative method. Another approximative solution is given by Hall and Kissenpfennig (1975). They distribute the horizontal stiffness along the vertical con-tact surfaces of embedded base, Fig. 15. The free-field dis-placement is applied through an additional two-mass-system. The author of this paper, however, finds it sufficient to apply the different free-field motions through the lateral springs of the model. Hall and Kissenpfennig (1975) compare the results obtained by this method to those obtained by a finite element analysis utilizing the computer program LUSH, (Lysmer et al. 1974). It can be seen in Fig. 16 that, even if the structure is deeply embedded, the responses of some charac-teristic points obtained by different analysis do not differ much for frequencies up to 10 cps.

Hadjian, Howard and Smith (1975) have investigated the eff of embedment on the stiffness of the interaction springs. Expe ments have been performed on a shaking table and, at larger scale, directly in the soil. Different methods of analysis ha been utilized for interpretation. Fig. 17 show the results: A long as the embedment depth does not exceed half the base dia-meter, embedment need not be considered, according to the authors. In cases of deeper embedment, simple models, that means the coupling method, can be utilized.

Veletsos and Verbic (1973) and Veletsos and Nair (1975) ha solved some vibrational problems of the elastic half space in-cluding viscous or hysteretic damping. The result are impeda functions which may be used with the coupling method.

5.5 Conclusions

To conclude the comparison of the two methods - coupling meth

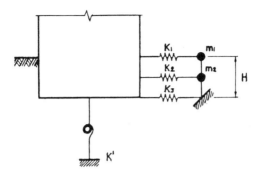

Fig. 15. Interaction model of embedded structure (From Hall and Kissenpfennig 1975).

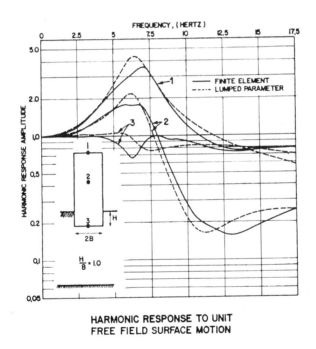

HARMONIC RESPONSE TO UNIT
FREE FIELD SURFACE MOTION

Fig. 16. Response of embedded structure (From Hall and Kissenpfennig 1975).

and direct method – it can be stated that both methods lead to reasonable good results in most practical problems if properly applied. The direct method seems to be the more flexible one, every practical case can be analyzed at least in principle, that is, if the analysis is not too expensive. The coupling method can more economically be utilized and actually most of the published parameter studies have been performed in this way. One should always keep in mind that the numerical method need not be more accurate than the input data. The dependency of soil stiffness and damping on strain is a widely scattering relationship according to published data, as mentioned by Hadjian et al. (1974).

6. SPECIAL INTERACTION EFFECTS

6.1 Base slab uplift

During strong earthquakes, the rocking motion may be so intens that part of the base slab looses contact with the soil, Fig.1 That is a nonlinear process because the contact area decreases with increasing overturning moment. Kennedy et al. (1975) take into account the uplift by modifying the interaction springs for horizontal translation and rocking according to

$$K_H' = K_H \cdot \sqrt{\frac{d'}{d}} \qquad (13)$$

$$K_R' = K_R \cdot \left(\frac{d'}{d}\right)^2$$

where d' is the part of the width d of the base still in contact with the soil and K_H' and K_R' are the modified horizontal and rocking stiffnesses, respectively. K_H and K_R are the stiff ness of full contact. It can be seen from Eq. (13) that the rocking stiffness decreases rapidly with increasing uplift.

Kennedy et al. (1975) analyzed a HTGR cointainment building with and without consideration of uplift. Soil accelerations of 0.3 and 0.5 g have been assumed and three different shear wave velocities, v_s = 1800,580 and 327 m/s have been considere The results of the analysis showed that the uplift hardly affects the vibrational behaviour though the structure soil interface has been reduced by uplift between 33 and 88 percent

The same problem has been investigated by Wolf (1975). He obtained similar results. The relative displacement increased and the stresses decreased somewhat if the uplift was consider

6.2 Wave propagation effects

In most seismic analyses, it is assumed that the seismic excitation propagates vertically upwards. Then, the free-field motion at the surface is synchronous. In reality, the seismic waves propagate in such a way that a propagation velocity depending on the properties of the site can be observed at the surface. This leads to phase differences of the free-field

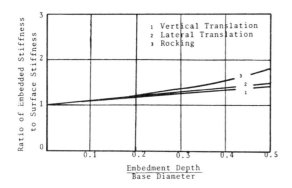

Fig. 17. Effect of embedment on stiffness of interaction springs (From Hadjian et al. 1975).

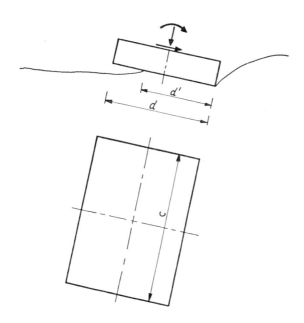

Fig. 18. Uplifted base due to strong-motion earthquake (After Kennedy et al. 1975).

displacement within the soil structure interface. This phenomenon is even more pronounced with traffic induced vibrations.
    Holzlöhner (1972) has shown how the free-field excitation
of a soil-structure interaction system can be calculated by an
averaging procedure from the free-field displacements. The
method leads to good results as has been shown in an example.
Scanlan (1975) assumed a rigid plate with distributed springs
which were passed by a seismic wave. His calculations lead to
the same modification factor as those previously mentioned.

6.3 Effects of adjacent structures

Savidis and Richter (1977) calculate the interaction between
two rigid rectangular footings placed at the surface of the
half space by using the analytic solution of Holzlöhner (1969)
Until now, the method of Savidis and Richter (1977) are
restricted to the cases that one or two footings are directly
loaded by a vertical force or a rocking moment. The vertical
displacements and the rotations about the horizontal axes are
calculated.
    If horizontal motion dominates as in seismic responses, the
direct method can be applied. A parametric study is being
undertaken in Germany at the BAM, Berlin.

7. REFERENCES

Bycroft, G.N. 1977, Soil-Structure Interaction at Higher Fre-
    quency factors, Earthq. Eng. and Struct. Dyn. 5: 235-248.
Chopra, A.K. and J.A. Gutierrez 1977, Earthquake Analysis of
    Structures Including Structure-Soil Interaction by a Sub-
    structure Method, 4th Int. Conf. on Struct. Mech. in Reac-
    tor Techn., San Francisco, Paper K 2/8.
Chu, S.L., P.K. Agrawal and S. Singh 1973, Finite element
    treatment of soil-structure interaction problem for nuclear
    power plant under seismic excitation, 2nd Int. Conf. on
    Structural Mechanics in Reactor Technology, Berlin, German
    Sept., Paper K 2/4.
Gutierrez, J.A. and A.K. Chopra 1975, Earthquake Analysis of
    Nuclear Reactor Buildings Including Foundation Interaction,
    Transact. of the 3rd Conf. on Struct. Mech. in Reactor
    Techn., London, Paper K 3/4.
Hadjian, A.H. 1976, Discussion, Journ. Geot. Eng. Div., ASCE,
    GT4: 380-384.
Hadjian, A.H., G.E. Howard and C.B. Smith 1975, A Comparison
    of Experimental and Theoretical Investigations of Embedmen
    Effects on Seismic Response, Transactions of 3rd Intern.
    Conf. on Structural Mechanics in Reactor Technology, Londo
    1-5 Sept., Paper K 2/5.
Hadjian, A.H., J.E. Luco and N.C. Tsai 1974, Soil Structure
    Interaction: Continuum or Finite Element? Nuclear Eng. and
    Design, Vol. 31, No.2: 151-167.
Hall, J.R.,Jr. and J.F. Kissenpfennig 1975, Special Topics on
    Soil-Structure Interaction, Paper U2/2, Extreme Load Con-
    ditions and Limit Analysis Procedures for Structural React

Safeguards and Containment Structures Seminar, Sept. 8-11, Berlin, Germany.

Holzlöhner, U. 1969, Schwingungen des elastischen Halbraums bei Erregung auf einer Rechteckfläche, Ing.Archiv, 38: 370-379.

Holzlöhner, U. 1972, A Contribution to Soil-Structure Inter-action, CREST Specialist Meeting on Antiseismic Design of Nuclear Power Plants, Pisa, Oct.

Holzlöhner, U. and W. Rücker, Structures Subjected to Traffic Induced Vibrations, will be published shortly.

Jennings, P.C. and J. Bielak 1972, Dynamics of building soil interaction,EERI 72-01, California Institute of Technology, Pasadena.

Johnson, G.R., P.P. Christiano and H.I. Epstein 1975, Stiffness Coefficients for Embedded Foundations, Journal of the Geo-technical Engineering Division, ASCE, Vol. 101, GT8, August.

Kausel, E., R.V. Whitman, F. Elsabee and J.P. Morray 1977, Dynamic Analysis of Embedded Structures, 4th Int. Conf. on Struct. Mech. in Reactor Techn., San Francisco, Paper 2/6.

Kennedy, R.P., S.A. Short, D.A. Wesley and T.H. Lee 1975, Non-linear Soil-Structure Interaction Due to Base Slab Uplift on the Seismic Response of a HTGR Plant, Transactions of the 3rd Intern. Conf. on Struct. Mech. in Reactor Techn., London, Paper K 3/5.

Lee, T.H. and D.A. Wesley 1971, Soil-foundation interaction of reactor structure subject to seismic excitation, 1st Int. Conf. on Structural Mechanics in Reactor Technology, Berlin Germany, Sept. Paper K 3/5.

Luco, J.E., A.H. Hadjian and H.D. Bos 1974, The Dynamic Modeling of the Half Plane by Finite Elements, Nuclear Engineering and Design, Vol. 31, No. 2: 184-194.

Lysmer, J. and R.L. Kuhlemeyer 1969, Finite Dynamic Model for Infinite Media, Journal of the Engineering Mechanics Division ASCE, Vol. 95, No. EM4, August.

Lysmer, J., T. Udaka, H.B. Seed and R. Hwang 1974, LUSH: A Com-puter Program for Complex Response Analysis of Soil Struc-ture Systems, Report Nr. EERC 74-4, Earthquake Engineering Research Center, University of California, Berkeley, April.

Lysmer, J., H.B. Seed, T. Udaka, R.N. Hwang and C.-F. Tsai 1975, Efficient Finite Element Analysis of Seismic Structure-Soil-Structure Interaction, Report No. EERC 75-34, Earthq. Eng. Res. Center, University of California, Berkeley, November.

Parmelee, R.A. 1967, Building-foundation interaction effects, J. Eng. Mech. Div. ASCE, 93 (EM 2) Apr: 131-152.

Parmelee, R.A. and J.H. Wronkiewicz 1971, Seismic Design of Soil-Structure Interaction Systems, Journ. of the Struct. Div., ASCE, ST 10: 2503-2517.

Parmelee, R.A. and J.H. Wronkiewicz 1973, Dynamic Coefficients for Evaluating the Seismic Response of Soil-Structure Inter-action Systems, 5th World Conf. on Earthq. Eng., Rome: 2606-2609.

Richart, F.E., Jr., J.R. Hall and R.D. Woods 1970, Vibrations of Soils and Foundations, Prentice-Hall, Inc., New Jersey.

Savidis, S.A. and T. Richter 1977, Dynamic Interaction of Rigid Foundations, 9th ICSMFE, Tokyo.

Scanlan, R.H. 1976, Seismic Wave Effects on Soil-Structure Interaction, Earthq. Eng. a. Struct. Dyn. 4: 379-388.

Seed, H.B., J. Lysmer and R. Hwang 1975, Soil-Structure Inter-
    action Analyses for Evaluating Seismic Response, Journal of
    the Geotechnical Engineering Division, ASCE, Vol. 101,
    No. GT5, May.
Seed, H.B., J. Lysmer and R. Hwang 1977, Discussion,GT4: 341-34
Veletsos, A.S. and V.V.D. Nair 1975, Seismic Interaction of
    Structures on Hysteretic Foundations, Journal of the Struc-
    tural Division, ASCE, Vol. 101, No. ST1, Proc. Paper 11460,
    Jan: 109-129.
Veletsos, A.S. and B. Verbic 1973, Vibrations of Viscoelastic
    Foundations, Earthquake Engineering and Structural Dynamics,
    2: 87-102.
Waas, G. 1972, Linear Two-Dimensional Analysis of Soil Dynamics
    Problems in Semi-Infinite Media, Thesis presented to the
    University of California, Berkeley, in partial fulfillment
    of the requirements for the degree of Doctor of Philosophy.
Whitman, R.V. and F.E. Richart, Jr. 1967, Design Procedures for
    Dynamically Loaded Foundations, Journal of the Soil Mechanic
    and Foundations Division, ASCE, Vol. 93, No. SM6, Paper 5569
    Nov.: 169-193.
Wolf, J.P. 1975, Approximate Soil-Structure Interaction with
    Separation of Base Mat from Soil (Lifting-off), Transact.
    3rd Intern. Conf. on Struct. Mech. in Reactor Techn.,
    London, Paper K 3/6.
Woods, R.D. 1977, Lumped Parameter Models, Advanced Study Inst.
    on Dynamical Methods in Soil and Rock Mechanics, Karlsruhe.

Dynamic interaction between adjacent structures

JOSÉ M. ROESSET
Massachusetts Institute of Technology, Cambridge, Mass., USA

JOSÉ J. GONZALEZ
INTECSA, Madrid, Spain

SYNOPSIS

The interaction, through the underlying soil, between two rigid masses or two structures idealized as simple one-degree-of-freedom systems is studied using a finite element type formulation based on the consistent boundary developed by Waas and extended by Kausel. Both the case of a harmonic external force applied at one of the masses and the case of a base motion are considered. It is found in both cases that the interaction effects increase when the two structures have the same natural frequency, when their masses increase and as the distance between them decreases. The most significant effect is in all cases a change in the natural frequencies of the soil structure system.

INTRODUCTION

The vibrations of soils and soil deposits and the dynamic interaction between soils and structures have received considerable attention in recent years. There are a variety of practical engineering problems associated with this general area: the design of foundations for vibrating machinery and the response of buried structures to blast loading were among the first ones to be studied; the effect of local soil conditions on the characteristics of earthquake motions, the stability of slopes and earth dams under seismic excitation (including the possibility of liquefaction) and the assessment of vibrations caused by transit systems were all considered later. In the last few years, and because of the need to perform a seismic analysis for all nuclear power plants, the earthquake response of structures accounting for the flexibility of the foundation (soil-structure interaction) has been a problem of considerable interest and research.

A substantial amount of work has been done in all phases of the soil-structure interaction problem, from the determination of the motions at the foundation level, before the structure is built, for different types of waves (of particular importance for embedded foundations) to the development of analytical or semi-analytical solutions for the foundation stiffnesses as a function of frequency (strip footings, circular foundations and rectangular foundations can now be studied on the surface of a horizontally stratified soil deposit), the derivation of simplified formulas from parametric studies, the consideration of nonlinear effects due to the soil behavior or to the possible separation of the mat from the soil, and the implementation of general computer programs to solve the complete problem for some idealized situations (normally with very simple models of the structure).

Two general approaches are used now for the solution of soil-structure interaction problems:

— A one-step or direct approach, in which the soil and the structure are analyzed together, modelling them through finite elements (or finite differences) and linear members. This procedure has a clear theoretical advantage if inelastic behavior, particularly in the soil, is to be accounted for through a step-by-step numerical integration of the equations of motion in the time domain. The advantage is hampered by the fact that the input motion must be specified at the base of the model, where it is not known a priori. When the design earthquake is specified at the free surface of the soil deposit, as is now normally the case, a deconvolution is necessary to obtain first a compatible motion at bedrock.

— A three-step approach also referred to as the substructure or spring method. In this case the first step is the determination of the seismic motion at the foundation level, considering a rigid but massless foundation (for an embedded structure the motion will have both translational and rotational components). The second step is the determination of the dynamic stiffnesses of the foundation, complex functions of the frequency. Finally the dynamic analysis of the structure resting on frequency dependent "springs" as obtained in the second step is carried out for the base motions computed in the first. This

128

procedure implies the validity of linear superposition and is therefore restricted in rigor to linear analyses or studies in which non-
linearities are simulated through an equivalent linearization. It
offers on the other hand considerably more flexibility in the way each
step is handled and it is particularly suited to parametric studies.

The advantages and disadvantages of each one of these two approaches
and the possibility of obtaining with either sensible results, when properly
implemented, has been extensively discussed, although some controversy seems
to persist, unfortunately, on the validity or adequacy of each method.

HISTORICAL REVIEW

It has long been recognized that no building stands alone, and that
the presence of neighboring structures may affect its dynamic response,
particularly under a seismic excitation, but very little work has been
.done to date to determine in a systematic way the importance of these
interactions. The problem is complicated by the large number of parameters involved: the number of structures, the relative position in space,
their size, mass and stiffness and the characteristics of the soil profile. The interaction of two structures through the underlying, or surrounding, soil is not only of interest in seismic analyses but also in
the assessment of potential damage to very sensitive structures due to
vibrations induced from adjacent buildings (vibrating machinery, wind
loads, etc.).

Richardson (1969)   studied the case of two rigid cylinders (with
the same dimensions) resting on the surface of an elastic half-space when
one of them was excited by external harmonic forces. This solution was
based on an analytical formulation using an averaging procedure to determine the motions of each cylinder from the free field displacements. With
the two cylinders at a distance of ten radii between centers, Richardson's
results indicate that the direct effects in the excited mass (vertical
displacement due to a vertical load) are only slightly affected by the
presence of the second mass, but that new effects appear which would not
have occurred had the mass been alone (horizontal displacements and rotations under a vertical load). Under this type of excitation the vertical
displacement of the second, passive mass, is about 20% of that of the
first with a similar variation with frequency. The horizontal displace-

ment and the rotation of the second mass are, however, much larger than those of the first. This result is quite logical if one takes into account that these effects are induced first in the passive mass and then transmitted, through a feedback mechanism, to the active one.

Chang Liang (1974) presented the results of a more extensive set of parametric studies on the interaction effects between two rigid masses resting on the surface of a soil layer of finite depth (underlain by rigid rock). A two-dimensional (plane strain) problem was considered and the solution was obtained using a finite element type formulation with the consistent boundary originally developed by Waas (1972). Both the case of a harmonic force (a horizontal force or a rocking moment) applied to one of the masses, and the case of a base motion affecting the two masses, and simulating a seismic type excitation, were studied.

For the first case and within the range of parameters studied (values of the masses, relative distances between their centers and stratum depths), Chang Liang concluded that the interaction effect between the two masses seemed to be less important for very shallow layers of soil (resting on rigid rock) than for layers of moderate thickness. In all cases the existence of a second mass at distances of 2.5 to 5 times the base width affected only slightly the direct response of the excited mass (although there were again vertical displacements that would not have occurred had the mass been alone). The significant part of this kind of study would thus be the determination of the vibrations induced in the second, unexcited mass. The horizontal displacement of this mass due to a horizontal force applied at the first is of the order of 50% that of the active mass when the stratum is relatively deep and the distance between the foundations is 2.5 times their base width; it reduces to about 20% if the stratum is very shallow or if the distance increases to five widths. This ratio is larger when the masses increase (particularly the second one). Similar results are obtained for the rotations induced by a rocking moment, but the ratios of the effects in the passive structure to those in the active one are slightly smaller (of the order of 30 to 40% for the deep stratum and the smaller distance). These observations are basically consistent with those of Richardson.

For the second case, when a base motion simulating an earthquake excitation (caused by a train of shear waves propagating vertically through the

130

soil at a specified frequency) was imposed on both masses, the main effect of the adjoining mass seemed to be a change in the natural frequencies of the system. As a consequence, if the displacements at the base of one of the masses (including the effect of the other) were divided by the corresponding result if the mass were alone, amplifications and deamplifications would take place at different frequencies. The amplifications could be of the order of 50 to 100% when the masses were heavy and the deamplifications were similarly of the order of 40 to 60%. The net effect under an actual earthquake would depend, however, on the frequency content of the specified motion. For a narrow band process interaction effects between the two masses could be very important: if most of the energy of the earthquake is around one of the frequencies where amplifications take place, a large increase in response could occur (the range of amplification is very narrow for the swaying frequency of the masses, but broader for the rocking frequency); on the other hand, if the energy was centered in the frequency range where the results show deamplifications, the effect of the adjoining mass would be beneficial. For a white noise or a wide band process, the main effect would be the appearance of small shifts in the frequencies at which the peaks of the response spectra occur. Whether the amplitude of these peaks would increase or decrease is not obvious from the results presented, but it appears that small amplifications might occur. In particular the presence of a larger and heavier structure seemed to have a detrimental effect on the response of a lighter one.

More recently Lysmer et al. (1975) have presented results for the seismic response of a nuclear power plant, including the effect of two auxiliary buildings. The formulation was based on a finite element solution by the one-step approach, using basically a two-dimensional model to which viscous dashpots are added on the lateral faces to increase the amount of damping and thus simulate radiation in the direction perpendicular to the plane. The model will have, however, the foundation stiffnesses and natural frequencies corresponding to a two-dimensional solution rather than those of the true three-dimensional problem. Furthermore, the buildings must be centered along a common axis, and their foundations must have the same width in the direction perpendicular to the plane (a serious limitation for the analysis of general three-dimensional situations).

In this study two equal auxiliary buildings were included, one on each side of the nuclear power plant (symmetrically placed). A layer of sand with a depth of 65 ft. resting on rigid rock was considered. The main building had a base width of about 2 times the stratum depth and was embedded to about one-third of this depth. The auxiliary buildings had base widths approximately equal to the stratum depth and were deeply embedded to about two-thirds of this depth. The control motion, with a maximum acceleration of 0.25g and a spectrum according to the AEC Regulatory Guide, seemed to be specified at the depth of the reactor foundation in the free field. The results presented were the acceleration response spectra for 2% oscillator damping at a point of the containment building at the level of the soil surface, considering only the main building (two-dimensional solution) and with the effect of the auxiliary buildings (for a two-dimensional and a pseudo three-dimensional solutions).

These results are particularly striking and show much stronger effects than those of previous studies. When the two-dimensional model is considered throughout, the peak of the response spectrum is amplified by a factor of almost 2.5, due to the presence of the auxiliary buildings. This would suggest that present practice, analyzing nuclear power plants under seismic excitation without due account for the presence of adjacent structures, could lead to very erroneous results and unsafe designs, a very disquieting thought. It would also indicate that the interaction of adjacent buildings is much more important than other effects which are now being studied.

The comparison of the two-dimensional solution without auxiliary buildings and the pseudo three-dimensional results shows an increase in the peak of the response spectrum of only 1.5 to 2 (still very significant), but it is not very meaningful because of the inconsistency in the models. The building alone should have been studied also with the pseudo three-dimensional model for a valid comparison.

Finally, the comparison between the normal two-dimensional solution and the one with additional viscous dashpots on the lateral faces (pseudo three-dimensional approach) shows a reduction in response which can be easily explained by the increase in damping provided in the latter.

An interpretation of these results is difficult without the complete data on the structures (stiffnesses and masses) and on the motion characteristics. It is possible on one hand that the interaction effects be so marked because the input at the base of the building has a large peak at a specific frequency, or because the parameters are not realistic for

132

typical structures (they would increase with very large masses). On the other hand, the very deep embedment of the auxiliary buildings may cause a box-type effect and decrease the effective radiation damping (it would be necessary to assess first what the soil-structure interaction effect is for the containment building alone and what is the amount of effective damping due to radiation). This is a case where a solution using the three-step approach would have greatly facilitated the identification of the factors contributing to the increase in response.

It seems, in any case, that because of these results, more studies are needed to improve present knowledge on the nature and magnitude of structure-soil-structure interaction effects.

SCOPE

In this work a formulation is presented for the general three-dimensional solution of structure-soil-structure interaction problems with the three-step approach. Any number of structures with foundations of arbitrary shape can be considered. The main two limitations are that the structures are founded on the surface (the method could be extended to embedded foundations, but it would become much more expensive computationally) and that a horizontally stratified soil deposit of finite depth (resting on much harder, rock-like material) must be considered. The formulation is essentially an extension of the procedure used in the studies of Chang Liang (1974) to the three-dimensional case.

The formulation is described first and evaluated by comparing the free field displacements due to concentrated loads on the surface of a layer of - finite depth to results of a semi-analytical solution suggested by Gazetas (2). The method proposed here furnishes results which are in good agreement with the analytical solution at considerably less computational expense. In addition, the dynamic stiffnesses of a square foundation are obtained and compared to those of an equivalent circular footing, and the convergence characteristics of the method are investigated as a function of the mesh size.

Results are then presented for the cases of two rigid masses and two simple structures resting on the surface of a homogeneous soil deposit with a depth equal to twice the foundation width. In order to reduce the number of parameters, both foundations are assumed to be square, of equal size, parallel and centered along a common horizontal axis. The values of

the masses, the natural frequencies and the distance between the foundations
are varied. Both the cases of a harmonic excitation applied to one of the
masses or structures and of a base motion (simulating an earthquake) are
considered. The results are presented in dimensionless form: the displacements are multiplied by the shear modulus of the soil and the side of the
foundation, and divided by the value of the applied force (they correspond
to unit values of these three quantities). They are plotted versus a dimensionless frequency which is the actual frequency in cycles per second (Hz)
multiplied by the size of the foundation and divided by the shear wave velocity of the soil. (Notice that this is not exactly the same dimensionless
frequency used in other studies where the frequency in radians per second
is multiplied by the radius of a circular foundation, or half the size of
a square one, and divided by the shear wave velocity. A factor of π should
be applied to the values used here to obtain the other).

FORMULATION

The starting point for the proposed formulation is the determination
of the displacements at any point on the surface of the soil deposit due
to a harmonic unit horizontal force x, a unit horizontal force y, and a unit
vertical force z applied at the origin, as a function of the frequency of
excitation. A semi-analytical type solution for this problem was suggested by
Gazetas (1975) using a double Fourier expansion in the x-y plane. In this
work the solution is based on a finite element type formulation using the
consistent boundary developed by Waas (1972), extended by Kausel (1974) to the
three-dimensional case. It is important to notice that contrary to what has
been sometimes reported (see, e.g. Lysmer, 1974) the use of this boundary matrix
is not restricted to the solution of axisymmetric problems. When part of the
soil profile or the structure are reproduced by axisymmetric (toroidal)
finite elements, then the geometry of the problem must be indeed axisymmetric,
although the loads or excitation can have any distribution expanding them
in a Fourier series along the circumference (the approach normally used
for the solution of shells of revolution under arbitrary loadings). This
restriction becomes meaningful when the soil-structure interaction problem
is to be solved in a single step but no longer applies if the three-step
solution is used. All that is necessary is to express the coordinates of
the points where displacements are desired in cylindrical coordinates and
to refer the radial and tangential displacement to orthogonal cartesian
components.

134

Figures 1 and 2 show the horizontal and vertical displacements at the free surface of a soil layer of finite depth due to a unit horizontal or vertical force at the origin, respectively. The abscissa is the distance to the point of application of the load divided by the stratum thickness H. To evaluate these results it must be taken into account that in the present solution a concentrated load is considered within the accuracy of the finite element method. In the semi-analytical solution, on the other hand, the load is expanded in a double Fourier series using a discrete Fourier transform. The solution corresponds then to a rectangular pulse, or a load uniformly distributed over a rectangle, with sides equal to 1/6 of the stratum depth for the case shown. One can expect therefore that in the immediate neighborhood of the point of application of the load, the results provided by the finite element model will be slightly larger (the actual condition of a point load is better modelled). A second parameter which affects the accuracy of Gazetas' solution is the number of points used for the double Fourier expansion. The results shown are for a mesh with 64 equally spaced points in each horizontal direction. Considering these differences the agreement in the results seems very good. The curves shown are for frequencies of 0.1 C_s/H and 0.4 C_s/H, where C_s is the shear wave velocity of the soil and H is the stratum depth. They are typical of those obtained for a number of frequencies studied.

For the problem at hand the foundation or foundations considered are discretized by a grid of equally spaced points in the x and y directions (Figure 3). (The method would accept of course unequal spacing of the points, but in this work only square foundations were treated). Assuming that each side of a foundation is divided into N equal segments, there are $(N+1)^2$ points for each foundation and a total of $2(N+1)^2$ points (if both foundations are equal). By considering a unit force applied at one of these points and determining the displacements u, v and w at all the others, one can form a column of a global flexibility matrix F of dimensions $6(N+1)^2$ by $6(N+1)^2$ (since there are three displacement components at each node), or in general $3N_1 + 3N_2$ if N_1 is the total number of points in the first foundation and N_2 the corresponding quantity in the second. It is important to notice that it is not necessary to repeat the computation for each point (or column of the flexibility matrix), since many of the coefficients can be derived from the others by a simple shift or by applying symmetry and antisymmetry conditions.

135

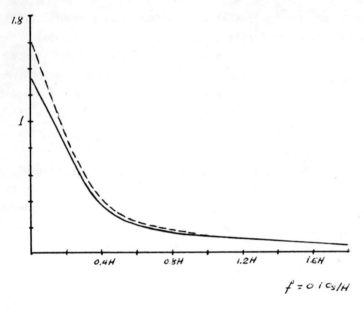

$f = 0 \ i \ c_s/H$

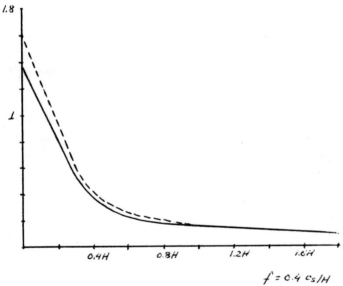

$f = 0.4 \ c_s/H$

Figure 1. Horizontal displacements due to a horizontal force.

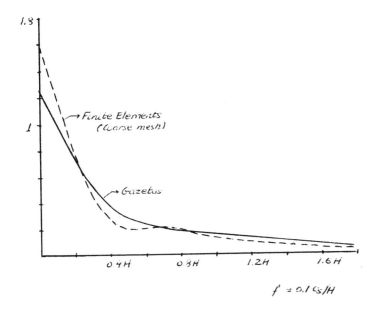

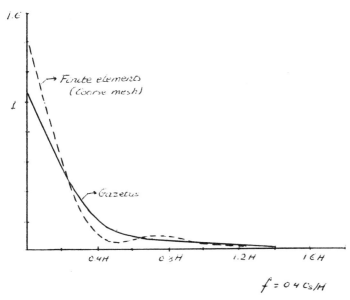

Figure 2.   Vertical displacements due to a vertical force.

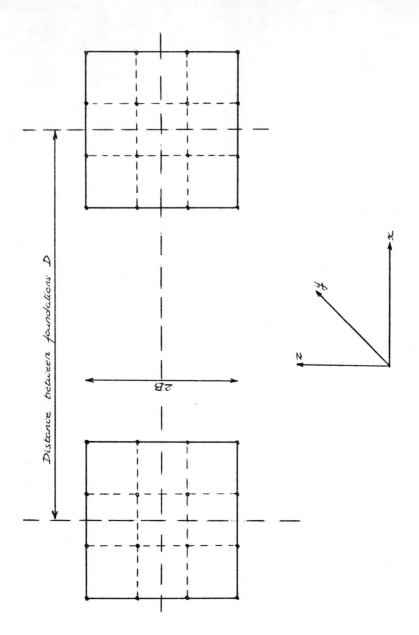

Figure 3. Basic geometry - plan view.

If P_1 represents a vector of forces (in the x, y and z directions) applied at the mesh points of the first foundation, U_1 is the vector of displacements at the same points, and P_2, U_2 are the corresponding variables for the second foundation.

$$U = \begin{Bmatrix} U_1 \\ U_2 \end{Bmatrix} = FP = F \begin{Bmatrix} P_1 \\ P_2 \end{Bmatrix}$$

Imposing now the condition of a rigid body motion for each foundation, the displacements of the mesh points can be related to those of the corresponding centroid by a transformation of the form

$$U_1 = T_1 \, u_1$$

$$U_2 = T_2 \, u_2$$

where U_1, U_2 have $3(N+1)^2$ components (or in general $3N_1$ and $3N_2$ respectively). T_1 and T_2 are matrices $3(N+1)^2$ by 6 (or in general $3N_1$ by 6 and $3N_2$ by 6), and u_1 and u_2 have six components each (three translational components along the x, y, and z axis and three rotations around each one of these axes).

The resultants of the forces applied at the mesh points of a foundation with respect to its centroid will consist of three forces and three moments. Denoting by p_1 and p_2 the vectors of these six force components,

$$p_1 = T_1^T \, P_1$$

$$\text{and} \quad p_2 = T_2^T \, P_2 \quad .$$

It is then possible to write

$$p = \begin{Bmatrix} p_1 \\ p_2 \end{Bmatrix} = \begin{bmatrix} T_1^T & 0 \\ 0 & T_2^T \end{bmatrix} F^{-1} \begin{bmatrix} T_1 & 0 \\ 0 & T_2 \end{bmatrix} \begin{Bmatrix} u_1 \\ u_2 \end{Bmatrix} = K u$$

K is then the dynamic stiffness matrix (with terms complex functions of frequency) of the system of two foundations on the soil layer. It should be noticed that in order to obtain K it is not necessary to invert the flexibility matrix F. It is enough instead to form directly the pro-

duct $F^{-1} \begin{bmatrix} T_1 & 0 \\ 0 & T_2 \end{bmatrix}$ by solving a system of equations with F as matrix of

coefficients and the columns of the transformation matrix as right-hand
side vectors. For the case of two foundations, matrix K will be of size
12 x 12. It can be partitioned into four submatrices of the form

$$K = \begin{bmatrix} K_{11} & K_{12} \\ K_{21} & K_{22} \end{bmatrix}$$

where K_{11} and K_{22} are symmetric and $K_{21} = K_{12}^T$.

It must be noticed that due to the interaction effects K_{11} and K_{22} are
not equal to the stiffness matrices of the individual foundation considered
alone. Furthermore, even if the two foundations are equal, K_{11} and K_{22} are
not identical.

When the two foundations are parallel and centered along the x axis
(as shown in fig. 3) in-plane and out-of-plane effects can be uncoupled.
The in-plane effects are represented by forces and displacements in the x
(horizontal) and z (vertical) directions, and moments and rotations
around the y axis. The out-of-plane effects consist of forces and displacements in the y direction, and moments and rotations around the x
axis (rocking) and around the z axis (torsion). For this situation the
in-plane matrices are of the form

$$K_{11} = \begin{bmatrix} K_{xx} & K_{xz} & K_{x\phi} \\ K_{xz} & K_{zz} & K_{z\phi} \\ K_{x\phi} & K_{z\phi} & K_{\phi\phi} \end{bmatrix} \qquad K_{22} = \begin{bmatrix} K_{xx} & K_{xz} & -K_{x\phi} \\ K_{xz} & K_{zz} & -K_{z\phi} \\ -K_{x\phi} & -K_{z\phi} & K_{\phi\phi} \end{bmatrix}$$

and

$$K_{21} = \begin{bmatrix} K'_{xx} & K'_{xz} & K'_{x\phi} \\ K'_{xz} & K'_{zz} & K'_{z\phi} \\ -K'_{x\phi} & -K'_{z\phi} & K'_{\phi\phi} \end{bmatrix} \qquad K_{12} = K_{21}^T$$

Similar relationships take place for the out-of-plane effects.

In order to check the accuracy of the solution procedure, the case
of a single foundation, rigid and square, was considered. Following the

approach used by Kausel (1974) the convergence of the solution with mesh
size was investigated by considering different grids and plotting the
values of the stiffnesses versus mesh size. In all cases it was considered
that the thickness of each sublayer of soil for the determination of the
consistent boundary matrix (and the computation of the surface displacements) was equal to the mesh size. As in Kausel's studies the values of
the stiffnesses for different meshes fell almost exactly on straight lines.
The extrapolated values of the static stiffnesses obtained from the study
were

$$K_{xx} = K_{yy} \simeq \frac{9.2 \ GB}{2-\nu} (1 + 0.6 \ \tfrac{B}{H}) \qquad \text{(horizontal)}$$

$$K_{zz} \simeq \frac{4.6 \ GB}{1-\nu} (1 + 1.6 \ \tfrac{B}{H}) \qquad \text{(vertical)}$$

$$K_{\phi\phi} \simeq \frac{4GB^3}{1-\nu} (1 + 0.11 \ \tfrac{B}{H}) \qquad \text{(rocking)}$$

$$K_{\theta\theta} \simeq 8.2 \ GB^3 (1 + 0.05 \ \tfrac{B}{H}) \qquad \text{(torsion)}$$

where G is the shear modulus of the soil, B is half the side of the
square foundation, and H is the stratum depth.

By comparison for a circular foundation, Kausel had suggested the
formulas (3)

$$K_{xx} = \frac{8GR}{2-\nu} (1 + 0.5 \ \tfrac{R}{H})$$

$$K_{\phi\phi} = \frac{8GR^3}{3(1-\nu)} (1 + 0.17 \ \tfrac{R}{H})$$

Taking an equivalent radius so as to obtain the same area for the
circular and the square foundation $R_e = \sqrt{4/\pi} \ B$ the first expression
would become

$$K_{xx} = \frac{9.1GB}{2-\nu} (1 + 0.56 \ \tfrac{B}{H})$$

Taking $R_e = \sqrt[4]{16/\pi} \ B$ to obtain the same moment of inertia, the second
expression yields

$$K_{\phi\phi} = \frac{11.9GB^3}{3(1-\nu)} (1 + 0.2 \ \tfrac{B}{H}).$$

These results indicate that the usual procedure of adapting the stiffnesses of a circular foundation to the case of a square footing by defining an equivalent radius provides an excellent approximation. Furthermore,

the variation of the stiffness coefficients with frequency was almost identical to the results of Kausel using again the equivalent radius and modifying accordingly the values of the dimensionless frequency.

It should be noted that the accuracy provided by a given mesh is not the same for all the terms of the stiffness matrix. Thus for instance for a coarse mesh with each side divided in three equal segments (a total of sixteen points under the foundation) the error in the terms K_{xx} K_{yy} or K_{zz} may be of the order of 15 to 20% (depending on Poisson's ratio) the error in the terms $K_{\phi\phi}$ or $K_{\theta\theta}$ is of the order of 50%. In all cases the results converge monotonically from above: that is to say, the computed values are larger than the actual ones. It is thus recommended, if the method is used for practical applications to use a mesh which is sufficiently fine, or even better, to compute results for two meshes (a coarse one and a medium one) and to obtain improved estimates using a linear extrapolation.

In order to study the interaction effects between two rigid masses, it is sufficient to form the inertia matrices of each mass. If M_i is the mass, I_{x_i} I_{y_i} and I_{z_i} are the mass moments of inertia with respect to axes parallel to the x, y and z directions passing through the center of gravity, and E_i is the height of this point with respect to the base, the inertia matrix referred to the base displacements is of the form

$$M_{b_i} = \begin{bmatrix} M_i & 0 & 0 & 0 & M_i E_i & 0 \\ 0 & M_i & 0 & -M_i E_i & 0 & 0 \\ 0 & 0 & M_i & 0 & 0 & 0 \\ 0 & -M_i E_i & 0 & I_{x_i} + M_i E_i^2 & 0 & 0 \\ M_i E_i & 0 & 0 & 0 & I_{y_i} + M_i E_i^2 & 0 \\ 0 & 0 & 0 & 0 & 0 & I_{z_i} \end{bmatrix}$$

The dynamic equations of motion for a steady state response in the frequency Ω are then

$$(K - \Omega^2 M)\, U = F$$

where

$$M = \begin{bmatrix} M_{b_1} & 0 \\ 0 & M_{b_2} \end{bmatrix}$$

and F is the vector of applied harmonic forces at the base of the masses.

For the case of a base motion, simulating an earthquake type excitation if $\ddot{U}$ are the absolute accelerations at the base of the masses (for a frequency Ω) and $\ddot{U}_G$ represents a vector of free field ground accelerations (with the specified amplitudes of acceleration in the appropriate degrees of freedom of each mass and zeroes for all other components), the corresponding equations are

$$(K - \Omega^2 M)\, \ddot{U} = K\, \ddot{U}_G$$

For two (or more) structures it is necessary to form first the dynamic stiffness matrix $D^i = K^i - \Omega^2 M^i$ of each structure. If this matrix is partitioned in the form

$$D^i = \begin{bmatrix} D^i_{11} & D^i_{12} \\ D^i_{21} & D^i_{22} \end{bmatrix}$$

where D_{11} corresponds to all the degrees of freedom above (and excluding) the foundation, D_{22} to the six degrees of freedom of the foundation (assumed to be rigid) and D_{12}, $D_{21} = D^T_{12}$ are the coupling terms, the equations of motion for a steady state condition are

$$A\,U = F$$

with

$$A = \begin{bmatrix} D^1_{11} & 0 & D^1_{12} \\ 0 & D^2_{11} & D^2_{12} \\ D^1_{21} & D^2_{21} & B+K \end{bmatrix}$$

$$B = \begin{bmatrix} D^1_{22} & 0 \\ 0 & D^2_{22} \end{bmatrix}$$

K is the stiffness matrix of the two foundations and the soil stratum as obtained from the proposed method of analysis. F is the vector of harmonic forces applied at any of the degrees of freedom of the structures and the two foundations. Partitioning it in a manner analogous to that used for matrix A,

143

$$F = \left\{ \begin{array}{c} F_1 \\ F_2 \\ F_b \end{array} \right\}$$

Finally, to determine the dynamic response of two (or more) structures to a base motion representing a seismic input at the free surface of the soil deposit (in the far field) obtaining the transfer functions of the absolute accelerations, the same equations can be used replacing the vector U by $\ddot{U}$ and the vector of applied forces F by a vector of the form

$$\left\{ \begin{array}{c} 0 \\ 0 \\ K\ddot{U}_G \end{array} \right\}$$

with the same partitioning used above and $\ddot{U}_G$ as previously defined.

INTERACTION BETWEEN TWO RIGID MASSES

In this section results are presented illustrating the interaction between two rigid masses on equal, square foundations and resting on the surface of a homogeneous soil deposit. The depth of the stratum (underlain by rigid rock) is equal to four times the side of each foundation (H = 4B); a Poisson's ratio of 1/3 and an internal hysteretic damping in the soil of 5% were assumed.

The masses are defined in terms of the dimensionless parameter $m = M/8\rho B^3$ where M is the actual value of the mass and ρ is the mass density of the soil. Values of m = 1 and 2 were studied; the first one may be typical of a heavy nuclear power plant; the second one would correspond to an extremely heavy building. These large values were selected in order to emphasize the interaction effects. It was assumed in all cases that the height of the center of gravity above the foundation was equal to 1/3 of the base size and that the radius of gyration (for the computation of the mass moments of inertia) was equal to half the base. Distances between the centers of the foundations of 3B (a clear spacing of half the foundation size), 4B, 5B, and 10B were considered. The results presented here are all for the case D = 3B where the interaction effects are more marked. They are plotted versus the dimensionless frequency $f_0 = f(2B/c_s)$ defined above.

Figure 4 shows the variation with frequency of the horizontal displacement at the base of the first mass when it is excited by a harmonic horizontal force (applied also at the base). The first, excited, mass has m = 1. It can be seen that the presence of the second mass originates shifts in the natural frequencies with amplification of the motion in certain ranges and deamplifications in others. Similar results are obtained when m_1 = 2 (figure 5), but it can be seen that the effect of interaction increases when the second mass is heavier (and also as both masses increase). To interpret these figures it is interesting to notice that the swaying rocking frequencies of the masses alone (frequencies at which the peaks occur) are approximately

for m = 1	0.125	0.21	0.375
for m = 2	0.125	0.16	0.34

The first peak is affected by the natural frequency of the soil deposit (at precisely 0.125). For the case of a half-space one would have expected to find only two peaks.

Figure 6 shows the horizontal displacements induced in the second mass. For a distance of 3B between the centers of the foundations, the displacement of the passive mass is about 0.6 times that of the excited one for m_2 = 1 and about 0.8 times for m_2 = 2. This ratio decreases obviously as the distance between the foundations increases, but it is still of the order of 0.25 to 0.30 for D = 10B. For the case (1,1) the variation with frequency of the motion of the second mass is similar to that of the first. For the other cases it tends to show only one peak. Notice that the results for (1,2) and (2,1) are almost identical (the difference in the scale of the figures should be taken into account).

Figure 7 shows the vertical displacements of the two masses for the two cases (1,1) and (2,2). It must be remembered that this displacement is entirely a result of the adjoining mass, since for a mass alone no vertical motion should take place (under a horizontal excitation). The displacement is almost zero below the fundamental frequency of the stratum, but it becomes significant in the neighborhood of the vertical frequencies of the masses (0.23 for m = 1 and 0.20 for m = 2). The effect ot the horizontal rocking frequency can still be seen when both masses are equal to 2, but is very small in the other instances.

145

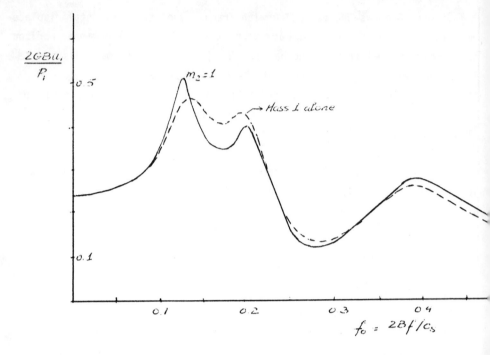

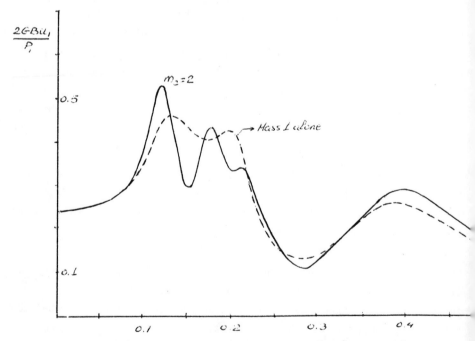

Figure 4. Displacements of excited mass for $m_1 = 1$. $f_0 = 2Bf/c_s$

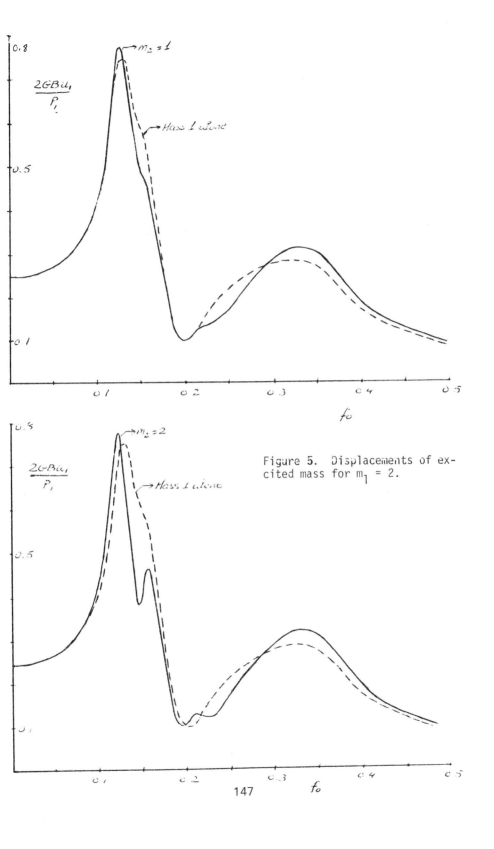

Figure 5. Displacements of excited mass for $m_1 = 2$.

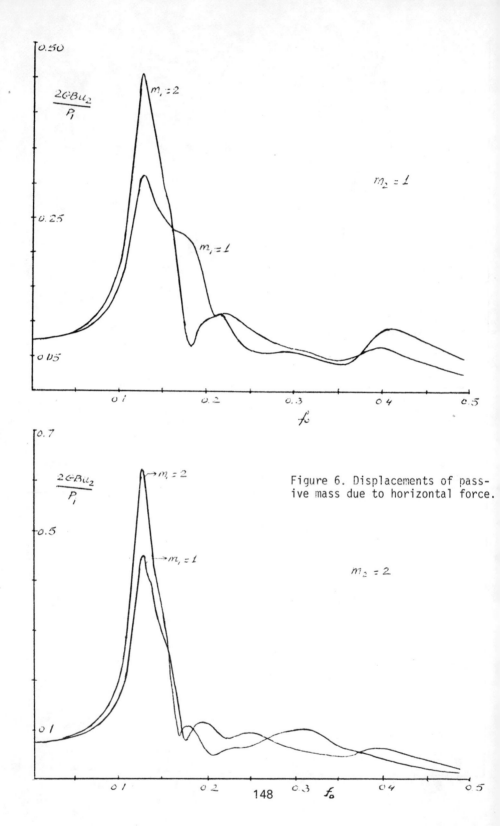

Figure 6. Displacements of passive mass due to horizontal force.

148

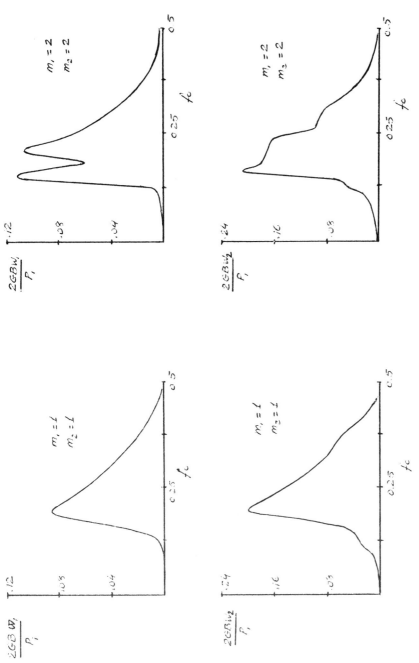

Figure 7. Vertical displacements of the masses due to a horizontal force at the base of the first one.

When the two masses are very close to each other (D = 3B) the vertical displacement induced in the active mass by the presence of the second can have a peak amplitude of the order of 1/5 to 1/8 of the peak horizontal displacement. The maxima of both effects occur, however, at different frequencies: if the excitation were to take place in the neighborhood of the vertical frequency, the vertical displacements could be almost as large as the horizontal ones (from 0.5 to 1 times, depending on the mass ratios). These vertical effects decay, though, very fast as the distance between the foundations increases. They are about half to two-thirds of the values shown at D = 4B, and only 1/20 of these values at D = 10B.

The vertical displacements of the passive mass have a frequency variation very similar to those of the first mass. It is important to notice that their amplitude is much larger (about 1.7 times when D = 3B and up to 10 times for D = 10B). This illustrates the fact that this displacement is induced first in the adjoining mass and then transmitted as a feedback to the active one.

The effect of the adjoining mass on the rotations caused by a unit horizontal force or a unit rocking moment is very similar to that described above for the horizontal displacement. Analogous conclusions are reached comparing vertical displacements due to vertical loads. The corresponding quantities in the passive mass are again of the order of 50 to 60% those of the primary mass for D = 3B and decrease to 10 to 20% for D = 10B. Finally, if a horizontal force is applied in the y direction to the first mass, torsional rotations appear in both. The behavior of these torsional motions is of the same nature as that of the vertical displacements caused by a horizontal force.

Figure 8 shows the horizontal accelerations at the base of the masses due to a ground acceleration specified at the free surface of the soil deposit (in the far field). The main effect of the adjoining mass is again the appearance of shifts in the frequences of the peaks. The values of the peaks seem to increase slightly (of the order of 5 to 20%), but whether an amplification or a deamplification will occur depends on the specific frequency of interest, or generally on the frequency content of the earthquake.

Figure 11 shows the torsional accelerations induced by the presence of the second mass when the ground motion takes place in the y direction.

150

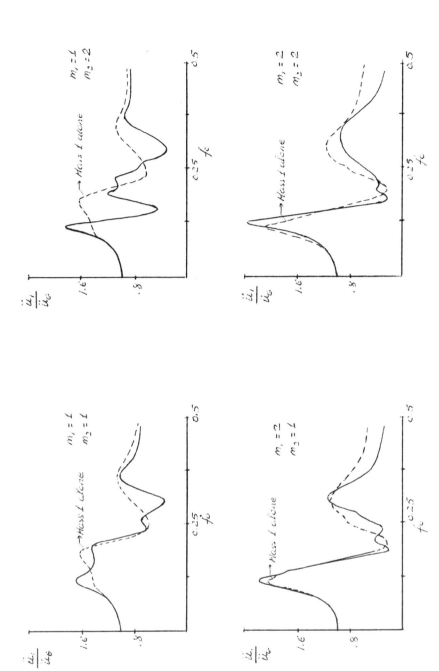

Figure 8. Horizontal accelerations at the base of the first mass due to a base motion.

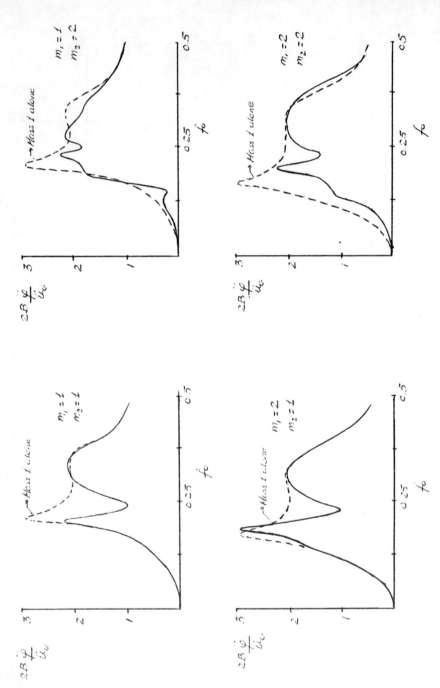

Figure 9. Rotational accelerations at the base of the first mass due to a horizontal base motion.

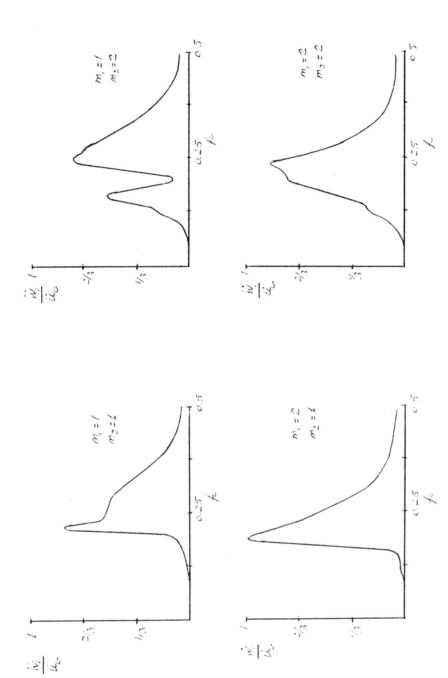

Figure 10. Vertical accelerations of first mass due to a horizontal ground motion.

153

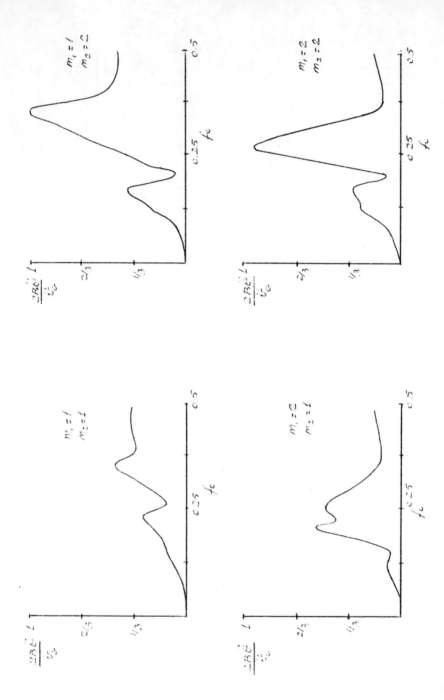

Figure 11. Torsional accelerations of first mass due to a horizontal ground motion.

154

The torsional frequencies of the masses on the soil stratum are 0.35 for m = 1 and 0.26 for m = 2. When both masses are equal to 1 the torsional acceleration shows two peaks: one at the second natural frequency in swaying rocking (0.21); the second one, larger, at the torsional frequency (0.35). The horizontal acceleration at the edge of the mass due to this torsional rotation would be about 10 to 20% of the acceleration of the centroid. When the second mass is heavier, the first peak increases slightly and shifts towards the frequency of the second mass; the second peak, on the other hand, increases by a factor of almost 2.5. In this case the lateral displacement of the edge of the mass due to torsion would be about 50% of the centroidal displacement (at that frequency). As the distance between the two foundations increases, the first peak changes very slowly; the second peak decreases faster and shifts to the left.

When the first mass is the heavy one (m_1 = 2) and the adjoining mass is smaller, peaks appear at frequencies of 0.125 (very small), 0.21 and 0.26 (torsional frequency of m = 2). The edge displacement caused by the torsion at 0.21 is about 50% of the centroidal displacement at that frequency, but is very small compared to the maximum horizontal displacement (at a different frequency). When both masses are equal to 2 the excited frequencies are 0.125, 0.16 (the two swaying-rocking frequencies) and 0.26 (the torsional frequency). The maximum response occurs again at the torsional frequency, but this peak decreases faster with distance.

INTERACTION BETWEEN TWO STRUCTURES

In order to study the interaction between two structures, introducing their own natural frequencies as parameters, two simple structures, idealized as single-degree-of-freedom systems, were used. Each structure was modelled as a single mass, lumped at a certain height, h, and a shear spring. The foundation was assumed to be rigid but massless. The structure was rigid in the vertical direction and had the same rotation at the base and at the top. The two foundations were still square, of equal size, and centered along the x axis so that in-plane and out-of-plane effects could be uncoupled. Considering only the in-plane effects, each structure alone, on the elastic foundation, had four degrees of freedom but only two dynamic ones.

155

In order to define the structural characteristics, the dimensionless frequency $k = K/2GB$ (K is the stiffness of the shear spring), the dimensionless mass $m = M/8\rho B^3$ (defined before), and the height h are used. In all cases it was assumed that $h = B$. A hysteretic type damping of 5% was employed. The soil profile was the same one of the previous studies with rigid masses. The results are shown again for a distance of 3B between the centers of the foundations.

The structures considered and their natural frequencies on a rigid base and on the elastic foundation were:

	$m=1$ $k=4$	$m=1$ $k=2$	$m=0.5$ $k=2$	$m=0.5$ $k=1$
for a rigid base	0.318	0.225	0.318	0.225
for an elastic foundation	0.170	0.154	0.220	0.175

It is interesting to notice that although structures 1 and 3 (and 2 and 4) have the same natural frequencies on a rigid base, the actual frequencies when considering the elastic foundation are different. For the soil structure system cases 1 and 4 have almost the same natural frequency. It is the frequency accounting for the flexibility of the foundation which is important in interpreting the shape of the response curves.

Figure 12 (a,b, and c) shows the displacements at the top of the two structures for various combinations of the structural models when a horizontal harmonic force is applied at the top of the first.

When the two structures are equal ($m=1$ $k=4$), the amplitude of the response at the resonant frequency (slightly shifted) increases by about 30 to 40% due to the presence of the second structure, but the peak is narrower. The displacement of the passive structure has a similar shape with a peak response about 50% of that of the first.

When $m_1=1$ $k_1=4$ and $m_2=1/2$ $k_2=2$ (same natural frequency on rigid base), the interaction effect is much smaller, with an increase in the peak value of about 10%. The response of the passive structure is still similar (controlled by the characteristics of the first) and of the order of 40%.

If $m_1=1$ $k_1=4$ and $m_2=1/2$ $k_2=2$ (same natural frequency approximately on elastic foundation), the peak displacement of the excited structure

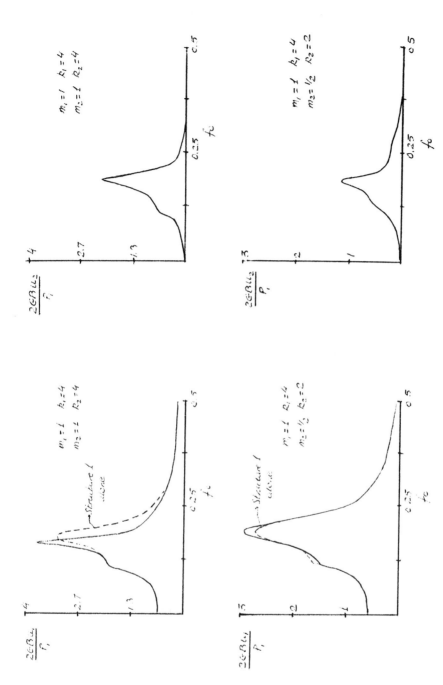

Figure 12a. Displacements at the top of the structures due to a horizontal force applied at the top of the first.

157

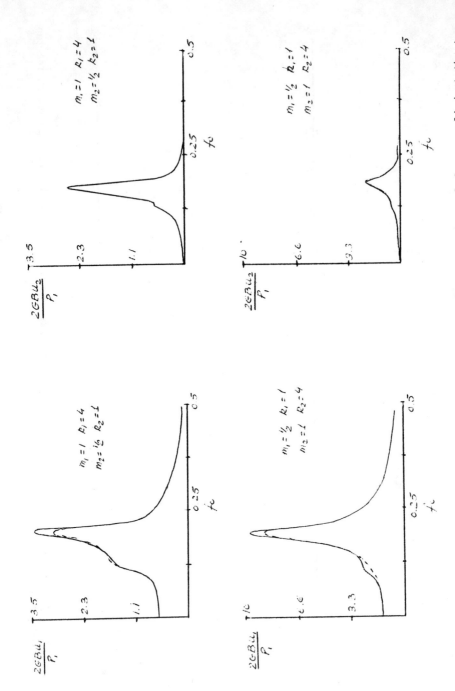

Figure 12b. Displacements at the top of the structures due to a horizontal force applied at the top of the first.

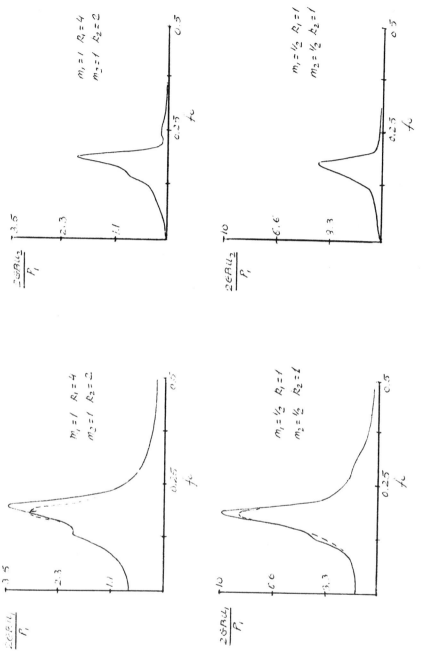

Figure 12c. Displacements at the top of the structures due to a horizontal force applied at the top of the first.

is increased by a factor of 1.2 to 1.3 over the value for the structure alone, a situation intermediate between the two previous ones. The response of the second structure becomes relatively larger, about 75% of that of the first. If the situation is reversed and the excited structure is the lighter one ($m_1=1/2$ $k_1=2$ $m_2=1$ $k_2=4$), the increase in the peak response of the excited structure is smaller (15 to 20%). The response of the passive structure is now much smaller (about 25% of that of the active one).

When the two structures have the same mass but different natural frequencies ($m_1=1$ $k_1=4$ $m_2=1$ $k_2=2$), the increase in the peak response is again of the order of 20 to 30%, and the response of the passive structure is half that of the active.

Finally, if both structures are equal but with smaller masses ($m=0.5$ $k=1$), the increase in the peak response of the excited structure is only of the order of 20%, and the response of the passive structure is less than half (about 40%).

It appears therefore that the interaction effects are more pronounced when both structures have the same natural frequency on the elastic foundation and when the masses are heavy. To illustrate better the reasons for the increase in the peak response due to the presence of the second structure, figure 13 shows the base displacements and rotations for the case when both structures are equal, with $m=1$ and $k=4$ (for the case of both structures and one structure alone). The main increase in response takes place in the rotation.

Figure 14 shows the horizontal accelerations at the top of the first structure when they are both excited by a ground acceleration in the x direction specified at the surface of the soil in the free field. In all cases the first structure has $m=1$ $k=4$; the second structure takes the values of each one of the four models. When the second structure is equal to the first, the interaction effect is more marked. Two peaks appear instead of one, and the peak response decreases by about 30%. When $m_2=0.5$ and $k_2=2$, the main effect is a shift in the frequency of the peak; the decrease in the amplitude of the peak is only of the order of 5%. If $m_2=0.5$ and $k_2=1$, the decrease is of the order of 15%. Finally, if $m_2=1$ and $k_2=2$, two peaks occur again and the amplitude of the larger is about 5 to 10% smaller than that of the structure alone.

160

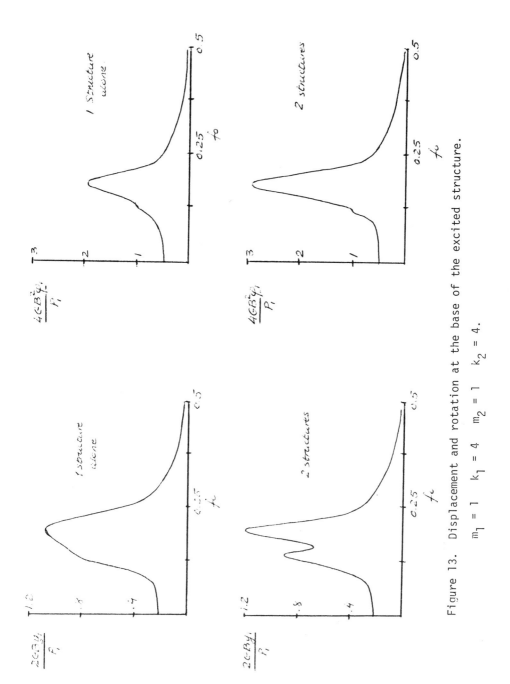

Figure 13. Displacement and rotation at the base of the excited structure.

$m_1 = 1$ $k_1 = 4$ $m_2 = 1$ $k_2 = 4$.

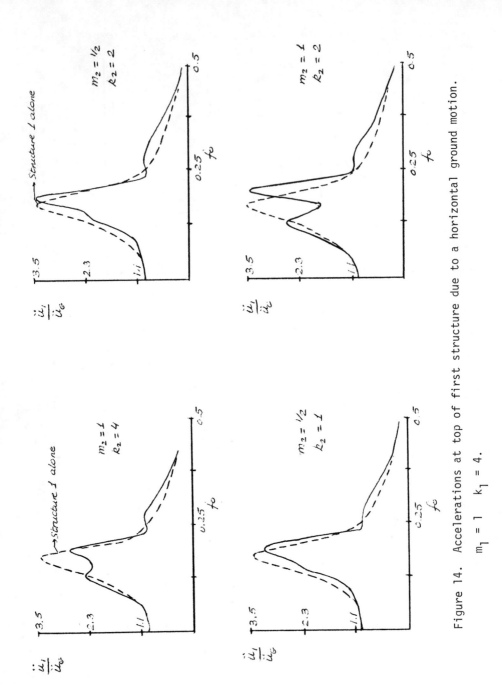

Figure 14. Accelerations at top of first structure due to a horizontal ground motion.
$m_1 = 1$ $k_1 = 4$.

Figure 15 shows the horizontal accelerations at the top of the second structure for the same situations. When $m_2=0.5$ and $k_2=2$, there is an increase in response at the frequency of the first structure and a slight decrease at its own frequency. In the second case, when $m_2=0.5$ and $k_2=1$ (both structures have approximately the same natural frequency on the elastic foundation), the reduction in the peak response is very marked, of the order of 40%. This implies that the fact that the other structure has a larger mass reinforces the interaction effect. In the last case ($m_2=1$ $k_2=2$), the interaction effect is relatively small.

The decrease in response at the peak is again primarily due to the base rotation. The results for a base excitation are in general terms, similar to those of the applied harmonic force, but the effect of the rotation is of opposite sign (it decreases the acceleration at the top for the case of a base motion and increases it in the other.

CONCLUSIONS

From the studies carried out and described here, it appears that the proposed formulation, based on a finite element type procedure with the consistent boundary developed by Kausel, can provide an excellent solution for truly three-dimensional problems and for the study of interaction effects between adjacent structures. In order to use the procedure in practice it would be necessary to use a finer mesh under each foundation or to obtain solutions for two different meshes and apply a linear extrapolation. (The results shown were obtained with a 3 x 3 mesh or a total of 16 points under each foundation for reasons of economy).

One of the main effects of an adjoining mass or structure is the excitation of modes of vibration which would not appear if the structure were alone (such as vertical vibrations under a horizontal force in the x direction and torsional rotations under a force in the y direction). When only one of the masses (or structures) is excited by an external force, these effects are due to a feedback from the passive structure, and they decay very rapidly with increasing distance between the foundations. When the two masses are excited (as in the case of a base motion), they become more significant and their rate of decay with distance is much slower.

163

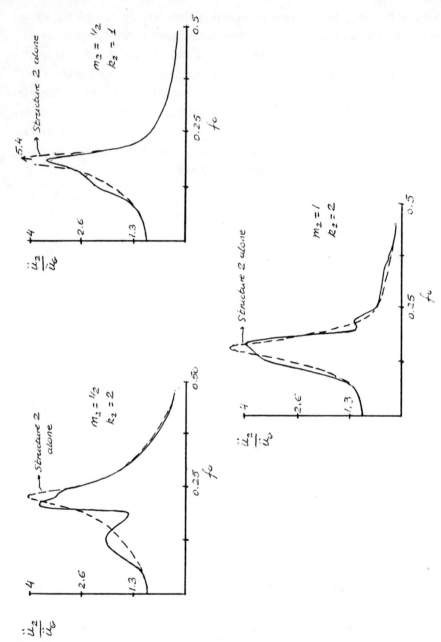

Figure 15. Accelerations at the top of the second structure due to a horizontal ground acceleration. $m_1 = 1$ $k_1 = 4$.

A second important effect is a change in the natural frequencies of the combined soil structure system. This change is more pronounced when dealing with two rigid masses, because their frequencies are affected only by the foundation stiffnesses. For the case of two structures, their own stiffness is an additional factor.

For the case of two rigid masses, when one of them is excited by a horizontal force, the effect of the adjoining mass on its horizontal displacement is only small. The effect is a little more pronounced in the rotations induced by a horizontal force or a rocking moment. The response of the passive mass is of the same order of magnitude as that of the excited one: from 30 to 80% when the masses are close (D=3B) and 10 to 30% when they are at a distance D=10B.

Under a base motion the existence of two masses tends to increase slightly the peaks of the base translation, but it reduces more importantly the rotation. The effect is more marked when both masses have the same natural frequency on elastic foundation and as their masses increase.

For the case of two structures, with one of them excited by a horizontal force at the top (where the mass is applied), the top displacement tended to increase in all cases due to the presence of the other structure, the increase being more noticeable when both of them had the same natural frequency on elastic foundation and as their masses increased. The increase in the peak response varied between 5 and 40% for the cases considered with the two foundations at a distance of 3B. The main cause for the increase lies in the base rotation.

Under a base motion, simulating a seismic input, the presence of the second structure tended instead to decrease the peak response. The decreases are of the same order as the increases found in the case of an exciting force and are also caused by the base rotation.

The results obtained are in general agreement with those reported by Richardson (1969) and by Chang Liang (1974). Because of the large number of parameters involved in this problem and the relatively small number of cases considered, it is difficult yet to present more specific conclusions. It appears, however, from these studies that simple rules could be derived to estimate the magnitude of the vibrations caused in a passive structure as a function of distance and feedback effects. These rules could then be extended to the case where both structures are excited.

In order to understand better the interaction effects, it would be convenient to isolate further translational and rotational effects (notice that in this work even for the case of rigid masses the horizontal force creates rotations due to the elevation of the center of mass). The rotational effects seem to be the most important ones. It would also be more convenient to work with a half-space rather than with a finite layer of soil to eliminate the effect of the stratum frequencies.

REFERENCES

Chang-Liang, V. 1974, Dynamic Response of Structures in Layered Soils, Research Report R74-10, Department of Civil Engineering, M.I.T.
Gazetas, G. 1975, Stiffness Functions for Strip and Rectangular Footings on Layered Media, S.M. Thesis, Department of Civil Engineering, M.I.T.
Kausel, E. 1974, Forced Vibrations of Circular Foundations on Layered Media, Research Report R74-11, Department of Civil Engineering, M.I.T.
Lysmer, J., Seed, H.B, Udaka, T., Hwang, R.N. and Tsai, C.F. 1975, Efficient Finite Element Analysis of Seismic Soil Structure Interaction. Report EERC 75-34, University of California, Berkeley.
Richardson, J.D. 1969, Forced Vibrations of Rigid Bodies on a Semi-Infinite Elastic Medium, Ph.D. Thesis, University of Nottingham, England.
Waas, G 1972, Linear Two-Dimensional Analysis of Soil Dynamics Problems in Semi-Infinite Layered Media, Ph.D. Thesis, University of California, Berkeley.

ACKNOWLEDGEMENTS

The results presented here are based on research conducted at the Civil Engineering Department of M.I.T. under a research grant from the National Science Foundation, Division of Advanced Environmental Research and Technology.

Dynamic interaction of a foundation
with viscoelastic halfspace

LOTHAR GAUL
Technische Universität, Hannover, Germany

SYNOPSIS   A mixed boundary value problem of soil-structure interaction is solved. The solution is based on a continuum approach for the viscoelastic halfspace. Results are given in terms of frequency dependent parameters of a simple halfspace model which has to be coupled with the model of the superstructure. The dynamic pressure distribution at the interface between base and soil is determined.

INTRODUCTION

Subject of the present paper is the dynamics of soil-foundation interactions. In calculating the dynamic response of footing-supported structures such as machine foundations this interaction has to be included.

Vertical and rocking vibrations of an arbitrarily shaped rigid base resting on the surface of a viscoelastic halfspace lead to mixed boundary value problems. These are treated by superposition based on the solution for a stress boundary value problem. Parameters of the dynamic interaction stiffness matrix are evaluated which determine the response of an excited rigid base. Actual pressure distributions at the interface between base and soil are presented.

FORCED VIBRATIONS OF A RIGID BASE ON THE SOIL

To illustrate the influence of soil-foundation interaction, an excited compact rigid footing (Figure 1) is regarded. From Newton's and Eulers's laws linearized equations of motion with respect to point A are obtained in the form

$$
\begin{bmatrix}
m & 0 & 0 & 0 & mz^c & -my^c \\
0 & m & 0 & -mz^c & 0 & mx^c \\
0 & 0 & m & my^c & -mx^c & 0 \\
0 & -mz^c & my^c & J_{xx} & J_{xy} & J_{xz} \\
mz^c & 0 & -mx^c & J_{xy} & J_{yy} & J_{yz} \\
-my^c & mx^c & 0 & J_{xz} & J_{yz} & J_{zz}
\end{bmatrix}^A
\begin{Bmatrix}
\ddot{u} \\ \ddot{v} \\ \ddot{w} \\ \ddot{\alpha} \\ \ddot{\phi} \\ \ddot{\gamma}
\end{Bmatrix}^A
=
\begin{Bmatrix}
R_x \\ R_y \\ R_z \\ M_x \\ M_y \\ M_z
\end{Bmatrix}^A
$$

$$[M]^A \{\ddot{U}\}^A = \{R\}^A \quad . \tag{1}$$

Here, $[M]^A$ is the inertia matrix, $\{\ddot{U}\}^A$ the vector composed of the second time derivatives of generalized coordinates, namely displacement coordinates u_i^A and angles φ_i , where $\varphi_i \ll 1$, $u_i^A \ll h$. The vector $\{R\}^A$ contains the forces P_i^A , F_i^B and moments T_i^A , M_i^B of excitation and halfspace reaction respectively according to Eq. 2:

$$
\begin{aligned}
R_i^A &= \qquad - F_i^B \qquad\qquad + P_i^A \\
M_i^A &= - (M_i^B + e_{ijk}\, x_j^o\, F_k^B) + T_i^A
\end{aligned}
\tag{2}
$$

For introducing the reactions of the halfspace, the soil-base plate interactions problem must be solved by intergrating the field equations of the halfspace. When only harmonic excitations are considered, the result can be summarized in a dynamic stiffness matrix $[\overline{K}]^B$, relating the complex halfspace reactions and the generalized coordinates, both at point B ,

$$\{F\}^B = [\overline{K}]^B \{\overline{U}\}^B \quad . \tag{3}$$

The complex valued elements of $[\overline{K}]^B$ are frequency dependent and represent lumped parameters of a simple halfspace model consisting of springs and dashpots in parallel. Including the halfspace reactions with the transformed stiffness matrix $[\overline{K}]^A$ and the complex vector $\{\overline{f}\}^A$ of excitation, we obtain from Eqs. 1 to 3

$$- \omega^2 [M]^A \{\overline{U}\}^A = \{\overline{R}\}^A = - [\overline{K}]^A \{\overline{U}\}^A + \{\overline{f}\}^A \quad , \tag{4}$$

where

$$[\overline{K}]^A = [T]^T [\overline{K}]^B [T] =$$

$$
= \begin{bmatrix}
\overline{c}_x & 0 & 0 & 0 & \overline{c}_x z^o & -\overline{c}_x y^o \\
 & \overline{c}_y & 0 & -\overline{c}_y z^o & 0 & \overline{c}_y x^o \\
 & & \overline{c}_z & \overline{c}_z y^o & -\overline{c}_z x^o & 0 \\
 & & & \overline{c}_\alpha + \overline{c}_y (z^o)^2 + \overline{c}_z (y^o)^2 & -\overline{c}_z x^o y^o & -\overline{c}_y x^o z^o \\
 & \text{symmetric} & & & \overline{c}_\phi + \overline{c}_z (x^o)^2 + \overline{c}_x (z^o)^2 & -\overline{c}_x y^o z^o \\
 & & & & & \overline{c}_\gamma + \overline{c}_x (y^o)^2 + \overline{c}_y (x^o)^2
\end{bmatrix}
$$

$$\tag{5}$$

The matrix $[\overline{K}]^B$ is diagonal.

Eq. 4 leads to a set of linear equations with complex coefficients

$$(- \omega^2 [M]^A + [\overline{K}]^A) \{\overline{U}\}^A = \{\overline{f}\}^A \tag{6}$$

for the unknown complex coordinates $\{\overline{U}\}^A$. From the real part of the steady state solution

$$\{U(t)\}^A = \text{Re}[\{\overline{U}\}^A \exp(i\omega t)] = \{U^R\}^A \cos \omega t - \{U^I\}^A \sin \omega t \tag{7}$$

the amplitudes and phase angles of the response can be evaluated.

INTERACTION BETWEEN RIGID BASE AND SOIL

The interaction between rigid base and soil leads to a mixed boundary value problem. The displacement boundary condition requires the displacement field to be linear in the interface (Figure 2) and surface stresses to vanish outside the interface. Compared with elastic halfspace theories based on a continuum approach, the theory of viscoelasticity presented here provides a better approximation of the rheological halfspace properties which are of considerable importance for rocking modes. The integration of the viscoelastic field equations is carried out in a way which is basically different from the mathematical treatment of the elastic problem. The method is applicable to vertical and rocking vibrations of arbitrarily shaped bases. The interface is discretised into r rectangular surface elements (Figure 2). The element number ℓ ($\ell = 1, .., r$) is loaded by a complex pressure $\overline{p}_\ell \exp(i\omega t)$, with $\overline{p}_\ell$ being constant over the area A_ℓ of the element. For the stress boundary value problem for the halfspace loaded by a single element of this kind a solution is given. It allows to calculate the complex compliance matrix of the interaction problem in the form

$$[\overline{h}_{k\ell}] = [f(x_{k\ell}, y_{k\ell}) + i\, g(x_{k\ell}, y_{k\ell})] \qquad k, \ell = 1, \ldots, r \quad . \tag{8}$$

This matrix is of importance also in the case, when the base plate is flexible and discretized by the finite element method.

For the complete base composed of r surface elements the complex displacements $\overline{w}_k$ and the rocking angles $\overline{\phi}_k$ are found by superposition

$$\overline{w}_k = \sum_{\ell=1}^{r} \overline{h}_{k\ell} \left(\frac{\overline{p}_\ell \, A_\ell}{G \, a_\ell}\right) \quad , \qquad \overline{\phi}_k = - \frac{\overline{w}_k}{x_k} \quad . \tag{9}$$

The displacement boundary conditions for the vertical and rocking mode

$$\overline{w}_k = \overline{w} \quad \text{and} \quad \overline{\phi}_k = \overline{\phi} \qquad k = 1, \ldots, r \quad , \tag{10}$$

respectively, are satisfied in the center of each element. Equivalence of halfspace reactions and resultants of the pressure distribution requires for the vertical mode

169

$$\bar{F}_z = \sum_{\ell=1}^{r} \bar{p}_\ell A_\ell \qquad \text{where} \qquad \bar{p}_\ell = \bar{k}_\ell \frac{\bar{F}_z}{A} \qquad (11)$$

and for the rocking mode

$$\bar{M}_y = - \sum_{\ell=1}^{r} x_\ell \bar{p}_\ell A_\ell \qquad \text{where} \qquad \bar{p}_\ell = - \bar{k}_\ell \left(\frac{\bar{M}_y}{I_y} x_\ell\right) \quad . \qquad (12)$$

In Eq. 12 I_y is defined as the moment of inertia of the base area A about the y axis. Eqs. 9 to 12 lead to a set of linear equations for the unknown complex pressure coefficients

$$\bar{k}_\ell = k_\ell^R + i\, k_\ell^I \qquad (13)$$

and for the unknown elements of the interaction matrix in Eq. 3:

$$\{\bar{F}\}^B = [\bar{K}]^B \{\bar{U}\}^B$$

$$\begin{Bmatrix} \bar{F}_z \\ \bar{M}_y \end{Bmatrix}^B = a\, G \left[\begin{array}{c|c} c_z + i\, a_o\, d_z & 0 \\ \hline 0 & a^2(c_\phi + i\, a_o\, d_\phi) \end{array} \right]^B \begin{Bmatrix} \bar{w} \\ \bar{\phi} \end{Bmatrix}^B \quad . \qquad (14)$$

These elements represent frequency dependent spring and damping coefficients. The frequency parameter a_o is defined as $a_o = \omega a / v_s$, $v_s = (G/\rho)^{1/2}$. The width of the base is $2\, a$.

STRESS BOUNDARY VALUE PROBLEM

In this section the solution of the stress boundary value problem leading to the compliance matrix in Eq. 8 is shown in some detail. The halfspace is loaded on one rectangular surface element by

$$\sigma_{zz}(x,y,0,t) = \begin{cases} - \bar{p}\, \exp(i\omega t) & \left|\frac{x}{a}\right|, \left|\frac{y}{b}\right| \leq 1 \ , \quad \text{where} \quad \bar{p} = \dfrac{\bar{F}}{4ab} \\[2ex] 0 & \left|\frac{x}{a}\right|, \left|\frac{y}{b}\right| > 1 \end{cases} \qquad (15)$$

In Table 1 the field equations of the viscoelastic continuum are presented. They are obtained by introducing alternatively constitutive relations of the hereditary integral type or of differential operator type. For harmonic excitations, the displacement field is specified as a harmonic function of time $u_i(x_i,t) = \bar{u}_i(x_i)\exp(i\omega t)$. Introducing this separation of variables condition into both types of field equations we arrive in the steady state at the following equation for the complex displacements $\bar{u}_i$:

VISCOELASTIC STRESS STRAIN CONSTITUTIVE RELATIONS

DEVIATORIC STATE	HYDROSTATIC STATE
$s_{ij} = 2 \int\limits_{-\infty}^{t} G(t-\tau) \dfrac{de_{ij}(\tau)}{d\tau} \, d\tau$	$\sigma_{kk} = 3 \int\limits_{-\infty}^{t} K(t-\tau) \dfrac{d\varepsilon_{kk}(\tau)}{d\tau} \, d\tau$
HEREDITARY INTEGRAL (RELAXATION)	
$\mathbb{P}(D)s_{ij} = \mathbb{Q}(D) e_{ij}$	$\mathbb{L}(D)\sigma_{kk} = \mathbb{M}(D)\varepsilon_{kk}$
z.B. $\mathbb{P}(D) = p_l D^l$, $D := \dfrac{d}{dt}$, $l = 0,..,N$	
DIFFERENTIAL OPERATOR FORM	

EQUATIONS OF MOTION OF THE VISCOELASTIC CONTINUUM

$\int\limits_{-\infty}^{t} \left[\lambda(t-\tau) + G(t-\tau)\right] \dfrac{\partial u_{j,ji}}{\partial \tau} \, d\tau + \int\limits_{-\infty}^{t} G(t-\tau) \dfrac{\partial u_{i,jj}}{\partial \tau} \, d\tau = \rho \dfrac{\partial^2 u_i}{\partial t^2}$
HEREDITARY INTEGRAL FORM , $\lambda(t) = K(t) - \dfrac{2}{3} G(t)$
$\left[\dfrac{\mathbb{M}(D)\,\mathbb{P}(D)}{3} + \dfrac{\mathbb{Q}(D)\,\mathbb{L}(D)}{6} \right] u_{j,ji} + \dfrac{\mathbb{Q}(D)\mathbb{L}(D)}{2} u_{i,jj} = \rho\,\mathbb{P}(D)\,\mathbb{L}(D)\dfrac{\partial^2 u_i}{\partial t^2}$
DIFFERENTIAL OPERATOR FORM

Table 1. Field equations of the continuum

$$[\lambda^*(i\omega) + 2G^*(i\omega)]\overline{u}_{j,ji} - G^*(i\omega)e_{ijk}(e_{klm}\overline{u}_{m,l})_{,j} + \rho\omega^2 \overline{u}_i = 0 . \quad (16)$$

The complex moduli $\lambda^*(i\omega)$, $G^*(i\omega)$ are related to the relaxation functions of shear $G(t)$ and of compression $K(t)$ and also to the differential
operators in Table 1. Introducing the dilatation $\overline{\varepsilon} = \overline{u}_{i,i}$ and the vector
of rotation $\overline{\omega}_k = \frac{1}{2} e_{klm} \overline{u}_{m,l}$ we can decouple Eq. 16 into the reduced wave
equations

$$\overline{\varepsilon}_{,jj} + (k_D^*)^2 \, \overline{\varepsilon} = 0 \qquad (\overline{\omega}_k)_{,ll} + (k_S^*)^2 \, \overline{\omega}_k = 0 \qquad (17)$$

with complex wavenumbers

$$k_D^* = \left[\frac{\rho\omega^2}{\lambda^*(i\omega) + 2G^*(i\omega)}\right]^{1/2} \qquad k_S^* = \left[\frac{\rho\omega^2}{G^*(i\omega)}\right]^{1/2} . \qquad (18)$$

171

Solutions of Eq. 17 lead to both the displacement field and the stress field, when $\bar{\omega}_{k,k} = 0$ is taken as constraint condition:

$$\bar{u}_i = - \frac{1}{(k_D^*)^2} \bar{\varepsilon}_{,i} + \frac{2}{(k_S^*)^2} e_{ijk} \bar{\omega}_{k,j}$$

(19)

$$\bar{\sigma}_{ij} = G^*(i\omega) \left[\delta_{ij} \left(\frac{1-2n^{*2}}{n^{*2}} \right) \bar{\varepsilon} - \frac{2}{(k_D^*)^2} \bar{\varepsilon}_{,ij} + \frac{2}{(k_S^*)^2} (e_{i\ell k} \bar{\omega}_{k,\ell j} + e_{j\ell k} \bar{\omega}_{k,\ell i}) \right]$$

where $n^* = k_D^*/k_S^*$.

The solution of the stress boundary value problem can be obtained by super-position of harmonic responses using the Fourier integral theorem. The travelling wave response of a vertical displacement wave

$$w(x,y,o,t) = \bar{w} \exp[i(\beta x + \gamma y + \omega t)] \quad , \quad (20)$$

which has been excited by a stress wave acting on the shearfree surface of the halfspace,

$$\sigma_{zz}(x,y,o,t) = - \bar{\sigma} \exp[i(\beta x + \gamma y + \omega t)]$$

$$\sigma_{zx}(x,y,o,t) = \sigma_{zy}(x,y,o,t) = 0 \quad , \quad (21)$$

is characterized by the complex compliance

$$\bar{H}_0(\beta,\gamma,\omega) = \frac{\bar{w}}{\bar{\sigma}} = \frac{\hat{w}}{\hat{\sigma}} \exp(i\varphi) \quad . \quad (22)$$

This compliance, evaluated from Eqs. 17 to 21, describes the amplitude and phase response. Using the Fourier integral theorem the solutions for distinc wavenumbers β and γ are now superimposed. The vertical displacement field at the halfspace surface which results from the dynamic pressure distribution given in Eq. 15 is thus obtained in the form

$$w(x,y,o,t) = \frac{\bar{F} \exp(i\omega t)}{(2\pi)^2} \int_{-\infty}^{\infty} \int_{-\infty}^{\infty} \bar{H}_0(\beta,\gamma,\omega) [I_A(\beta,\gamma) + I_B(\beta,\gamma) + i \ I_C(\beta,\gamma)] d\beta \ d\gamma$$

(23)

where

$$I_A(\beta,\gamma) = \frac{1}{4(\beta a)(\gamma b)} \{ \sin[\beta(a-x)] \sin[\gamma(b-y)] + \sin[\beta(a+x)] \sin[\gamma(b-y)]$$
$$+ \sin[\beta(a-x)] \sin[\gamma(b+y)] + \sin[\beta(a+x)] \sin[\gamma(b+y)] \}$$

$$I_B(\beta,\gamma) = - \frac{\sin(\beta a) \ \sin(\beta x) \ \sin(\gamma b) \ \sin(\gamma y)}{(\beta a)(\gamma b)} \quad ,$$

$$I_C(\beta,\gamma) = \frac{\sin(\beta a) \ \sin(\gamma b)}{(\beta a)(\gamma b)} [\cos(\beta x) \sin(\gamma y) + \sin(\beta x) \cos(\gamma y)] \quad .$$

Under the integral appears the compliance $\bar{H}_0$, which is of fundamental importance in this solution procedure. Figure 3 shows the absolute value of the dimensionless compliance $|\bar{H}| = (\hat{w}G\zeta)/\hat{\sigma}$ plotted versus the ratio v/v_S (v = stress wave velocity, v_S = velocity of shear waves and $\zeta = (\beta^2+\gamma^2)^{1/2}$). For an elastic halfspace the well-known Rayleigh pole of first order occurs, so that the integral of Eq. 23 becomes improper with respect to the integrand. Solutions of this case were given for the center deflection (Kobori 1962) and for the displacement field (Holzlöhner 1969) by choosing Cauchy's principal value of the integral and by performing numerical contour integration in the complex plane. An application of the elastic-viscoelastic correspondence principle does not offer any advantage because the solutions for the elastic case are not available in a closed form. The viscoelastic halfspace leads to a finite resonance peak instead of a pole (Figure 3). Thus, the integral in Eq. 23 is no longer improper with respect to the integrand and can be integrated directly. Results are presented for example for two viscoelastic two-parameter models, namely the constant hysteretic model and the Kelvin-Voigt model. In the expression for the complex shear modulus $G^*(i\omega) = G[1+i\eta(\omega)]$, the former leads to $\eta(\omega) \equiv \eta$ wheras for the latter $\eta(\omega)$ depends on frequency according to $\eta(\omega) = a_0 \cdot \xi$.

Experimental data for soils indicate that the absorbed energy in one cycle of vibration - the area under the stress strain hysteresis loop - and thus the damping factor $\eta(\omega)$ are approximately frequency independent (Crandall, Kurzweil & Nigam 1971, Krizek & Franklin 1970). This is valid only in a limited range of frequencies. In a larger frequency range, the actual frequency dependence should be taken into account. This is possible in the present solution. A viscoelastic two-parameter model might be insufficient if the frequency range of interest is not known in advance. If the resonance frequencies of a system are known, the parameters of simple viscoelastic models can be fitted to the corresponding actual damping (Mahrenholtz & Gaul 1977). This provides good results because the response outside the resonance closely matches the solutions using more sophisticated viscoelastic models.

VERTICAL DISPLACEMENT FIELD $w(x,y,o,t) = w(x,y) \exp(i\omega t)$

$$\frac{w(x,y)aG}{\bar{F}} = f(x,y) + ig(x,y) = \frac{2}{(2\pi)^2} \int_{K_v=0}^{\infty} A_0 K_v \, \bar{H}(K_v) \left[\int_{\theta=0}^{\pi/2} T(A_0 K_v, \theta) \, d\theta \right] dK_v$$

$$\bar{H}(K_v) = \frac{\bar{w} G k_v}{\bar{\sigma}} = \frac{U_D\left[-(1-\eta^2(\omega)) \cos\frac{\varphi_D}{2} - 2\eta(\omega)\sin\frac{\varphi_D}{2} + i\left\langle -(1-\eta^2(\omega))\sin\frac{\varphi_D}{2} + 2\eta(\omega)\cos\frac{\varphi_D}{2}\right\rangle\right]}{(1+\eta^2(\omega))\left[(2K_v^2-1)^2 - \eta^2(\omega) - 4K_v^2 U_S U_D \cos\left(\frac{\varphi_S+\varphi_D}{2}\right) + i\left\langle 2\eta(\omega)(2K_v^2-1) - 4K_v^2 U_S U_D \sin\left(\frac{\varphi_S+\varphi_D}{2}\right)\right\rangle\right]}$$

$$a_0 = \frac{\omega a}{v_S} \qquad U_D = \left[(K_v^2-n^2)^2 + (n^2 \eta(\omega))^2\right]^{1/4} \qquad \varphi_D = \arctan\frac{n^2 \eta(\omega)}{K_v^2-n^2}, \qquad 0 \le \varphi_D < \pi$$

$$A_0 = \frac{a_0}{[1+\eta^2(\omega)]^{1/2}} \qquad U_S = \left[(K_v^2-1)^2 + \eta^2(\omega)\right]^{1/4} \qquad \varphi_S = \arctan\frac{\eta(\omega)}{K_v^2-1}, \qquad 0 \le \varphi_S < \pi$$

$$T(A_0 K_v, \theta) = \sum_{i,k=1}^{2} (-1)^{i+k} \frac{\sin(a_i A_0 K_v \cos\theta) \sin(b_k b_0 A_0 K_v \sin\theta)}{(A_0 K_v \cos\theta)(b_0 A_0 K_v \sin\theta)} \qquad n^2 = \frac{1-2v}{2(1-v)}$$

$$b_0 = b/a \qquad a_{1,2} = (x/a \pm 1) \qquad b_{1,2} = (y/b \pm 1)$$

Table 2. Solution of the stress boundary value problem

173

Considerable algebraic manipulations lead to the solution for the dynamic displacement field given in Table 2. The result, which has to be integrated numerically, is presented as complex compliance with real and imaginary parts f and g, respectively. The functions f and g do also describe the phase lag of the displacement with respect to the resultant of the exciting pressure distribution $\overline{F}$. The general form of the solution allows investigating the influence of several parameters:

$$w = w[x, y, \frac{b}{a}, a_0 = \frac{\omega a}{v_S}, n(\nu), n(\omega)] \tag{24}$$

DYNAMIC SOIL PRESSURE DISTRIBUTION

As a result of the mixed boundary value problem, the static and dynamic pressure distributions at the interface between a rigid square base and the halfspace are presented. The pressure distributions in Figures 4 and 5 are caused by moments. The fields of the real and imaginary parts of the complex pressure coefficients $\overline{k}_\ell$ from Eq. 12 shown in these figures determine the actual pressure distribution:

$$p_\ell(t) = - \frac{\hat{M}_y}{I_y} x_\ell (k_\ell^R \cos \omega t - k_\ell^I \sin \omega t) \quad . \tag{25}$$

Comparison between the static case (Figure 4) and the dynamic case (Figure 5) reveals that with increasing frequency parameter a_0 the real part of the pressure distribution becomes more nearly linear and tends to the equivalent linear pressure distribution. Thus, the singularity at the edge becomes sharper and more localized. The equivalent linear pressure distribution $-(\hat{M}_y/I_y) x_\ell$ is indicated by a dashed line. The dynamic pressure distribution associated with vertical excitation (Figure 6) is determined from Eq. 11:

$$p_\ell(t) = \frac{\hat{F}}{A} (k_\ell^R \cos \omega t - k_\ell^I \sin \omega t) \quad . \tag{26}$$

LUMPED PARAMETERS OF A HALFSPACE MODEL

The elements of the dynamic interaction stiffness matrix $[\overline{K}]^B$ in Eq. 14 are plotted in Figures 7, 8, 9, 10 versus the frequency parameter a_0. The real parts c and the imaginary parts d represent dimensionless equivalent spring and damping coefficients, respectively. They correspond to vertical (c_z, d_z) and rocking vibrations (c_ϕ, d_ϕ).

The Kelvin-Voigt model and the constant hysteretic model are used for example.

An increase of the viscosity coefficient ξ and the damping factor η from 0.05 to 0.2 causes the damping coefficients d to increase and the spring coefficients c to decrease. The damping coefficients d calculated for light material damping $(\xi, \eta = 0.05)$ result primarily from geometrical radiation of wave energy, referred to as geometrical damping. For rotational modes geometrical damping is considerably smaller than for the vertical translational mode. The damping coefficients d_ϕ in the low frequency range

174

of Figures 9 and 10 indicate the strong influence of viscous halfspace behaviour especially for rotational modes.

The question remains into how many elements should the interface be divided in order to obtain the parameters of the halfspace model with reasonable accuracy. To find an answer, the halfspace compliances corresponding to static tilt and to vertical loading are plotted in Figure 11 versus the number of elements. The results are compared with approximate solutions of integral equations with Boussinesq kernel (Gorbunov-Possadov & Serebrjanyi 1961). The convergence behaviour indicates that a small number of elements is sufficient.

With the values obtained for the halfspace parameters Eq. 7 can be used to determine the response of an excited base. Figure 12 shows the response of a base of mass m, which is excited in a purely vertical mode by a rotating mass $(\hat{f}_z = m_u e \omega^2)$. The dashed resonance curves describe the center deflection of a base with zero flexural rigidity (uniform pressure distribution). The curves drawn in solid lines indicate that the resonance peaks for a rigid base are considerably smaller.

For values of the mass ratio $b = m/(\rho a^3)$ which are typical for practical situations the resonance peaks are small even for light material damping. This is a consequence of the strong geometric damping. This statement is only correct for the vertical mode, however. As Figures 9 and 10 indicate, this is different for rocking vibrations. Such vibrations occur with high resonance peaks which are typical for systems with low damping (Richard, Hall & Woods 1970).

CONCLUSIONS

A mixed boundary value problem of soil-structure interaction is solved. Solutions are given in terms of frequency dependent parameters of a simple halfspace model, which has to be coupled with the model of the superstructure. Using integral transform techniques transient motions can be determined in addition to the presented steady-state response. Based on the solution for a dynamic stress boundary value problem also halfspace interactions with a flexible base plate can be determined by using finite element analysis.

Finally, the problem of interactions between separate structures resting on a viscoelastic halfspace can be solved.

REFERENCES

Crandall, S.H., L.G. Kurzweil & A.K. Nigam 1971, On the measurement of Poisson's ratio for modeling clay, Experimental Mechanics: 402-407.

Gorbunov-Possadov, M.I. & R.V. Serebrjanyi 1961, Design of structures on elastic foundations, Proc. 5th Int. Conf. Soil Mech. and Found. Eng., Paris 1: 643-648.

Holzlöhner, U. 1969, Schwingungen des elastischen Halbraumes bei Erregung auf einer Rechteckfläche, Ing.-Archiv 38: 370-379.

Kobori, T. 1962, Dynamical response of rectangular foundations on an elastic halfspace, Proc. Japanese National Symp. on Earthquake Eng.: 81-86.

Krizek, R.J. & A.G. Franklin 1970, Energy dissipation in soft clay, Proc. Int. Symp. on Wave Propagation and Dynamic Properties of Earth Materials, Univ. of Mexico: 797-807.

Mahrenholtz, O. & L. Gaul 1977, Dämpfungsfragen, VDI-Bildungswerk, Lehr-gangshandbuch 32-12-03, BW 2950.

Richard, F.E.Jr., J.R. Hall, Jr. & R.D. Woods 1970, Vibrations of soils and foundations. Englewood Cliffs, New Jersey, Prentice-Hall.

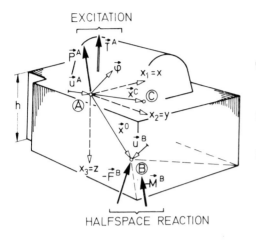

EXCITATION

Ⓐ ORIGIN OF
 BODY-FIXED BASE
 COORDINATES x_i

Ⓒ CENTER OF MASS
 MASS m

Ⓑ POINT OF ACTION
 IN INTERFACE

HALFSPACE REACTION

F_i^B, M_i^B FORCE, MOMENT OF HALFSPACE REACTION

P_i^A, T_i^A FORCE, MOMENT OF EXCITATION

      COORDINATES OF RIGID BODY MOTION

u_i^A, u_i^B DISPLACEMENTS (REFERENCE: STAT. EQUIL. POSITION)

φ_i ANGLES ($\varphi_i \ll 1$, $u_i^{A,B} \ll h$)

Figure 1. Rigid footing on the soil

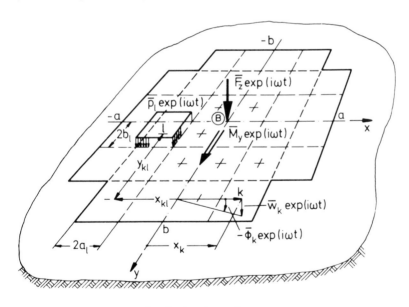

Figure 2. Discrete model of the interface

177

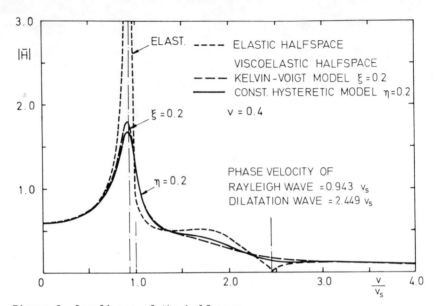

Figure 3. Compliance of the halfspace.
Excitation by a stress wave

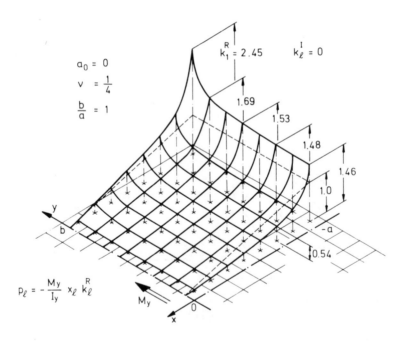

Figure 4. Static pressure distribution.
Rigid base loaded by a moment M_y

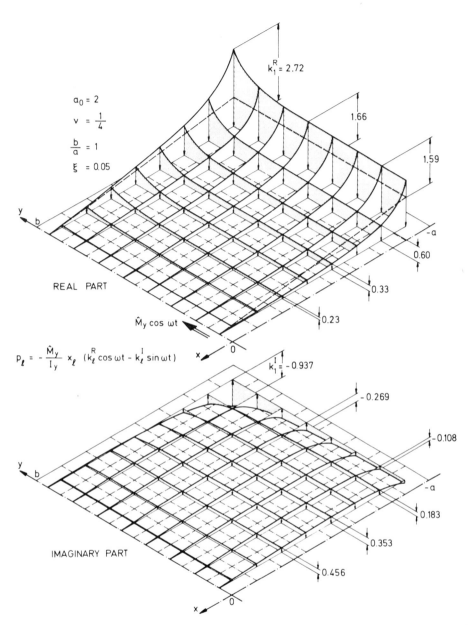

Figure 5. Dynamic pressure distribution.
Rocking vibration of a rigid base

179

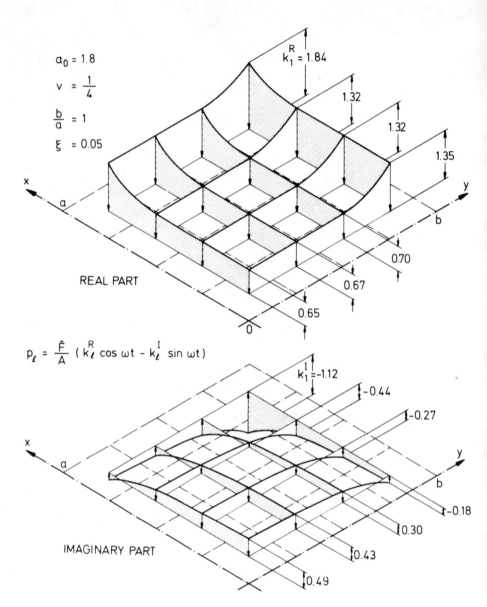

$$p_\ell = \frac{\hat{F}}{A} \left(k_\ell^R \cos \omega t - k_\ell^I \sin \omega t \right)$$

Figure 6. Dynamic pressure distribution.
Vertical vibration of a rigid base

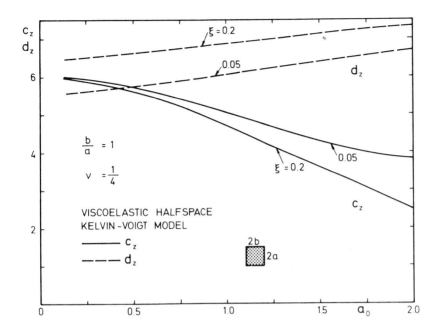

Figure 7. Equivalent spring and damping coefficient.
Vertical vibration of a rigid base

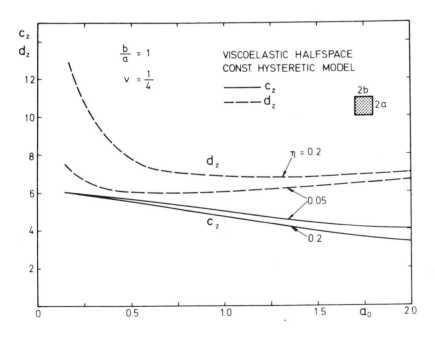

Figure 8. Equivalent spring and damping coefficient.
Vertical vibration of a rigid base

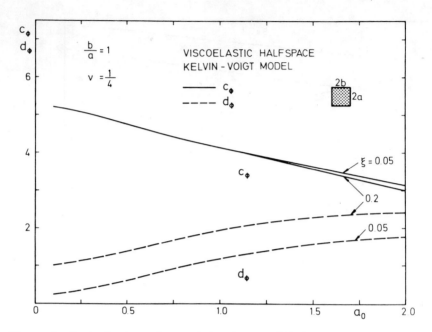

Figure 9. Equivalent spring and damping coefficient.
Rocking vibration of a rigid base

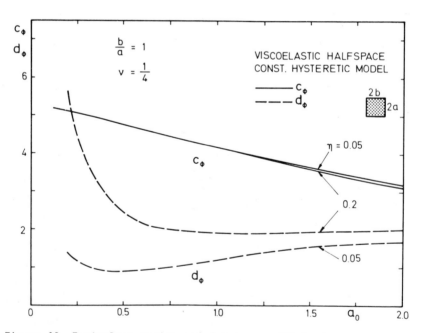

Figure 10. Equivalent spring and damping coefficient.
Rocking vibration of a rigid base

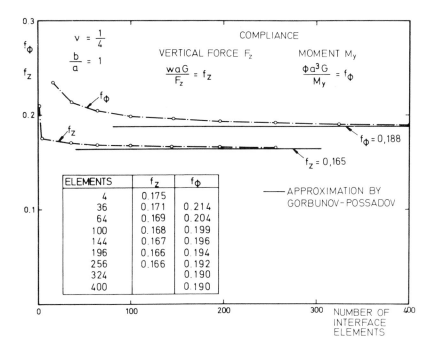

Figure 11. Convergence of halfspace compliances

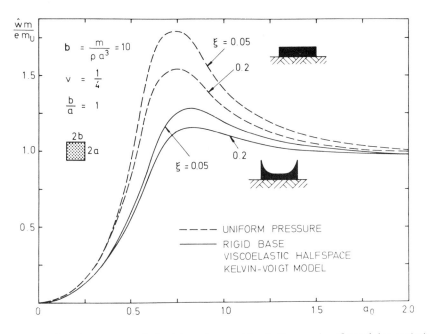

Figure 12. Response of footing to vertical force developed by rotating mass exciter

183

Effects of piles on dynamic response
of footings and structures

M. NOVAK
University of Western Ontario, London, Canada

1 INTRODUCTION

The last few years have seen a dramatic increase in interest in dynamic behavior of pile foundations. Many research papers have appeared describing
a number of theoretical and experimental studies into various aspects of
soil-pile-structure interaction. Many papers on the subject were presented
recently at the Sixth World Conference on Earthquake Engineering in New
Delhi and the Ninth International Conference on Soil Mechanics and Foundation Engineering in Tokyo.
　This paper is not intended as a state-of-the-art report; rather, it is
confined to a brief description of some results of research conducted at The
University of Western Ontario. This research concerns the theoretical approaches to impedance functions of embedded piles, experimental verification
of the theory and inclusion of piles into the analysis of footings and
structures.

2 IMPEDANCE FUNCTIONS OF EMBEDDED PILES

Impedance functions are complex, frequency dependent expressions that relate
harmonic exciting forces applied to the pile head and the resulting displacements at the same level. The real and imaginary parts of the impedance
functions define, respectively, the discrete stiffness and damping constants
of the soil-pile system.
　These constants are a result of interaction between the piles and surrounding soil. Several approaches to the study of this interaction have been
used, the main ones being: the discrete model of lumped masses (Penzien
1970), the continuum approach (Tajimi 1969, Kobori et al. 1977, Nogami &
Novak 1976, 1977, Novak 1974b, 1977, Novak & Nogami 1977, Novak & Aboul-Ella
1978) and the finite element method (Blaney et al. 1976, Kuhlemeyer 1976,
Shimizu et al. 1977, Tajimi & Shimomura 1977). Each of these approaches has
some advantages. The lumped mass model can approximately incorporate nonlinearity, the continuum approach offers fundamental insight into the problem and the finite element method can handle nonhomogeneity and, at a cost,
even nonlinearity. The work reported here was based on continuum considerations. Two approaches were developed and will be referred to as the more
rigorous theory and the simpler theory.

2.1 The More Rigorous Theory

In all approaches, the key to the solution of the interaction between the elastic pile and the soil is the description of the resistance of the soil to the motion of the pile. In continuum approaches, this soil resistance can be established from the equations of the viscoelastic medium. Their solution, however, imposes some rather limiting assumptions: the pile has to be vertical, of circular cross section, end bearing and perfectly bonded to the soil; the soil must be homogeneous, isotropic and linearly elastic or viscoelastic. Despite these idealizations, the solutions are useful.

The equations of motion u, v, w of the viscoelastic medium with hysteretic type damping, can be obtained from the equations of an elastic medium by complimenting the Lamé's constants, λ, G, with their imaginary (out of phase) components, λ', G'. Thus, the equations of motion of the viscoelastic media in cylindrical coordinates are

$$[(\lambda+2G)+i(\lambda'+2G')] \frac{\partial\Delta}{\partial r} - \frac{2(G+iG')}{r} \frac{\partial\omega_z}{\partial\theta} + 2(G+iG') \frac{\partial\omega_\theta}{\partial z} = \rho\frac{\partial^2 u}{\partial t^2}$$

$$[(\lambda+2G)+i(\lambda'+2G')] \frac{\partial\Delta}{r\partial\theta} - 2(G+iG') \frac{\partial\omega_r}{\partial z} + 2(G+iG') \frac{\partial\omega_z}{\partial r} = \rho\frac{\partial^2 v}{\partial t^2} \qquad (1)$$

$$[(\lambda+2G)+i(\lambda'+2G')] \frac{\partial\Delta}{\partial z} - \frac{2(G+iG')}{r} \frac{\partial}{\partial r}(r\omega_\theta) + \frac{2(G+iG')}{r} \frac{\partial\omega_r}{\partial\theta} = \rho\frac{\partial^2 w}{\partial t^2}$$

in which relative volume change:

$$\Delta = \frac{1}{r} \frac{\partial}{\partial r}(ru) + \frac{1}{r} \frac{\partial v}{\partial\theta} + \frac{\partial w}{\partial z} \qquad (2)$$

and components of rotational vector:

$$\omega_r = \frac{1}{2}(\frac{1}{r} \frac{\partial w}{\partial\theta} - \frac{\partial v}{\partial z})$$

$$\omega_\theta = \frac{1}{2}(\frac{\partial u}{\partial z} - \frac{\partial w}{\partial r}) \qquad (3)$$

$$\omega_z = \frac{1}{2}(\frac{1}{r} \frac{\partial(rv)}{\partial r} - \frac{1}{r} \frac{\partial u}{\partial\theta})$$

The soil resistance, p(z) (Fig. 1), to be derived from Eqs. 1, acts on the pile as described by the governing equation of the pile motion:

$$E_p I \frac{\partial^4}{\partial z^4} (u\ e^{i\omega t}) + m \frac{\partial^2}{\partial t^2} (u\ e^{i\omega t}) = -p(z)e^{i\omega t} \qquad (4)$$

in which $E_p I$ = bending stiffness of the pile, m = mass of the pile per unit length and p(z) = amplitude of the soil resistance to the harmonic motion of the pile.

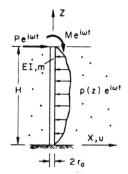

Fig. 1 External forces and soil resistance to motion of pile

Eqs. 1 and 4 describe the soil-pile interaction problem. It is very difficult to satisfy all the boundary conditions at the soil surface, pile surface and the surface of the underlying bedrock. To facilitate the solution of this problem, Kobori et al. (1977) considered an infinitely long pile and Tajimi (1966) suggested that the vertical component of the motion be ignored when solving horizontal response. This idea was exploited by Nogami and the writer (1977) to derive the soil resistance in the form of an infinite series:

$$p(z) = \alpha_h U(z) = \sum_{n=1}^{\infty} \alpha_{hn} U_n \sin(h_n z) \tag{5}$$

where α_{hn} = the horizontal resistance factor in the nth mode. It is the horizontal resistance of the soil generated around the circle $r = r_0$ by a unit horizontal displacement of the pile in the nth mode.

The resistance factor α_h can also be written with the dimensionless factor $\overline{\alpha}_h$ as

$$\alpha_{hn} = \pi G \overline{\alpha}_{hn} \tag{6}$$

where $\overline{\alpha}_{hn} = \overline{r}_0 [(1 + iD_s)\overline{h}_n^2 - \overline{a}_0^2]\overline{T}_n \tag{7}$

and

$$\overline{T}_n = \frac{4K_1(\overline{q}_n \overline{r}_0)K_1(\overline{s}_n \overline{r}_0) + \overline{s}_n \overline{r}_0 K_1(\overline{q}_n \overline{r}_0)K_0(\overline{s}_n \overline{r}_0) + \overline{q}_n \overline{r}_0 K_0(\overline{q}_n \overline{r}_0)K_1(\overline{s}_n \overline{r}_0)}{\overline{q}_n K_0(\overline{q}_n \overline{r}_0)K_1(\overline{s}_n \overline{r}_0) + \overline{s}_n K_1(\overline{q}_n \overline{r}_0)K_0(\overline{s}_n \overline{r}_0) + \overline{q}_n \overline{s}_n \overline{r}_0 K_0(\overline{q}_n \overline{r}_0)K_0(\overline{s}_n \overline{r}_0)} \tag{8}$$

Here, $K_{0,1}$ = modified Bessel functions of the second kind of complex arguments and

$$\left.\begin{array}{l} \overline{h}_n = Hh_n = \frac{1}{2}\pi(2n-1), \quad \overline{r}_0 = r_0/H \\[2mm] \overline{q}_n = Hq_n = \sqrt{\left\{\dfrac{(1+iD_s)\overline{h}_n^2 - \overline{a}_0^2}{n^2 + i[(n^2-2)D_v + 2D_s]}\right\}} \\[4mm] \overline{s}_n = Hs_n = \sqrt{\left[\dfrac{(1+iD_s)\overline{h}_n^2 - \overline{a}_0^2}{1+iD_s}\right]} \end{array}\right\} \tag{9}$$

Constants $D_V = \lambda'/\lambda$ and $D_S = G'/G$ represent material damping associated with the volumetric and shear strains respectively. Other notation used for D_S is

$$\tan\delta = \frac{G'}{G} = D_S \qquad (10)$$

where δ = the loss angle; finally, $\bar{a}_o = H\omega/V_S$ and $\eta = V_p/V_S$.

An example of the resistance factor is shown in Fig. 2. The sharp minima and upswings are due to layer resonances.

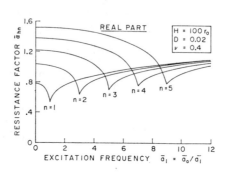

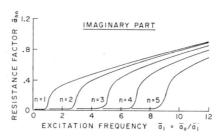

Fig. 2 Variation of resistance factor with frequency for the first five modes of layer.

With the soil reaction described by Eq. 5, the soil-pile interaction problem can be solved from Eq. 4.

The complex stiffnesses are defined as end forces of the pile corresponding to the boundary conditions

$$\left.\begin{array}{ll} u(z = H) = 1, & \psi(z = H) = 0 \\[4pt] u(z = H) = 0, & \psi(z = H) = 1 \end{array}\right\} \qquad (11)$$

A very efficient direct solution was developed that always reduces to a four by four problem of establishing four integration constants. Hence, the computing costs are very low and extensive parametric studies can be conducted (Novak & Nogami 1977).

An example of the complex impedance function, K, for horizontal translation is shown in Fig. 3. The effect of dimensionless frequency ($a_0 = r_0\omega/V_S$), soil layer resonances and material damping can be seen. The effect of material damping on real stiffness (Re K), damping (Im K) and the constant of the equivalent viscous damping (Im K/ω) is further illustrated by Fig. 4.

Marked differences can be seen in the effects of viscous material damping
and hysteretic material damping.

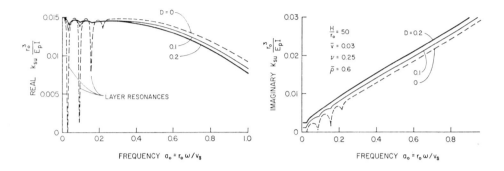

Fig. 3   Effect of material damping on impedance function of embedded
         pile

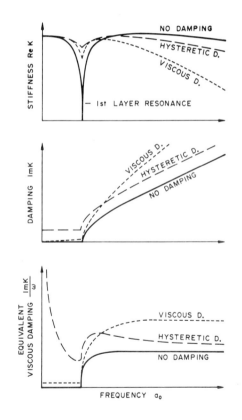

Fig. 4   The effect of material damping on stiffness, total damping and
         equivalent viscous damping of pile (schematic)

189

The response of short and long piles is shown in Fig. 5. The layer resonances and the system resonance may or may not appear depending on material
damping, pile slenderness, tip condition and the stiffness ratio. Long
piles without load are usually overdamped and hence, not very suitable for
dynamic testing. ($\bar{v} = V_s$ in soil/V_p in pile).

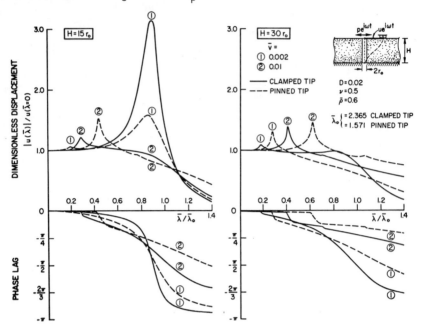

Fig. 5 Response of pile head to horizontal excitation and
corresponding phase lag.

2.2 The Simpler Approach

In the simpler approach, the soil reactions are derived from Eqs. 1 under
the simplifying assumption that only horizontally propagating waves are accounted for. This assumption is equivalent to the assumption of plane
strain which means that the soil reaction is mathematically accurate for an
infinitely long embedded pile undergoing uniform harmonic motions. Then,
the soil resistance to the motion of a unit length of the cylinder can be
defined by complex stiffness constants that are, for vertical direction

$$k_{ws} = G[S_{w1}(a_0, D_s) + i\ S_{w2}(a_0, D_s)]$$ (12)

for horizontal direction

$$k_{us} = G[S_{u1}(a_0, v, D) + i\ S_{u2}(a_0, v, D)]$$ (13)

and for torsion

$$k_{\zeta s} = Gr_0^2[S_{\zeta 1}(a_0, D_s) + i\ S_{\zeta 2}(a_0, D_s)]$$ (14)

In these formulae, v = Poisson's ratio and D_s, D = damping ratios. Parameters $S_{u1,2}$ are obtained as real and imaginary parts of the expression

190

$$k_{us} = \pi G \, a_0^2 T \tag{15}$$

where the dimensionless factor T is (Novak et al. 1977)

$$T = -\frac{4K_1(b_0^\star)K_1(a_0^\star) + a_0^\star K_1(b_0^\star)K_0(a_0^\star) + b_0^\star K_0(b_0^\star)K_1(a_0^\star)}{b_0^\star K_0(b_0^\star)K_1(a_0^\star) + a_0^\star K_1(b_0^\star)K_0(a_0^\star) + b_0^\star a_0^\star K_0(b_0^\star)K_0(a_0^\star)} \tag{16}$$

and $K_{0,1}$ are the modified Bessel functions of complex frequencies $a_0^\star$ and $b_0^\star$. Similar expressions define S_w and S_ζ. Parameters S are shown in Figs. 6 to 8.

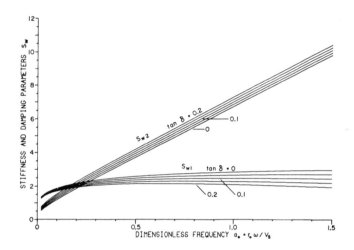

Fig. 6 Parameters S_w

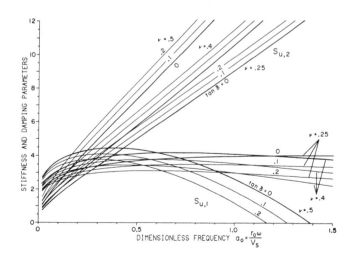

Fig. 7 Parameters S_u

191

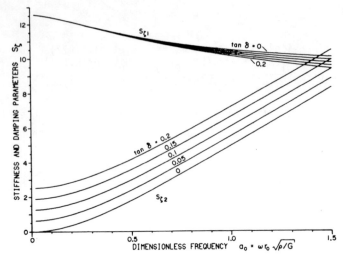

Fig. 8 Parameters S_ζ

With these parameters the horizontal soil resistance in Eq. 4 is

$$p(z) = k_{us} u(z) \tag{17}$$

For the soil reactions defined in this way, close form solutions for displacements and impedance functions can be obtained for all vibration modes.

The comparison of the simpler theory with the more rigorous theory is shown in Fig. 9. With increasing frequency the simpler solution approaches the more rigorous solution as the waves propagate more and more horizontally in agreement with the main assumption of the simpler theory. With frequency a_0 approaching zero, the stiffness from the simpler solution diminishes. Therefore, the stiffness needs to be adjusted in this region as indicated in Fig. 9.

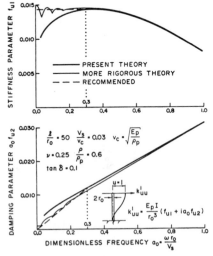

Fig. 9 Comparison of simpler theory with more rigorous theory.

192

The damping is adjusted similarly in order to avoid its overestimation at low frequencies. The comparison with the finite element method is also quite favorable (Kuhlemeyer 1976, Blaney et al. 1976).

The method is much simpler and in some aspects more versatile than the more rigorous approach. E.g., it can be used for friction (floating) piles. Fig. 10 indicates that friction piles yield lower stiffness but higher damping than end bearing piles.

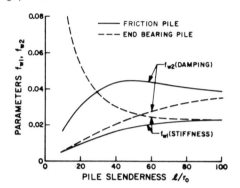

Fig. 10 Dimensionless stiffness and damping of floating (friction) and end bearing piles in the vertical direction ($a_o = 0.3$)

Using the impedance functions of individual piles, the impedance functions of the pile cap or a footing can be calculated. A correction for the effect of pile interaction (group efficiency) may be needed with closely spaced piles. Then, the response to various loads can be predicted. Examples of the response of a rigid footing in different modes are shown in Figs. 11 and 12. The analysis can be formulated in a way analogous to the approach to shallow foundations and individual types of foundations can be compared.

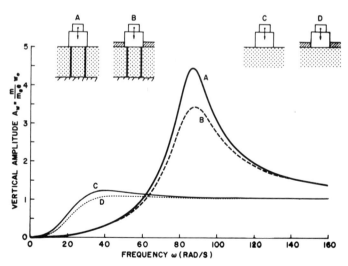

Fig.11 Vertical response of (A) pile foundation, (B) embedded pile foundation, (C) shallow foundation, and (D) embedded shallow foundation

193

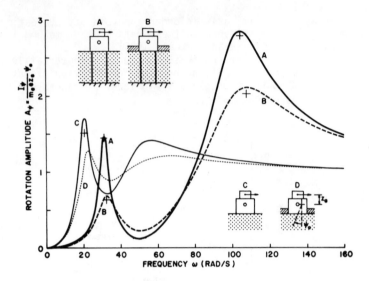

Fig. 12 Rocking component of coupled footing response to horizontal
         load. (A) pile foundation, (B) embedded pile foundation,
         (C) shallow foundation, and (D) embedded shallow foundation.

The simpler approach has the advantage that the tip condition can be re-
laxed and soil layering considered (Fig. 13). This can be achieved by defining the soil reactions using Eqs. 12 to 14 in each soil layer, which may
have different properties, and analyzing the pile in terms of the matrix
stiffness method like in the finite element method.

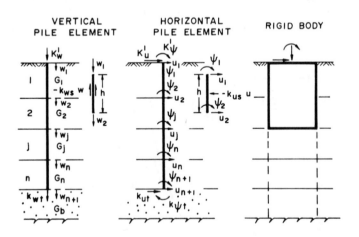

Fig. 13 Pile and rigid body embedded in layered media

From Eq. 4, the element stiffness matrix is obtained and the structure
stiffness matrix, $[K_u]$, assembled. The impedance functions related to the
horizontal translation, u, and rotation, ψ, follow from equation

194

$$
\begin{array}{c}
H_1 : \\
M_1 : \\
H_2 : \\
M_2 : \\
\vdots \\
M_{n+1}
\end{array}
\begin{bmatrix}
\overset{1}{K_{uu}^1} & \overset{2}{K_{u\psi}^1} \\
K_{\psi u}^1 & K_{\psi\psi}^1 \\
0 & 0 \\
0 & 0 \\
\vdots & \vdots \\
0 & 0
\end{bmatrix}
\begin{bmatrix}
\quad \\ \quad \\ \quad \\ \quad \\ \quad \\ \quad
\end{bmatrix}
=
K_u
\begin{bmatrix}
\quad \\ \quad \\ \quad \\ \quad \\ \quad \\ \quad
\end{bmatrix}
\begin{bmatrix}
\overset{1}{u_1 = 1} & \overset{2}{u_1 = 0} \\
\psi_1 = 0 & \psi_1 = 1 \\
u_2 & u_2 \\
\psi_2 & \psi_2 \\
\vdots & \vdots \\
\psi_{n+1} & \psi_{n+1}
\end{bmatrix}
\tag{18}
$$

Analogous equations hold for the other modes. The computing costs associated with the calculation of the impedance functions are negligible. With this method, described in detail in (Novak & Aboul-Ella 1978), arbitrary soil profiles and tip conditions can be considered. The inclusion of these factors greatly increases the versatility of the approach and facilitates good agreement between the theory and field experiments in the range of small amplitudes. Rigid bodies can be treated in the same way if the effects of shear and rotatory inertia are included in Eq. 4. The method is similar to that of Tajimi and Shimomura 1976, 1977.

3 PILE SUPPORTED STRUCTURES

When analyzing pile supported structures, the case of horizontal excitation is of major interest. With the impedance functions available, any approach of structural analysis can be applied. The notable exception may be the case of strong earthquakes where the joint model for soil, piles and structure with the inclusion of nonlinearity may be preferable.

However, in many cases, linearity may be acceptable and modal analysis is sufficient. With horizontal excitation, horizontal translation, u, and rotation (rocking), ψ need to be considered. Then, the pile foundation generates reactions

$$
\left\{ \begin{array}{c} R_x \\ R_\psi \end{array} \right\}
=
\left[\begin{bmatrix} k_{xx} & k_{x\psi} \\ k_{x\psi} & k_{\psi\psi} \end{bmatrix}
+ i\omega
\begin{bmatrix} c_{xx} & c_{x\psi} \\ c_{x\psi} & c_{\psi\psi} \end{bmatrix} \right]
\left\{ \begin{array}{c} u \\ \psi \end{array} \right\}
\tag{19}
$$

in which $i = \sqrt{-1}$, ω = circular frequency and the subscripts indicate the directions involved. The stiffness and damping coefficients, k and c, are frequency dependent but can often be taken approximately as independent of frequency over a range of interest.

When the modal analysis is used, the effect of the foundation on the modal properties of the structure must be established first. This effect can be viewed as three separate actions: (1) Reduction of the natural frequencies and modification of the modal shapes due to foundation flexibility; (2) Generation of damping in the structure through energy dissipation in the soil; (3) The modification (usually reduction) of the original structural damping, which the structure would have for the case of a rigid foundation, due to the flexibility of the foundation.

The foundation effect on the structural modes and frequencies is established using stiffnesses k (Eqs. 19) by standard procedures. The damping ratio generated in the structure due to energy dissipation in soil in mode j is

$$D_j = \frac{1}{2\omega_j M_j} (c_{xx} u_{1j}^2 + c_{\psi\psi} \psi_{1j}^2 + 2c_{x\psi} u_{1j} \psi_{1j})$$ (20)

in which generalized mass

$$M_j = \sum_{i=1}^{n} (m_i u_{ij}^2 + I_i \psi_{ij}^2)$$ (21)

Here, c = equivalent damping constants appearing in Eq. 19, u_{ij}, ψ_{ij} = modal displacements taken to arbitrary scale, u_{1j}, ψ_{1j} = modal displacements of the cap and ω_j = the jth natural frequency of the structure on flexible foundation; m_i, I_i = mass and mass moment of inertia and n = number of masses (Fig. 14).

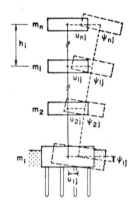

Fig. 14 Structure vibrating in j-th mode.

Denoting all the magnitudes corresponding to the rigid foundation by a bar, the structural damping ratio for the structure on a flexible foundation is

$$D_j^s = \overline{D_j^s} \beta_j$$ (22)

where

$$\beta_j = \frac{\overline{\omega}_j}{\omega_j} \frac{\overline{M}_j}{M_j} (\sum_{i=1}^{n-1} m_{i+1} (u_{i+1,j} - u_{ij} - h_i \psi_{1j})^2) / (\sum_{i=1}^{n-1} m_{i+1} (\overline{u}_{i+1,j} - \overline{u}_{ij})^2)$$ (23)

Coefficient β_j describes the modification of structural damping due to foundation flexibility, usually a slight reduction. The total damping for mode j of a structure on a flexible foundation is

$$D_j^t = D_j^s + D_j$$ (24)

196

These expressions were obtained from energy consideration (Novak 1974a).
Examples of the effect of foundation on modal properties of structures
are shown in Figs. 15 and 16. In these figures, the damping generated by
soil, D_j, and the coefficient of structural damping modification, β_j, are
given for two slender structures having different foundations.

With the modal properties of the structure established, the analysis of
the structure can proceed in the same manner as with shallow foundations.

The resultant effect of foundation flexibility on structural response depends on the properties of the structure and its foundation and also on the
type of excitation. With a given seismic input, the effect of foundation
flexibility is, in most cases, to reduce structural stresses. This was
found by many authors but should not be confused with the possible modification or amplification of the seismic excitation due to local site conditions.
With wind loading, the effect of foundation flexibility is favorable for all
types of aerodynamic excitation. This favorable effect increases with decreasing stiffness of the foundation and can be exploited practically in
controlling the often detrimental vortex induced oscillation of slender
structures if they are founded on soft soil. This is shown in more detail
in (Novak 1977b).

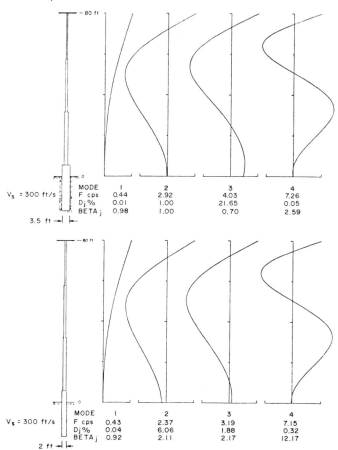

Vs = 300 ft/s	MODE	1	2	3	4
	F cps	0.44	2.92	4.03	7.26
	Dj%	0.01	1.00	21.65	0.05
	BETA j	0.98	1.00	0.70	2.59

3.5 ft

Vs = 300 ft/s	MODE	1	2	3	4
	F cps	0.43	2.37	3.19	7.15
	Dj%	0.04	6.06	1.88	0.32
	BETA j	0.92	2.11	2.17	12.17

2 ft

Fig. 15 Modal properties of lighting pole founded on deep foundation

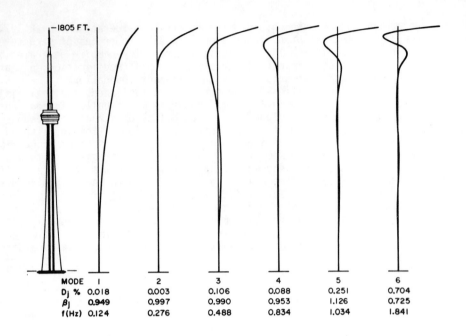

MODE	1	2	3	4	5	6
D_j %	0.018	0.003	0.106	0.088	0.251	0.704
β_j	0.949	0.997	0.990	0.953	1.126	0.725
f(Hz)	0.124	0.276	0.488	0.834	1.034	1.841

Fig. 16   Modal properties of Toronto CN Tower founded on shallow
foundation.

4   FURTHER RESEARCH

Further research is needed, particularly into the effect of interaction of
piles (grouping), nonlinearity, batter piles and effects of strong earthquakes. Some information on these aspects of pile behavior can be found in
Poulos 1971, Poulos and Madhav 1971 and Tajimi 1977.

5   ACKNOWLEDGEMENTS

The research was funded by the National Research Council of Canada. The
assistance of T. Nogami, J.F. Howell and F. Aboul-Ella is gratefully acknowledged.

6   REFERENCES

Blaney, G.W., E. Kausel and J.M. Roesset, 1976, Dynamic Stiffness of Piles,
Proc. of the Second Int. Conf. on Numerical Methods in Geomechanics,
Virginia Polytechnic Institute and State University, Blacksburg, Virginia,
U.S.A., June, 1976, ASCE, New York, pp. 1001-1012.

Bielak, J. and V.J. Palencia, 1977, Dynamic Behaviour of Structures With
Pile Supported Foundations, Proc. of the Sixth World Conf. on Earthquake
Engrg., New Delhi, Vol. 4, pp. 115-120.

Kobori, T., R. Minai and K. Baba, 1974, Torsional Vibration Characteristics of the Interaction Between a Cylindrical Elastic Rod and Its Surrounding Viscoelastic Stratum, Proc. of the 24th Japan National Congress for Appl. Mech., Theor. and Appl. Mech., Vol. 24, Univ. of Tokyo Press, pp. 197-205.

Kobori, T., R. Minai and K. Baba, 1977a, Dynamic Interaction Between An Elastic Cylinder and Its Surrounding Visco-Elastic Half-Space, Proc. of the 25th Japan National Congress for Appl. Mech., 1975, Theor. and Appl. Mech., Vol. 25, Univ. of Tokyo Press, pp. 215-226.

Kobori, T., R. Minai and K. Baba, 1977b, Dynamic Behavior of a Laterally Loaded Pile, Ninth Intern. Conf. on Soil Mech. and Found. Engrg., Spec. Session No. 10, Tokyo, p. 6.

Kuhlemeyer, R.L. 1976, Static and Dynamic Laterally Loaded Piles, Res. Rep. No. CE 76-9, Dept. Civ. Engrg., Univ. Calgary, Calgary, Alta., Canada, p. 48.

Nogami, T. and M. Novak, 1976, Soil-Pile Interaction in Vertical Vibration, Intern. J. Earthquake Engrg. and Struct. Dyn., Vol. 4, No. 3, pp. 277-293.

Nogami, T. and M. Novak, 1977, Resistance of Soil to a Horizontally Vibrating Pile, Int. J. Earthquake Engrg. and Struct. Dyn., Vol. 5, No. 3, pp. 249-262.

Novak, M. 1974a, Effect of Soil on Structural Response to Wind and Earthquake, Int. J. Earthquake Engrg. and Struct. Dyn., Vol. 3, No. 1, pp. 79-96. Additional Note in Vol. 3, 1975, pp. 312-315.

Novak, M. 1974b, Dynamic Stiffness and Damping of Piles, Canadian Geotechnical J., Vol. 11, No. 4, pp. 574-598.

Novak, M. 1977a, Vertical Vibration of Floating Piles, J. Engrg. Mech. Div. ASCE, Vol. 103, No. EM1, February, pp. 153-168.

Novak, M. 1977b, Soil-Structure Interaction Under Wind Loading, in Recent Advances in Engineering Science, ed. G.C. Sih, Lehigh Univ. Publication, pp. 1099-1110.

Novak, M. and T. Nogami, 1977, Soil Pile Interaction in Horizontal Vibration Int. J. Earthquake Engrg. and Struct. Dyn., Vol. 5, No. 3, pp. 263-282.

Novak, M. and R.F. Grigg, 1976, Dynamic Experiments With Small Pile Foundations, Canadian Geotechnical J., Vol. 13, No. 4, pp. 372-385.

Novak, M. and J.F. Howell, 1977, Torsional Vibration of Pile Foundations, J. Geotech. Engrg. Div., ASCE, Vol. 103, No. GT4, pp. 271-285.

Novak, M., T. Nogami and F. Aboul-Ella, 1977, Dynamic Soil Reactions for Plane Strain Case, Res. Report, No. BLWT-1-77, Fac. of Engrg. Sci., Univ. of Western Ontario, pp. 1-26.

Novak, M. and F. Aboul-Ella, 1978a, Impedance Functions for Piles in Layered Media, J. Engrg. Mech. Div. ASCE.

Novak, M. and F. Aboul-Ella, 1978b, Stiffness and Damping of Piles in Layered Media, ASCE Specialty Conf. on Earthquake Engrg. and Soil Dynamics, Pasadena, Calif., p. 16.

Penzien, J. 1970, Soil-Pile Foundation Interaction, Ch. 14 in Earthquake Engrg. (Ed. R.L. Wiegel et al), Prentice-Hall, pp. 349-381.

Poulos, H.G. 1971, Behaviour of Laterally Loaded Piles: II - Pile Groups, Soil Mech. and Foundation Div., ASCE, 97 (SM5), pp. 733-751.

Poulos, H.G. and M.R. Madhav, 1971, Analysis of the Movements of Battered Piles, Proc. 1st Australia-New Zealand Conf. on Geomechanics, Melbourne, Vol. 1, pp. 268-275.

Shimizu, N., S. Yamamoto, Y. Koori and N. Minowa, 1977, Three-Dimensional Dynamic Analysis of Soil-Structure Systems by Thin Layer Element Method, (Part 2), Trans. of the Architectural Inst. of Japan, No. 254, pp. 39-47.

Tajimi, H. 1966, Earthquake Response of Foundation Structures, Report of the Fac. Sci. Engrg., Nihon Univ., 1966.3, pp. 1.1-3.5. (In Japanese).

Tajimi, H. 1969, Dynamic Analysis of a Structure Embedded in an Elastic Stratum, Proc. 4th WCEE (Chile).

Tajimi, H. and Y. Shimomura, 1976, Dynamic Analysis of Soil-Structure Interaction by the Thin Layer Element Method, Trans. Architectural Inst. of Japan, No. 243, pp. 41-51. (In Japanese).

Tajimi, H. and Y. Shimomura, 1977, Dynamic Analysis of a Single Pile Embedded in Horizontally Layered Soils, private communication, p. 32.

Tajimi, H. 1977, Seismic Effects on Piles, State-of-the-Art Report, Spec. Session 10, Ninth Int. Conf. Soil Mech. and Found. Engrg., Tokyo, p. 12.

Effective shear modulus beneath dynamically loaded foundations on sand

KENNETH H. STOKOE, II
University of Texas, Austin, Tex., USA

Discussion Contribution to the Workshop:
"Lumped Parameter Models"

1 DISCUSSION

For low-amplitude dynamic loadings such as those generated by machine vibrations, the machine-foundation-soil system is typically idealized in the analytical model as a rigid circular cylinder resting on or partially embedded in an elastic half space (Richart et al 1970). The key soil property used to represent the elastic half space is the shear modulus, G. For foundations on a soil with a varying shear modulus, the question arises as to the appropriate value or values of G to use in the analytical model.

Several experimental studies have been reported in which the effective shear modulus for use in a lumped-parameter model was determined for dynamically loaded foundations resting on sand (Drnevich and Hall 1966, Richart et al 1970, Stokoe 1972, and Erden 1974). The shear modulus of the sand under these low-amplitude dynamic loadings could be approximated by (Hardin and Drnevich 1972):

$$G = C \frac{(2.973 - e)^2}{1 + e} \sqrt{\bar{\sigma}_o} \qquad (1)$$

in which C equals a constant which may vary with sand type and is usually taken as 1230, e is the void ratio, $\bar{\sigma}_o$ is the effective mean principal stress in psi, and G is the shear modulus in psi. Equation 1 clearly shows how the shear modulus of the sand varied with stress level beneath the foundations in these experimental studies.

To use the lumped-parameter model, a single value of G, hence an effective shear modulus, had to be determined for the sand beneath the foundations. The effective shear modulus was calculated by: 1. determining the minimum vertical effective stress beneath the outside edge of the foundation, $\bar{\sigma}_{v(min)}$, following the procedure outlined by Richart et al 1970 with the theoretical solution developed by Prange 1965, and 2. substituting $\bar{\sigma}_{v(min)}$ or some fraction of $\bar{\sigma}_{v(min)}$ for $\bar{\sigma}_o$ in Equation 1.

Stokoe 1972 and Erden 1974 have shown with model studies that the effective shear modulus varies with mode of vibration. For vertical excitation of cast-in-place concrete foundations with circular bases, the effective shear modulus can be determined from Equation 1 by substituting in $\bar{\sigma}_{v(min)}$ for $\bar{\sigma}_0$. However, for the same foundations excited in torsional or first-mode coupled rocking and sliding motion, the effective shear modulus should be determined by substituting $0.67\,\bar{\sigma}_{v(min)}$ for σ_0 in Equation 1.

Erden 1974 has also shown that the effective shear modulus varies with pressure distribution beneath the foundation base. This effect is most pronounced in vertical motion and for all practical purposes can be neglected in torsional and first-mode coupled rocking and sliding motions. In vertical motion, as weight is added to the foundation after casting, the pressure distribution beneath the base changes which results in a change in the effective shear modulus. This change in G can be accounted for by a change in the fraction of $\bar{\sigma}_{v(min)}$ which is used in Equation 1. In Erden's study, as the weight added to the model footings varied from zero to a weight equal to the original cast-in-place weight of the models, the fraction of $\bar{\sigma}_{v(min)}$ used in Equation 1 varied about linearly from 1.0 to 0.67.

Erden 1974 has shown that this approach for determining an effective shear modulus of sand is also valid for foundations with rectangular bases. The rectangular base is approximated by an equivalent circular base using the relationships presented by Richart et al 1970. The weight of the rectangular foundation is then used with the equivalent circular base to determine $\bar{\sigma}_{v(min)}$.

It is interesting to note that in these studies $\bar{\sigma}_{v(min)}$ occurred at a depth beneath the foundation edge about equal to the radius of the circular foundations for cast-in-place foundations with no added weight. As weight was added to the circular foundations, the depth of $\bar{\sigma}_{v(min)}$ increased. For rectangular foundation with no added weight $\bar{\sigma}_{v(min)}$ occurred at depths less than one equivalent radius. This depth also increased as weight was added.

Finally, it should be noted that the results by Stokoe 1972 and Erden 1974 were generated for surface model footings with small imposed stresses on the sand, stresses well below those required to cause bearing capacity failure. As the imposed stress level increases, the effective modulus may change. In addition, all footings were initially cast-in-place. It was found that if the foundations were not cast-in-place the effective shear modulus varied rather erratically and was always less than the effective shear modulus determined by using $0.67\,\bar{\sigma}_{v(min)}$.

2 REFERENCES

Drnevich, V.P. and J.R. Hall, Jr., 1966, Transient Loading Tests on a Circular Footing, J. Geotechn. Engrg. Div. Proc. ASCE 92 (SM6): 153-167

Erden, S.M. 1974, Influence of Shape and Embedment on Dynamic Foundation Response, Ph.D. Dissertation, Univ. of Massachusetts, Amherst.

Hardin, B.O. and V.P. Drnevich, 1972, Shear Modulus and

Damping: Design Equations and Curves, J. Soil Mech. Found
   Div., Proc. ASCE 98 (SM7): 667-692.
Prange, B. 1965, Ein Beitrag zum Problem der Spannungsmessung
   im Halbraum, Ph.D. dissertation, Technischen Hochschule
   Karlsruhe, Germany, O. Berenz, Karlsruhe.
Richart, F.E., Jr., J.R. Hall, Jr. and R.D. Woods 1970,
   Vibration of Soils and Foundations, Englewood Cliffs,
   New Jersey, Prentice-Hall, Inc.
Stokoe, K.H., II, 1972, Dynamic Response of Embedded Founda-
   tions, Ph.D. Dissertation, Univ. of Michigan, Ann Arbor.

Determination of spring constants in SSI by a finite element method

H. WERKLE
University of Karlsruhe, Karlsruhe, Germany

Discussion Contribution to the Workshop:
"Lumped Parameter Models"

The classical methods for the determination of soil spring constants are based on the theory of elastic halfspace. Embedded foundations and different soil properties in the environment of the foundation can be considered in a numerical analysis by the finite element method. A dynamic finite element analysis in soil mechanics differs from a finite element analysis of two- or three-dimensional structures by the following:

The vibrations of a foundation cause waves propagating in the surrounding soil. When these waves in a finite element analysis reach the boundary of the finite element net they will be reflected at the boundary and the energy of these waves will remain within the soil region defined in the finite element model. Therefore the net must be large enough to keep the influence of the reflections at the boundary small (Vaish et al., 1974). This causes extended finite element models. A possibility to prevent the reflection at the model boundary is to use finite boundary elements with the stiffness and damping properties of the infinitely extended layered soil. Since stiffness matrices of these boundary elements are frequency-dependent, the analysis is performed in the frequency domain.

Finite boundary elements for two-dimensional finite element analysis were given by Waas (Waas, 1972). The soil in the boundary element is arbitrarily layered. A similar boundary element of axisymmetric structures with non-axisymmetric deformation was given by Kausel (Kausel, 1974).

For the nuclear power plant given in Figure 1, a calculation of the soil spring constants was made. The building has a circular foundation with a diameter of 22 m and is embedded in the depth of 13 m. The soil in the immediate surroundings of the power plant was filled and compacted below the later constructed building which resulted in the complicated soil conditions given in Figure 1. For the determination of the frequency-dependent stiffness and damping constants a finite element analysis of an axisymmetric model according to Kausel (Kausel, 1974) was performed.

The soil properties and the finite element net used in the analysis are given in Figure 2. As no defined bedrock exists, the depth of the finite element net must be chosen large enough so that its influence on the soil spring constants can be neglected. For comparison purposes a less deep soil model, as given in Figure 3, was analysed in addition.

A comparison of the soil spring constants resulting from the analysis of the two soil models shows that only the vertical spring constant is significantly influenced by the depth of the two finite element models. For rocking, horizontal and torsional vibrations, only a small difference can be noted between the spring constants of the two models (Figures 4-7).

The spring constants are given as complex numbers, in dependence of the excitation frequency. Instead of introducing a complex spring constant $\hat{k}$, it is possible to use a real spring constant k and a damping constant c, where

$$k = \text{real part } (\hat{k})$$

$$c = \frac{\text{imaginary part } (\hat{k})}{\Omega}$$

and $\Omega = 2\pi f$ as circular frequency of excitation.

The differential equation for a hysteretically damped one-degree-of-freedom-system with mass m and spring constant k excited by a force P with frequency Ω is:

$$m \cdot \ddot{x} + 2\frac{k}{\Omega}\, \xi \cdot \dot{x} + k \cdot x = P \cdot e^{i\Omega t} \qquad (2)$$

The value $\xi = \dfrac{c \cdot \Omega}{2 \cdot k}$ is the hysteretic damping constant. (3)

As k and c are frequency-dependent, the damping value also depends on the excitation frequency f. In the resonance case, the hysteretic damping constant equals the viscous damping constant and gives an idea of the amount of damping at different frequencies.

At low frequencies the damping constant ξ equals the material damping of soil (1%). At higher frequencies the constant increases because of the radiation damping added to material damping and reaches values of more than 50%. For horizontal and vertical vibrations, as expected, the effect of radiation damping was particularly great.

The analysis was performed as part of the HDR-Safeguard Research Program (KFK-Fachbericht, 1978; Müller-Dietsche et al. , 1977). During this programme extensive soil investigations and vibration measurements at the building were made by other workers (KFK-Fachbericht, 1978; Gundy et al. , 1977). A comparative vibration analysis of the building was performed with computer code SAP IV using a beam model. The lowest eigenfrequencies were found to be influenced mainly by the value of the soil spring constants and far less by a variation of the structural properties of the building. In order to compare the measured values with the calculated ones, the soil spring constants were varied until the lowest calculated eigenfrequencies equalled the lowest measured eigenfrequencies. The soil spring constants obtained are also given in Figures 4-7. They accord well with the values obtained independently by the finite element analysis.

REFERENCES

Gundy, W. E. , Howard, G. E. , Ibanez, P. , Keowen, R. S. , Smith, C. B. , Spencer, R. B. , Taylor, G. B. and W. B. Walton. A comparison of vibration tests and analysis on nuclear power plant structures and piping Transactions of the 4th International Conference on Structural Mechanics in Reactor Technology, San Francisco, California, August 1977. Paper K 8/6

Kausel, E. Forced Vibrations of Circular Footings on Layered Media. Research Report R74-11, MIT, Cambridge, Massachusetts, January 1974.

KFK-Fachbericht, Kernforschungszentrum Karlsruhe, Abteilung PHDR, Postfach 3640, 7500 Karlsruhe 1, Germany. 1978. In preparation.

Müller-Dietsche, W. , Jehlicka, P. and H. Steinhilber. Objectives of seismic tests in the HDR Safeguards Research Program. Transactions of the 4th International Conference on Structural Mechanics in Reactor Technology, San Francisco, California, August 1977. Paper K 8/5.

Vaish, A. K. and A. K. Chopra. Earthquake Finite Element Analysis of Structure - Foundation Systems, ASCE, EM 6, December 1974.

Waas, G. Analysis Method for Footing Vibrations through Layered Media. Technical Report S-71-14 US Army Engineer Waterways Experiment Station. Vicksburg, Mississippi, September 1972.

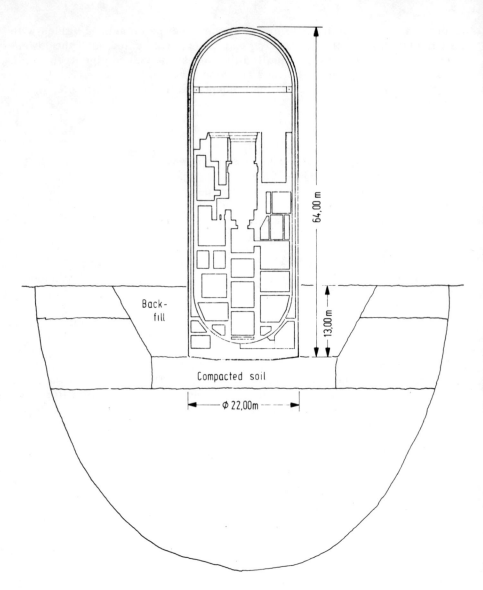

Fig.1   Embedded Nuclear Power Plant

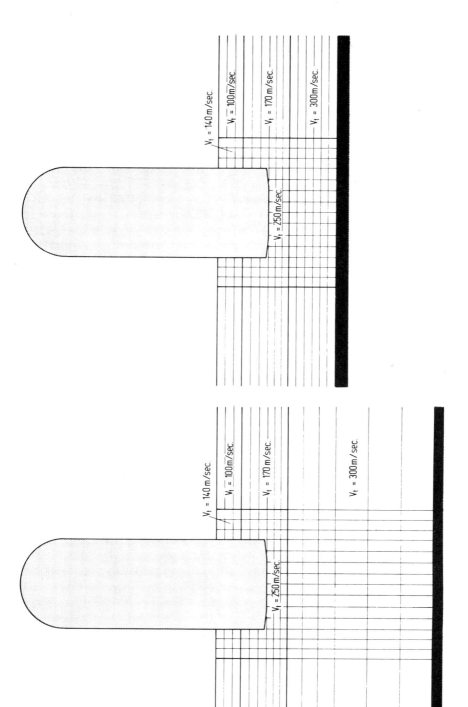

Fig. 2   Finite element idealization   Model 1

Fig. 3   Finite element idealization   Model 2

209

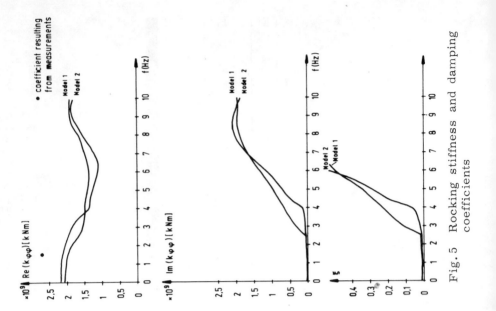

Fig.5 Rocking stiffness and damping coefficients

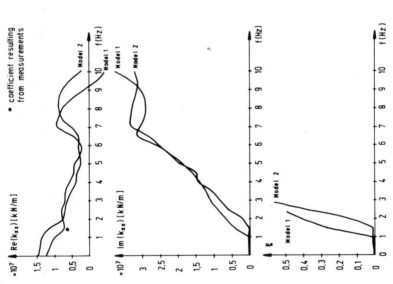

Fig.4 Horizontal stiffness and damping coefficients

210

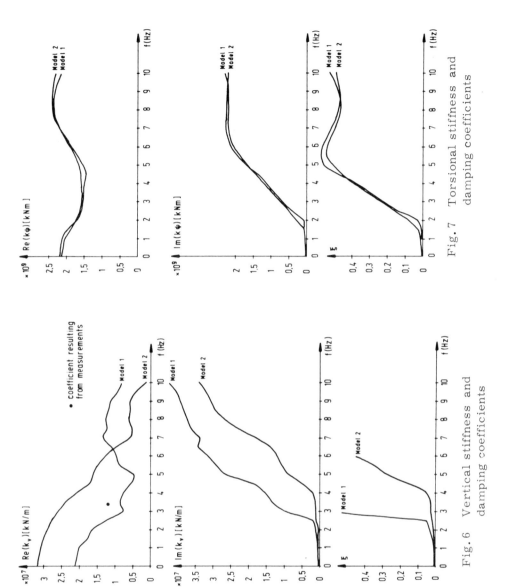

Fig. 6 Vertical stiffness and damping coefficients

Fig. 7 Torsional stiffness and damping coefficients

211

Dynamic behavior of embedded foundations

EDUARDO KAUSEL
Stone & Webster Engineering Corporation, Boston, Mass., USA

Discussion Contribution to the Workshop:
"Lumped Parameter Models"

With the advent of fast digital computers, and particularly
following the development of finite element techniques, many
difficult problems involving the dynamic response of structures
to a seismic excitation, accounting for foundation interaction
effects, became amenable to calculation. In earlier times, it
was only possible to analyze idealized cases of structures resting on circular or rectangular rigid plate foundations underlain by elastic, homogeneous, isotropic half spaces. More recently, however, analytical or closed-form solutions have also
been developed for (geometrically) more complicated plates
welded to viscoelastic, layered half spaces. Nevertheless most,
if not all, of these newer solution disregard embedment, and
are thus strictly applicable to surface foundations only. This
embedment has in many cases considerable effect on the dynamic
response of the structure, both in terms of relative frequency
contents and amplitude of the resulting motions. While rigorous
analytical solutions for embedded foundations are nonexistent
at present, it is possible to use approximations and corrections
to the available results for surface foundations, to account
for the increase in stiffness (Elsabee 1975, Kausel et al. 1977,
Novak 1973) and the scattering of the incident seismic waves by
the embedded walls (Kausel et al. 1977, Morray 1975). These
approximate solutions are particularly useful in cases where
direct analyses with finite elements are not practical or feasible, either when the ultimate use of the analysis does not
warrant the added cost or refinement, or when the geometry -
fully three dimensional - cannot adequately be modeled with
available programs based on plane-strain axisymmetric formulations.

KINEMATIC INTERACTION

Soil structure interaction effects arise on the one hand by relative motions of the foundation with respect to the surrounding soil - the free field - as a result of inertial forces in
the structure being transmitted to the compliant soil. This
effect can be defined as Inertial Interaction (Ad hoc group
ASCE). On the other hand, interaction effects in the absence

of inertial forces can also arise because the stiffer structural foundation cannot conform to the distortions of the soil genera- ted by the travelling waves. For ideally rigid, massless found- ations, this effect depends only on the geometry of the found- ation, the soil configuration, and the travel path of the seis- mic excitation across the soil-structure interface. It is ge- nerally referred to as Wave Passage Effects, Scattering of the Seismic Waves by a Rigid Foundation or Kinematik Interaction (Ad hoc group ASCE). For embedded foundations, this effect is always present, irrespective of the seismic environment pro- scribed, and is one of the chief sources of discrepancy between traditional lumped spring interaction analyses and more involved procedures using finite elements.

It can be shown that in order to obtain consistent results using both approaches, it would be necessary to prescribe at the base of the soil "springs and dashpots" a support motion consisting of translations and rotations (Kausel, Roesset,1974). The components of this support motion vector U_b are related to the displacements U and internal forces P along the soil- structure interface in the free field (i.e., before any struc- ture has been built), and to the subgrade impeance matrix X for the foundation under consideration. This matrix represents the forces acting on the soil along the soil-structure interface, necessary to produce unit harmonic displacements of these same points when there is no seismic excitation and the soil has been excavated. For a structure embedded in a viscoelastic, hori- zontally layered half space or stratum, the solution to the free field problem U , P can be obtained analytically or numerically for any train of waves, and in particular, for any combination of plane body and surface waves. However, obtaining the sub- grade impedance matrix X is not an easy task for the stated conditions except for the case of zero embedment (Gazetas, Roesset 1976).

In most cases, only the global subgrade stiffness matrix $K_0 = T^T X T$ (where T is a rectangular rigid body transformation matrix) containing the frequency dependent soil "springs" is available, either from an analysis with finite elements or from published approximate solutions. This matrix K_0 is not suffi- cient to solve the free field problem, that is, to determine the support motion vector (Novak 1973)

$$U_b = K_0^{-1} T^T (P + XU)$$ (1)

because the inverse transformation, from stiffness matrix to im- pedance matrix, is not defined. Note, however, that the sub- grade impedance matrix X and the subgrade stiffness matrix K_0 are independent of the seismic environment, except for possible nonlinear effects which are not being considered here. There- fore, parametric studies can easily be conducted once any of these matrices is available.

Most of the research efforts for embedded foundations have in the past been directed towards determining approximate so- lutions of the stiffness matrix K_0 (Elsabee 1975, Novak 1973 etc.). More recently, the problem of kinematic interaction, expressed formally by equation 1 above, has also received some attention (Kausel et al. 1977, Morray 1975) although much

214

remains to be studied. It is true that some earlier papers have investigated the problem of wave passage on soil-structure interaction, but their models were generally simple (Winkler or Boussinesq foundation impedances; sinusoidal wave train etc.) and did not account for embedment. While these solutions have a theoretical interest, they are inadequate in a practical situation involving deep embedment.

A point of concern is the simulation of material nonlinear effects in the soil when the impedance (spring or lumped parameter) method is used, since the method is based on the principle of superposition. In particular, the components of the subgrade stiffness matrix K_0 and support motion vector U_b are sensitive to a detachment of the lateral walls with the surrounding backfill. Admittedly, this situation is rarely given attention in direct finite element analysis as well, although it can have an effect as important as that of the embedment itself. Some debate has also centered around the use of one single wave pattern to define the free field motion (vertically propagating shear waves; Raleigh waves; Love waves, etc.) which could result in unrealistic spation variations of the seismic motions with depth. The resulting deamplification (or lack of it) would lead to unconservative estimates for the motions of the structure, or greatly exaggerate wave passage effects at selected frequencies. Under such conditions, it might be better to regard the design motion as an average motion in the vicinity of the structure and use it directly at the support of the springs, without correction for kinematic interaction. It is hoped that more work will be done to provide relatively simple "engineering" solutions to the problems posed by embedded foundations so that a greater confidence can be placed in the resulting dynamic analyses, and realize savings in design through smaller factors of safety.

REFERENCES

"Analyses for Soil-Structure Interaction Effects for Nuclear Power Plants", Report by the Ad Hoc Group on Soil— Structure Interaction, Nuclear Structures and Materials Committee of the Structural Division of ASCE

Elsabee, F. 1975, Static Stiffness Coefficients for Circular Foundations Embedded in an Elastic Medium, Thesis presented to the Massachusetts Institute of Technology, at Cambridge, Massachusetts, in 1975, in partial fulfillment of the requirements for the degree of Master of Science

Gazetas, G.C., J.M. Roesset 1976, Forced Vibrations of Strip Footings on Layered Soils, Proceedings of the National Structural Engineering Conference, ASCE, held August 1976 in Madison, Wisconsin, Vol. 1, pp. 115-131

Kausel, E., J.M. Roesset 1974, Soil-Structure Interaction Problems for Nuclear Containment Structures, Electric Power and the Civil Engineer, ASCE, Proceedings of the Power Division Specialty Conference held in Boulder, Colorado, August 1974

Kausel, E., R.V. Whitman, J.P. Morray, F. Elsabee 1977, The Spring Method for Embedded Foundations, paper submitted for possible publication, Cambridge, Massachusetts, 1977

215

Morray, J.P. 1975, The Kinematic Interaction Problem of Embedde
    Circular Foundations, Thesis presented to the Massachusetts
    Institute of Technology, at Cambridge, Massachusetts, in 1975
    in partial fulfillment of the requirements for the degree of
    Master of Science
Novak, M. 1973, Vibrations of Embedded Footings and Structures,
    presented at the ASCE National Structural Meeting, held
    April 1973 in San Francisco, California, Meeting Preprint
    No. 2029

A simple lumped parameter model
for foundation interaction

B. PRANGE
University of Karlsruhe, Karlsruhe, Germany

Discussion Contribution to the Workshop:
"Lumped Parameter Models"

SYNOPSIS

Based on the well known lumped parameter models of SSI a simple
model is presented to calculate the interaction of a number of
foundations subjected to vibrations of the same angular frequency . The total system is described as a multi-degree-of-
freedom system with system coupling coefficients and transfer
coefficients.

INTRODUCTION

In cases of foundations subjected to vibrations and belonging
to either the same superstructure or to adjacent structures, in
addition to the structure-soil-interaction the effects of
structure-soil-structure interaction have to be considered.
This consideration in general leads to a very complex multi-
degree-of-freedom system analysis if the actual geometry and
the flexibility of such foundations have to enter the calculations. In this paper, a simple lumped parameter model shall
be presented in analogy to the lumped parameter models of rigid
foundations on the homogeneous, elastic, isotropic halfspace.
The only parameters needed in addition to the well known lumped
parameters of foundations are the properties of the free wavefields generated by the vibrating foundations.
  Otherwise, the same assumptions made for lumped parameter
models will be made here with the necessity to prove their
validity.

VISCO-ELASTIC SUPPORT

The motion of a structure on a wavefield can be described by a
visco-elastic support as seen in fig.1 for the example of one
degree of freedom (vertical translation).
  The node $\bar{k}$ defines the contact point of the freefield with
the harmonic motion

$$y_{\bar{k}}(t) = A_{\bar{k}} \, e^{i\varepsilon_{\bar{k}}} \, e^{i\omega t} \qquad (1)$$

with $A_{\bar{k}}$ = Amplitude at $\bar{k}$
                $\varepsilon_{\bar{k}}$ = Phase angle at $\bar{k}$
                ω = Angular frequency

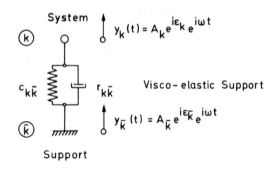

Fig.1   Structure excitation due to support motion
of visco-elastic support

The node k defines the contact point of the spring reaction $c_{k\bar{k}}$
and the damping reaktion $r_{k\bar{k}}$ with the structure, i.e. the
distorted wavefield.
It has the motion

$$y_k(t) = A_k \, e^{i\varepsilon_k} \, e^{i\omega t} \tag{2}$$

with the notations above.
   The indices k and $\bar{k}$ have to be used because with respect to
the geometry of the superstructure it is one and the same system
point k, the spring- and damping reaction models having a fictitious length k-$\bar{k}$.
   Phase angles $\varepsilon_{\bar{k}}$ have to be introduced because the freefield
exhibits different phase relations at the nodes $\bar{k}$, $\bar{i}$, ...
   The force acting on node k due to the rigid support motion
is then

$$S_k(t) = A_{\bar{k}} \, e^{i\varepsilon_{\bar{k}}} \, (c_{k\bar{k}} + i_{k\bar{k}} \, \omega) \, e^{i\omega t} \tag{3}$$

with the force amplitude

$$|S_k| = A_{\bar{k}} \, \sqrt{c_{k\bar{k}}^2 + (\omega \, r_{k\bar{k}})^2} \tag{4}$$

and the phase angle

$$\phi_k = \varepsilon_{\bar{k}} + \varepsilon_{k\bar{k}} = \varepsilon_{\bar{k}} + \text{atan} \, r_{k\bar{k}}\omega / c_{k\bar{k}} \tag{5}$$

If we apply the force of equ.3 to node k, the rigid support $\bar{k}$
is now formally rest and we end up with a structure loaded by
external forces rather than external vibrations.
   If the spring $c_{k\bar{k}}$ and the dashpot $r_{k\bar{k}}$ would not deform at
the presence of freefields, the node k would vibrate

synchronously with the freefield motion of node $\bar{k}$. If, however, we have a vectorial difference in the motions of k and $\bar{k}$,

$$y_{k\bar{k}}(t) = y_k(t) - y_{\bar{k}}(t) = (A_k \, e^{i\varepsilon_k} - A_{\bar{k}} \, e^{i\varepsilon_{\bar{k}}}) \, e^{i\omega t} \tag{6}$$

an additional interference-wavefield will be generated at k with waves propagating to other surface points, for example to $\bar{i}$.

TRANSFER COEFFICIENTS

Since we have replaced the support-excitation by external forces (or moments), the structure under consideration consists now of a multi-degree-of-freedom system with localized masses, springs and dashpots with the respective system-coupling-coefficients and the so called transfer-coefficients describing the interference wavefields propagating from k to i.

The interference wavefield at i due to non-synchronous motion (equ.6) at k can be described by

$$y_{ik}(t) = \bar{y}_k(t) \; g_{ik} \; d_{ik} \; e^{i\varepsilon_{ik}} = \bar{y}_k(t) \; t_{ik} \tag{7}$$

where

$$
\begin{aligned}
\bar{y}_k(t) &= \text{source vibration at } k \\
g_{ik} &= \text{attenuation due to geometric damping} \\
d_{ik} &= \text{attenuation due to material damping} \\
\varepsilon_{ik} &= \text{phase difference between } i \text{ and } k \\
t_{ik} &= \text{transfer coupling coefficient}
\end{aligned}
$$

If the source vibration is identified with the difference in motion, equ.6, i.e. the interference wavefield generated at k, we find

$$y_{ik}(t) = (A_k \, e^{i\varepsilon_k} - A_{\bar{k}} \, e^{i\varepsilon_{\bar{k}}}) \, t_{ik} \, e^{i\omega t} \tag{8}$$

The freefield vibration at i and the interference wavefield vibration at i (propagating from k) have to be added as vectors.

The system of differential equations for a multi-degree-of-freedom system is represented by the following equation for degree of freedom i:

$$i: \quad \sum_{k=1}^{n} \left[m_{ik} \, \ddot{y}_k(t) + r_{ik} \, \dot{y}_k(t) + c_{ik} \, y_k(t) \right] + S_i(t) = 0 \tag{9}$$

with

$$
\begin{aligned}
m_{ik} &= \text{inertia effect in } i \text{ due to unit acceleration at } k \\
r_{ik} &= \text{damping effect in } i \text{ due to unit velocity at } k \\
c_{ik} &= \text{spring effect in } i \text{ due to unit displacement at } k
\end{aligned}
$$

The solution for harmonic motion

$$y_k(t) = A_k \, e^{i\omega t} \, e^{i\varepsilon_k} \tag{10}$$

219

yields

$$i: \sum_{k=1}^{n} [A_k \, e^{i\varepsilon_k} \, s_{ik}] + S_i \, e^{i\varphi_i} = 0 \tag{11}$$

where $\quad s_{ik} = (-m_{ik} \, \omega^2 + i \, r_{ik} \, \omega + c_{ik})$

and

$\quad \varphi_i$ = phase angle of the external exciting quantity $S_i(t$

If we transform the support excitation (freefield and interference field) into external exciting quantities (forces or
moments resp.) as seen in equ.3, equ.11 has to be supplemented
with regards to equ.8 as follows:

$$i: \sum_{k=1}^{n} [A_i \, e^{i\varepsilon_k} \, s_{ik} + A_{\bar{k}} \, e^{i\varepsilon_{\bar{i}}} \, (c_{i\bar{i}} + i \, \omega \, r_{i\bar{i}})] +$$

$$+ \sum_{k=1}^{n} (A_i \, e^{i\varepsilon_k} - A_{\bar{k}} \, e^{i\varepsilon_{\bar{k}}}) \, t_{ik} \, (c_{i\bar{i}} + i \, \omega \, r_{i\bar{i}}) +$$

$$+ S_i \, e^{i\varphi_i} = 0 \tag{12}$$

for one angular frequency ω.

Rearranging equ.12 we arrive at

$$i: \sum_{k=1}^{n} A_k \, e^{i\varepsilon_k} [s_{ik} + t_{ik} \, (c_{i\bar{i}} + i \, \omega \, r_{i\bar{i}})] +$$

$$+ \sum_{k=1}^{n} A_{\bar{k}} \, e^{i\varepsilon_{\bar{k}}} [\bar{t}_{ik}(c_{i\bar{i}} + i \, \omega \, r_{i\bar{i}})] + S_i \, e^{i\varphi_i} = 0 \tag{13}$$

with $\quad \bar{t}_{ik} = \begin{array}{l} -t_{ik}; \quad i \neq k \\ 1 \quad ; \quad i = k \end{array}$

or in vector form

$$\{A\}^T \, [S,T] + \{\bar{A}\}^T \, [\bar{T}] + \{S\} = 0 \tag{14}$$

where $\quad \{A\}$ = n-dimensional vibration vector of the
multi-degree-of-freedom system
$\{\bar{A}\}$ = j-dimensional vibration vector of the free-
field at the nodes $\bar{k} = 1...j$, $j \leqslant n$
$\{S\}$ = m-dimensional vector of external
forces (or moments) $m \leqslant n$

For ω, the set of differential equations equ.13 can be solved
by already available computer programs.

Calculation of Transfer Coefficients

In the following, the transfer coefficients will be calculated for only some selected but most common vibration modes:
   Vertical and horizontal translation of foundations as exciting modes
   Vertical, radial and tangential translation of foundations as coupling modes

We have thus not considered the remaining modes rocking and torsion. We know at least in the case of rocking, that the damping ratio is small due to a much smaller amount of radiation. This may be the justification not to investigate these coupling effects in this simple lumped parameter model of structure-soil-structure interaction.

The transfer coefficients calculated below (after Bycroft, 1956) are valid for circular foundations, but similar result may readily be obtained by putting the values for rectangular foundations into the relevant equations 17 and 19 (e.g. Richart, Hall and Woods, 1970).

We first have to establish the relationship between the translation amplitude at k and the exciting force:

$$S_k(t) = A_k(c_{k\bar{k}} + i \; r_{k\bar{k}} \; \omega) \; e^{i\omega t} \tag{15}$$

with the phase angle between force and displacement

$$\varepsilon_k = \text{atan} \; \omega \; r_{k\bar{k}}/c_{k\bar{k}} \tag{16}$$

We can write the argument in the following form (e.g. Richart, Hall and Woods, 1970):

$$\omega \; r_{k\bar{k}}/c_{k\bar{k}} = \omega \; \frac{f_{jr}}{f_{jc}} \; \frac{f(\nu)}{f(\nu)} \; \frac{r_o^2 \; \sqrt{\rho G}}{r_o \; G} = \omega \; \frac{r_o}{v_s} \; \frac{f_{jr}}{f_{jc}} = a_o \; f_{jr}/f_{jc} \tag{17}$$

where f_{jc} = normalized spring constant as function of a_o
       f_{jr} = normalized damping constant as function of a_o
       $f(\nu)$ = dependency of f_{jc} and f_{jr} upon Poisson's ratio ν

and hence

$$\varepsilon_k = \text{atan} \; a_o \; f_{jr}/f_{jc} \tag{18}$$

(j being the translation index, vertical or horizontal resp.). This value of the phase difference between force and displacement is plotted in fig.2 with the values of f_{jc} and f_{jr} taken from Bycroft (1956) for circular contact areas.
The resulting force magnitude is then

$$S_k = A_k \; \sqrt{c_{k\bar{k}}^2 + \omega^2 \; r_{k\bar{k}}^2}$$

$$S_k = A_k \; r_o \; G \; f(\nu) \; \sqrt{f_{jc}^2 + a_o^2 \; f_{jr}^2} = A_k \; r_o \; G \; f(\nu) \; F_{jk} \tag{19}$$

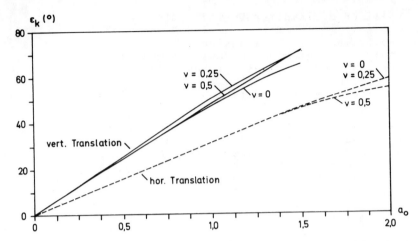

Fig.2 Phase-angle ε_k between translation amplitude
at k and exciting force

The values F_{jk} are given in fig.3 and fig.4, again valid for
circular contact areas:

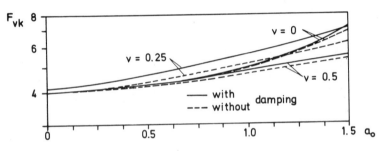

Fig.3 Coupling-Function F_{vk} for vertical translation

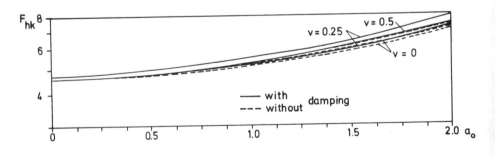

Fig.4 Coupling-Function F_{hk} for horizontal translation

Since we have calculated the vector of the force necessary to generate $A_k e^{i\omega t}$ at k, we can easily determine the respective wavefield due to A_k (taking advantage of the fact that wavefield data due to unit exciter forces are available in the literature, e.g. Cherry 1962, Prange 1978, et al.).

If the normalized wavefield displacement amplitude due to unit exciter force is denoted y_i^* (vertical, radial or tangential) and the Rayleigh-wavelength λ_r taken as reference, we can write for the displacement amplitude y_i at i:

$$y_i = \frac{S_k}{G\lambda_r} y_i^* \tag{20}$$

and hence with equ.19

$$y_i(t) = A_k \frac{r_0}{\lambda_r} f(\nu) F_{jk} y_i^* e^{i(\omega t + \varepsilon_k - \varepsilon^*)} \tag{21}$$

where ε^* describes the phase angle due to wave propagation and r_0 = radius of contact area. y_i^* includes attenuation due to geometric damping as well as interferences in the nearfield (Prange 1978) and is therefore identical to g_{ik} of equ.7. If we want to include material damping (d_{ik} of equ.7), we have to multiply equ.21 by the term

$$e^{\pi\tan\phi(r_0-r)/\lambda_r}$$

where $\tan\phi = \omega\mu/G$
     μ = coefficient of viscosity of Rayleigh-wave

The transfer coefficients t_{ik} of equ.7 are consequently

$$t_{ik} = T_{ik} e^{i\varepsilon_{ik}} = \frac{r_0}{\lambda_r} f(\nu) F_{jk} e^{\pi\tan\phi(r_0-r)/\lambda_r} y_i^* e^{i(\varepsilon_i - \varepsilon^*)} \tag{22}$$

where the displacements components y_i^* and the phase angles are taken from wavefield data and the other values from fig.2-4 as well as the table 1 below.

CONCLUSIONS

If two foundations are assumed to be rigid and their dimensions small in comparison to the Rayleigh-wavelength ($a_0 < 2$), their interaction via the subsoil can be calculated for a given angular frequency quite straightforward taking advantage of already known values of lumped parameter models and wavefields.

Considering only translational interaction the corresponding effects for rocking and torsion are neglected, which may be justified at least by the small intensity of the wavefield generated in the rocking mode.

The accuracy of the structure-soil-structure interaction model is of the same order as the SSI lumped parameter models it is based upon.

Table 1. Complex Transfer Coefficients

| Excitation at k | Transfer Coefficient Ampl. T_{ik} |
Displacement at i	Transfer Coefficient Phase ε_{ik}
vertical vertical	$T_{ikvv} = R_0 \ 1/(1-\nu)F_{vk} \ e^{\pi\tan\phi(R_0-R)} \ w_i^*$ $\varepsilon_{ikvv} = \varepsilon_{vk} - \varepsilon_w^*$
vertical horizontal	$T_{ikhv} = R_0 \ 1/(1-\nu)F_{vk} \ e^{\pi\tan\phi(R_0-R)} \ q_i^*$ $\varepsilon_{ikhv} = \varepsilon_{vk} - \varepsilon_q^*$
horizontal vertical	$T_{ikvh} = R_0 \ 7(1-\nu)/(7-8\nu)F_{hk} \ e^{\pi\tan\phi(R_0-R)} q_{zi}^* \sin\alpha$ $\varepsilon_{ikvh} = \varepsilon_{hk} - \varepsilon_z^*$
horizontal radial	$T_{ikrh} = R_0 \ 7(1-\nu)/(7-8\nu)F_{hk} \ e^{\pi\tan\phi(R_0-R)} q_{ri}^* \sin\alpha$ $\varepsilon_{ikrh} = \varepsilon_{hk} - \varepsilon_r^*$
horizontal tangential	$T_{ikth} = R_0 \ 7(1-\nu)/(7-8\nu)F_{hk} \ e^{\pi\tan\hat{\phi}\hat{k}(R_0-R)} q_{ti}^* \cos\alpha$ $\varepsilon_{ikth} = \varepsilon_{hk} - \varepsilon_t^*$

with $\hat{k} = v_r/v_s$; $R_0 = r_0/\lambda_r$; $R = r/\lambda_r$
    and α angle between horizontal exciter force at k and the
        normal to line i-k
(see also Prange 1978)

REFERENCES

Bycroft, G.N. 1956, Forced Vibrations of a Rigid Circular
    Plate on a Semiinfinite Elastic Space and on an Elastic
    Stratum. Phil.Trans.Royal Soc. London Ser.A, Vol. 248
Cherry, J.T. 1962, The Azimuthal and Polar Radiation Patterns
    Obtained from a Horizontal Stress Applied at the Surface of
    an Elastic Halfspace. Bull.Seism.Soc.America Vol.52 No.1
Prange, B. 1978, Primary and Secondary Interferences in Wave-
    fields. Proc.Dynamical Methods in Soil and Rock Mechanics,
    DMSR 77, Balkema, Rotterdam
Prange, B. 1978a, On the Interaction Between a Body and the
    Wavefield. Publ. Institute of Soil and Rock Mech., Uni-
    versity of Karlsruhe
Richart, F.E., Hall, J.R. and Woods, R.D. 1970, Vibrations of
    Soils and Foundations. Prentice Hall, Englewood Cliffs, N.J.

Proceedings of DMSR 77 / Karlsruhe / 5-16 September 1977 / Volume 1

Analytical methods for the computation of wavefields

STAVROS A. SAVIDIS
Technische Universität, Berlin, Germany

SYNOPSIS Analytical methods of calculating wave fields and the response of dynamically loaded foundations are presented. The soil is treated as an elastic isotropic and homogeneous half-space. The kinematics and generation of waves, the method of integral transforms for solving boundary value problems and the contour integration are discussed. Numerical results are given for the case of one and two dynamically loaded rigid rectangular footings, including interaction effects.

1 INTRODUCTION

The calculation of wave fields in a soil mass caused by dynamic forces is of very great importance to Engineering sciences in general. Scientists from many different fields such as seismology, geology, applied physics and soil dynamics etc. are all interested in solving this problem as it affects them.
Whilst in seismology we are interested in the far-wave field and the energy transmission of the waves caused by earthquakes, the aim of soil dynamics is the calculation of the displacements at or near their sources, these being dynamically loaded foundations.
Until recently the calculation of wave fields could only be made analytically. In order to come to any numerial results at all, the real system had to be simplified. During the last decade, due to the quick development of digital computers, new powerful numerical methods such as the finite element method or the method of characteristics etc., have become available.
These methods allow us to treat more complicated models such as multilayered and inhomogeneous systems with rather complicated boundary conditions.
The assumptions made in these methods and the correctness of their results can be tested by comparing them with known analytical solutions of simple systems. Only when this comparison is positive we can use these methods to calculate more sophisticated systems.

In the course of this paper we will be concerned with analytical methods of solving problems in soil dynamics. We shall

treat the soil as an elastic medium. This is a resonable
assumption, if the dynamic strain is somewhere in the region of
10^{-5} or even smaller.
We will further assume an isotropic and homogeneous material.
We shall discuss first the generation and the kinematics of
waves in an unbounded medium and afterwards we will be concerned with the free and forced vibrations of a layer and of
the half space.
The method of integral transforms for solving boundary value
problems and the contour integration will also be presented,
and finally we will give some numerical results.

2 KINEMATICS AND GENERATION OF WAVES

The theory of elasticity is concerned with the deformation of
a material under stress. Here we shall only discuss the most
elementary topics which are prerequisites of the study of the
generation and propagation of waves.

2.1 The analysis of stress

For non Cosserat-media the state of stress at a point P is
completely defined by the symmetric stress tensor Σ

$$\Sigma = \begin{pmatrix} \sigma_{11} & \sigma_{12} & \sigma_{13} \\ \sigma_{21} & \sigma_{22} & \sigma_{23} \\ \sigma_{31} & \sigma_{32} & \sigma_{33} \end{pmatrix} \cong \sigma_{ij} \quad i,j = 1,2,3 \qquad (1)$$

By knowing Σ it is possible to calculate every force vector
$\underline{t}$ dA acting on a small surface dA with the unit normal $\underline{n}$ and
passing through P. (Figure 1)

$$\underline{t} = \Sigma \cdot \underline{n} = \sigma_{ij} (\underline{e}_j \cdot \underline{n}) \underline{e}_i \qquad (2)$$

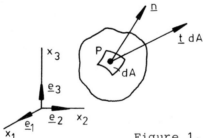

Figure 1. Stress vector $\underline{t}$ at
point P.

2.2 Analysis of deformation

The term deformation refers to a change in the shape of the
continuum between some initial (undeformed) configuation and a

226

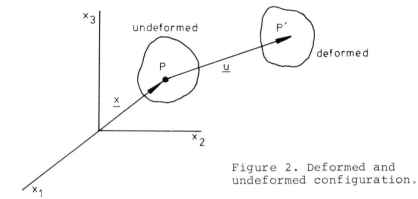

Figure 2. Deformed and undeformed configuration.

subsequent (deformed) configuation (Figure 2). To describe the state of deformation of a deformable medium a displacement vector $\underline{u}$ is used, which is defined for every point P of the continuum. In this analysis of wave propagation the so called Lagrangian formulation, i.e. the displacement vector $\underline{u}$ is a function of the time t and of the position vector $\underline{x}$ of the initial (undeformed) configuation, will be used. Moreover, small deformations, i.e. the displacement gradient will be small compared to unity, will be assumed. In this case the state of deformation around any point P is completely determined by the nine coefficients of the deformation tensor $\mathbf{D}$

$$
\mathbf{D} = \begin{pmatrix} \dfrac{\partial u_1}{\partial x_1} & \dfrac{\partial u_1}{\partial x_2} & \dfrac{\partial u_1}{\partial x_3} \\[2ex] \dfrac{\partial u_2}{\partial x_1} & \dfrac{\partial u_2}{\partial x_2} & \dfrac{\partial u_2}{\partial x_3} \\[2ex] \dfrac{\partial u_3}{\partial x_1} & \dfrac{\partial u_3}{\partial x_2} & \dfrac{\partial u_3}{\partial x_3} \end{pmatrix} \triangleq \dfrac{\partial u_i}{\partial x_j} \tag{3}
$$

The deformation tensor can be divided into two parts:

$$
\mathbf{D} \triangleq \frac{\partial u_i}{\partial x_j} = \frac{1}{2}(\frac{\partial u_i}{\partial x_j} + \frac{\partial u_j}{\partial x_i}) - \frac{1}{2}(\frac{\partial u_j}{\partial x_i} - \frac{\partial u_i}{\partial x_j}) \tag{4}
$$

The first quantity is the infinitesimal symmetric strain tensor e_{ij}

$$
\mathbf{E} \triangleq e_{ij} = \frac{1}{2}(\frac{\partial u_i}{\partial x_j} + \frac{\partial u_j}{\partial x_i}) \tag{5}
$$

The second quantity is the rotation tensor ω_{ij}

$$
\mathbf{\Omega} \triangleq \omega_{ij} = \frac{1}{2}(\frac{\partial u_j}{\partial x_i} - \frac{\partial u_i}{\partial x_j}) \tag{6}
$$

which is antisymmetric. Hence, the tensor ω_{ij} has only three independent components. From such an antisymmetric tensor a dual vector can always be built:

$$\omega_k = \frac{1}{2}1_{kij}\omega_{ij} \qquad \text{i.e.} \qquad \underline{\omega} = \frac{1}{2} \text{ curl } \underline{u} \tag{7}$$

2.3 Geometric interpretation of strain and rotation components.

The normal strain component e_{ii} simply represents the change in length $ds_i - d\bar{s}_i$ per unit of the initial length $d\bar{s}_i$ of a line element lying initially in the x_i direction (Figure 3).

$$e_{ii} = \frac{\partial u_i}{\partial x_i} = \frac{ds_i - d\bar{s}_i}{d\bar{s}_i} \tag{8}$$

The shearing strain component e_{ij} simply represents one-half of the change in the angle between two line elements lying initially in the directions x_i and x_j

$$e_{ij} = \frac{1}{2}(\frac{\partial u_i}{\partial x_j} + \frac{\partial u_j}{\partial x_i}) = \frac{1}{2}(\alpha_j + \alpha_i) \tag{9}$$

A positive or negative value of e_{ij} denotes that the initial right angle becomes, after deformation, an acute or an obtuse angle respectively.
The component ω_k of the rotation vector, when positive, represents the counterclockwise rotation of the bisector of the angle between two line elements initially pointing in the x_i and x_j directions.

$$\omega_k = \frac{1}{2}(\frac{\partial u_j}{\partial x_i} - \frac{\partial u_i}{\partial x_j}) = \frac{1}{2}(\alpha_i - \alpha_j) \tag{10}$$

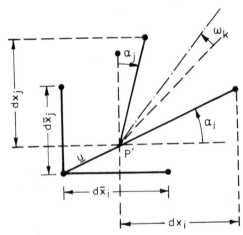

Figure 3. Deformation in two dimensions.

As in the case of the stress tensor, the strain tensor can be separated into two parts, a spherical part v_{ij} and a deviator part ε_{ij}. The spherical part is given by:

$$v_{ij} = \frac{1}{3} \Delta \, \delta_{ij} \tag{11}$$

where the dilatation $\Delta = e_{11} + e_{22} + e_{33}$ represents the change in volume per unit of initial volume.

2.4   Stress-strain relationship, equation of motion and the wave equations

Let us now examine an elastic, isotropic and homogeneous body. The stress-strain relationship for such a material is given by

$$\sigma_{ij} = \lambda \, \Delta \, \delta_{ij} + 2\mu \, e_{ij} \tag{12}$$

λ and μ are known as Lamé's elastic constants. The equation of motion by absence of body forces is

$$\rho \, \frac{\partial^2 u}{\partial t^2} = (\lambda+\mu) \nabla (\Delta) + \mu \nabla^2 u \tag{13}$$

∇ represents the operator $\nabla = \frac{\partial}{\partial x_i} e_i$ and $\Delta = \nabla \cdot u = \mathrm{div}\, u$ is the dilatation.
It is well known from Helmholtz's theorem that any analytic vector field u can be expressed in the form

$$u = \nabla \phi + \nabla \times \psi \tag{14}$$

where ϕ is a scalar function and ψ is a vector field with an arbitrary divergence. We may impose $\nabla \cdot \psi = 0$, which provides the necessary additional condition, to uniquely determine the three components of the vector field u from the four components of ϕ and ψ. A physical interpretation of the potentials ϕ and ψ can be obtained by performing the operation of divergence and curl on Eq.(14).

$$\mathrm{div}\, u = \nabla \cdot u = \Delta = \nabla \cdot (\nabla \phi) = \mathrm{divgrad}\, \phi \tag{15a}$$
$$\mathrm{curl}\, u = \nabla \times u = 2\, \omega = \nabla \times (\nabla \times \psi) = \mathrm{curlcurl}\, \psi \tag{15b}$$

The above results show that the dilatation Δ is related to the scalar potential ϕ and that the rotation ω can be expressed by the vector potential ψ.
Substituting Eq.(14) in Eg.(13) gives

$$\nabla \left[(\lambda+2\mu) \nabla \phi - \rho \, \frac{\partial^2 \phi}{\partial t^2} \right] + \nabla \times \left(\mu \nabla^2 \psi - \rho \frac{\partial^2 \psi}{\partial t^2} \right) = 0 \tag{16}$$

This equation will be satisfied if each vector vanishes, thus giving

$$\nabla \left[(\lambda+2\mu)\nabla\phi - \rho\frac{\partial^2\phi}{\partial t^2}\right] = \underline{0} \tag{17a}$$

$$\nabla \times \left(\mu\nabla^2\underline{\psi} - \rho\frac{\partial^2\underline{\psi}}{\partial t^2}\right) = \underline{0} \tag{17b}$$

Sternberg and Guttin (1962) have established that the complete solution of Eq.(17a,b) ist given by the solution of the following two wave equations

$$\rho\frac{\partial^2\phi}{\partial t^2} = (\lambda+2\mu)\nabla^2\phi \tag{18a}$$

$$\rho\frac{\partial^2\underline{\psi}}{\partial t^2} = \mu\nabla^2\underline{\psi} \tag{18b}$$

It is often more advantageous in soil dynamics to introduce by Eq.(14) the potentials ϕ and $\underline{\psi}$ and solve the wave equations (18), than to solve the equation of motion (13).

From the above analysis it can be seen that an infinite elastic, isotropic and homogeneous medium can sustain two kinds of waves which are propagated at two different velocities:

$$v_P = \sqrt{\frac{\lambda+2\mu}{\rho}} \quad \text{and} \quad v_S = \sqrt{\frac{\mu}{\rho}} \tag{19}$$

Eq.(18a) denotes that a change in volume, or dilatational disturbance, will propagate at the velocity v_P. Eq.(18b) shows that a rotation in the medium propagates at the velocity v_S. When the rotation $\underline{\omega}$ is zero, then only Eg.(18a) is valid. That means that an irrotational disturbance propagates at the velocity v_P. When the dilatation Δ is zero, only Eq.(18b) is valid, meaning that an equivoluminal disturbance propagates at a velocity v_S.
The wave propagating at the velocity v_P is referred to as a dilatational wave (primary wave, P-wave, compression wave, irrotational wave), and the wave at the velocity v_S is called a distortional wave (secondary wave, S-wave, shear wave, equivoluminal wave).

2.5 Plane waves

A very important general class of wave is that of the so called plane wave (Figure 4) which are represented by

$$\underline{u} = \underline{A}\, f(\underline{x}\cdot\underline{n} - ct) \tag{20}$$

Here $\underline{A}$ is the amplitude of the wave;
       $\overline{f}$ is an arbitrary scalar function which is called
         the wave form;

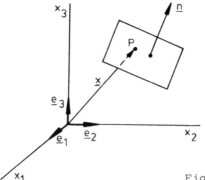

Figure 4. Plane wave.

$\underline{n}$ is a unit normal, indicating the direction of wave
  propagation;
c is the wave velocity;
$\underline{x} \cdot \underline{n}$ - ct is called the phase, and represents the equation
  of a plane which builds the wave front and passes
  through the point P with the position vector $\underline{x}$.

Not every c and every amplitude vector $\underline{A}$ are feasible solutions
of the equation of motion Eq.(13). Substitution shows that only
when

$$(\lambda + \mu)(\underline{A} \cdot \underline{n})\underline{n} + (\mu - \rho c^2)\underline{A} = 0 \qquad (21)$$

does the form given by Eq.(20) represent plane waves in an
elastic isotropic and homogenenous solid.
The Eq.(21) is satisfied if

$$\underline{A} = |\underline{A}|\underline{n} \quad \text{and} \quad c = \sqrt{\frac{\lambda + 2\mu}{\rho}} = v_P \qquad (22a)$$

$$\text{or} \quad \underline{A} \cdot \underline{n} = 0 \quad \text{and} \quad c = \sqrt{\frac{\mu}{\rho}} = v_S \qquad (22b)$$

The first is a plane P-Wave with the displacement vector parallel to the wave normal; the second is a plane S-Wave where the
displacement vector is always at right angles to the wave normal. Instead of Eq.(20) and the condition Eq.(21) we can also
represent plane elastic waves in terms of displacement potentials. A P-Wave is given by:

$$\phi = \phi_0 f(\underline{x} \cdot \underline{n} - v_P t)$$
$$\psi = 0 \qquad (23)$$

Plane S-Waves can be represented by:

$$\phi = 0$$
$$\underline{\psi} = \underline{\psi}_0 f(\underline{x} \cdot \underline{n} - v_S t) \qquad (24)$$

2.6 Harmonic Waves

A function of special interest for the wave form is the harmonic function $\exp(i\omega t)$ where ω is the angular frequency. Thus for Eq.(20) we now have

$$\underline{u} = \underline{A}\, \exp\{ik(\underline{x}\cdot\underline{n} - ct)\} \tag{25}$$

where $k = \frac{\omega}{c}$ is the wave number. For $\underline{k} = k\underline{n}$ Eq.(25) takes the form

$$\underline{u} = \underline{A}\, \exp\{i(\underline{x}\cdot\underline{k} - \omega t)\} \tag{26}$$

By superpositioning of plane harmonic waves we can construct cylindrical or spherical waves, which supply other important solutions for the wave equation. For cylindrical waves e.g. we have (Figure 5a):

$$u_r = \int_0^{2\pi} \exp\{ik(x\cdot\cos\theta + y\cdot\sin\theta) - i\omega t\}d\theta$$

$$= \int_0^{2\pi} \exp\{ikr\cdot\cos(\theta - \phi) - i\omega t\}d\theta = 2\pi\exp(-i\omega t)\cdot J_0(kr) \tag{27}$$

Thus, a rotationally symmetric solution of the wave equation, a so called cylindrical wave is given by the Bessel function of first kind J_0. This solution is regular at the origin $r=o$. Another solution which is singular at the origin, and corresponds to a radiation process with a source at the origin, can be constructed by the superposition of plane harmonic improper or damped waves, which possess a complex angle of incidence. We consider the complex path L of integration illustrated in Figure 5b and form the complex integral

$$u_r = \int_L \exp\{ikr\cdot\cos(\theta - \phi) - i\omega t\}d\theta = \pi\exp(-i\omega t)H_0^1(kr) \tag{28}$$

where $H_0^1(kr)=J_0(kr)+iY_0(kr)$ denotes the Hankel function of first order. This function is always a part of the displacement amplitude of rotationally symmetric problems in soil dynamics.

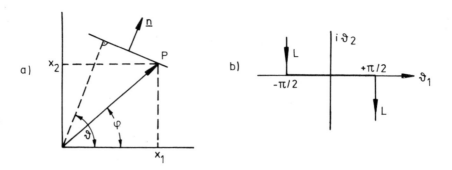

Figure 5. Superposition of plane harmonic waves.

2.7 Harmonic and transient input

The equation of motion Eq.(16), or the wave equations Eq.(18), can be presented as

$$\mathcal{L}\{\underline{u}(\underline{x},t)\} = \rho\underline{\ddot{u}}(\underline{x},t) \qquad (29)$$

where $\mathcal{L}$ is a linear differential operator. It might be noted here, that we have not mentioned any external force. The motion of the elastic medium will depend on the nature of the external force $\underline{q}(\underline{x},t)$. The applied force may arise:

1. as an internal or surface source; then the governing equation becomes

$$\mathcal{L}\{\underline{u}(x,t)\} = \rho\underline{\ddot{u}}(\underline{x},t) - \underline{q}(\underline{x},t) \qquad (30)$$

where $q(\underline{x},t)$ describes the source density. Eq.(30) must be accompanied by a homogeneous boundary condition.

2. in the form of boundary and initial conditions. In this case, which is most important for soil dynamic problems, the governing equation is the same as Eq.(29) i.e. a homogeneous equation. However, the boundary conditions, associated with the problem, now become inhomogeneous.

A fundamental property for the linear differential Eq.(29) is such that, if $\underline{u}^1, \underline{u}^2, \ldots, \underline{u}^n$ are solutions to this equation, then $\underline{u} = \sum_i c_i \underline{u}^i$, where c_i are arbitrary constants, is also a solution of Eq.(29).

We mentioned that the motion $\underline{u}(\underline{x},t)$ of the medium is caused by externally applied forces $\underline{q}(\underline{x},t)$, either as sources or as boundary conditions. In both cases the applied forces are called the input, and the consequent motion of the elastic medium is the response or output.

With regard to time-dependence of the response, any elastic body is assumed to follow two basic conditions (Pao, Mow, 1973)
1. The condition of causality, i.e. if the input

$$\underline{q}(\underline{x},t) = \begin{cases} 0 & t < 0 \\ q(\underline{x},t) & t > 0 \end{cases}$$

is causal, the output $\underline{u}(\underline{x},t)$ is also causal, i.e.

$$\underline{u}(\underline{x},t) = \begin{cases} 0 & t < 0 \\ u(\underline{x},t) & t > 0 \end{cases}$$

2. The condition of time invariance, i.e. if $\underline{u}(\underline{x},t)$ is the response for the source $\underline{q}(\underline{x},t)$, then the response to $\underline{q}(\underline{x},t-t_1)$ is $\underline{u}(\underline{x},t-t_1)$ where t_1 is any constant. This condition is true for an elastic medium with time-independent material constants.

As a function of time, the response can be classified as either

233

steady-state or transient.
By steady-state we mean a response that is harmonic in time. It is represented by

$$\underline{u}(\underline{x},t) = \underline{U}(\underline{x},\omega) \exp(\pm i\omega t) \tag{31}$$

Harmonic motion of an elastic medium can arise either during free or forced vibrations under a harmonic source, or harmonic forces applied at the boundaries:

$$\underline{q}(\underline{x},t) = \underline{Q}(\underline{x},\omega) \exp(\pm i\omega t) \tag{32}$$

When the source has a magnitude of unity, i.e.

$$\underline{q}(\underline{x},t) = \underline{1}\cdot\exp(\pm i\omega t),$$

the coefficient of $\exp(\pm i\omega t)$ in the steady-state response is called the admittance $\chi(\underline{x},\omega)$ (Pao, Mow, 1973).

For $\underline{Q} = \underline{1}$: $\quad \underline{u}(\underline{x},t) = \chi(\underline{x},\omega)\cdot\exp(\pm i\omega t)$

for $\underline{Q} \neq \underline{1}$: $\quad \underline{u}(\underline{x},t) = \underline{Q}(\underline{x},\omega)\cdot\chi(\underline{x},\omega) \exp(\pm i\omega t)$ $\qquad(32a)$

Knowing the admittance $\chi(\underline{x},\omega)$ of an elastic body and the magnitude $\underline{Q}$ of the external source, the steady-state response is simply given by Eq.(32a).

For steady-state motion the wave equations Eq.(18) reduce to the so called Helmholtz's equations:

$$(\nabla^2 + k_1^2)\Phi = 0$$
$$(\nabla^2 + k_2^2)\underline{\Psi} = 0 \tag{33}$$

where

$$\phi(\underline{x},t) = \Phi(\underline{x},\omega) \exp(-i\omega t)$$
$$\underline{\psi}(\underline{x},t) = \underline{\Psi}(\underline{x},\omega) \exp(-i\omega t) \tag{34}$$

and $k_1 = \dfrac{\omega}{v_P}$, $k_2 = \dfrac{\omega}{v_S}$ are the wave numbers for P- and S-waves resp. ,

To find the transient response due to an input $q(\underline{x},t)$ we analyse $q(\underline{x},t)$ into its simple harmonic components by means of the Fourier transform

$$\underline{q}(\underline{x},t) = \frac{1}{2\pi} \int_{-\infty}^{\infty} \underline{Q}(\underline{x},\omega) \exp(-i\omega t)d\omega \tag{35a}$$

with

$$\underline{Q}(\underline{x},\omega) = \int_{-\infty}^{\infty} q(\underline{x},t) \exp(i\omega t)dt \tag{35b}$$

Eq.(35b) supplies the Fourier transform of $\underline{q}$ i.e. $\mathcal{F}\{q\} = \underline{Q}$. Having solved the steady-state problem to obtain the admittance, $\chi(\underline{x},\omega)$, we superimpose the components to obtain the

234

response of the system resulting from the original transient input $\underline{q}(\underline{x}, t)$

$$\underline{u}(\underline{x}, t) = \frac{1}{2\pi} \int_{-\infty}^{\infty} \chi(\underline{x}, \omega) \, \underline{Q}(\underline{x}, \omega) \, \exp(-i\omega t) \, d\omega \tag{36}$$

For a unit impulse, represented by the Dirac delta function $\delta(t)$

$$\begin{cases} \delta(t) = 0 & |t| > 0 \\ \int_{-\infty}^{\infty} \delta(t) \, dt = 1 \end{cases} \tag{37}$$

the Fourier transform of $\delta(t)$ is $\mathcal{F}\{\delta(t)\} = 1$, and the impulse response of the system becomes

$$\underline{u}^{\delta} = \frac{1}{2\pi} \int_{-\infty}^{\infty} \chi(\underline{x}, \omega) \, \exp(-i\omega t) \, d\omega \tag{38}$$

In any case, whether a steady state or a transient input exists or not, the governing equations are still Helmholtz's equations Eq. (33). Our aim is to find solutions satisfying these equations and the boundary conditions of any given problem.

We shall first discuss some simple boundary value problems for free vibrations of an elastic medium and afterwards we will pass on to forced vibration problems.

3 FREE VIBRATIONS

3.1 Free vibrations of an elastic half-space

This is the oldest boundary value problem of elastodynamics, which was solved by Lord Rayleigh in 1885. The results led to the now well known Rayleigh surface wave which is described in detail in many standard textbooks (Ewing, W.M. et al. 1957, Kolsky, H. 1963, Richart, F.E. et al. 1970).

3.2 Propagation of SH-Waves in a surface layer

Let us consider a homogeneous layer of the thickness d and

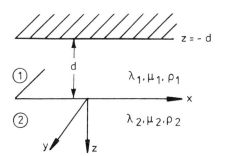

Figure 6. SH-Waves in a surface layer

235

material properties λ_1, μ_1, ρ_1 attached to the surface of an elastic, isotropic and homogeneous half-space, with the material constants λ_2, μ_2, ρ_2 (Figure 6). We will consider a plane problem with displacements u_{yj} only in the y-direction, so that

$$u_{yj} = u_{yj}(x,z,t), \quad u_{xj} = u_{zj} = 0 \quad j = 1,2 \qquad (39)$$

The index j = 1 refers to the layer and j = 2 to the half-space. Due to Eq. (39) the governing equation of motion reduce to

$$\nabla^2 u_{yj} = \frac{1}{v_{Sj}} \frac{\partial^2 u_{yj}}{\partial t^2} \qquad j = 1,2 \qquad (40)$$

This is the wave equation for horizontally polarized S-waves, the so called SH-waves. The non-trivial boundary conditions are:

$$(\tau_{yz})_1 = 0 \quad \text{for } z = -d \qquad (41a)$$

$$(\tau_{yz})_1 = (\tau_{yz})_2 \quad \text{for } z = 0 \qquad (41b)$$

$$u_{y1} = u_{y2} \quad \text{for } z = 0 \qquad (41c)$$

For solutions of Eq. (40) we consider harmonic plane waves of the form

$$u_{yj}(x,z,t) = U_{yj}(z) \exp\{ik(x-ct)\} \quad j=1,2 \qquad (42)$$

Substituting this in to Eq. (40) we obtain

$$\frac{d^2 U_{y1}}{dz^2} + \beta_1^2 U_{y1} = 0 \qquad \frac{d^2 U_{y2}}{dz^2} - \beta_2^2 U_{y2} = 0 \qquad (43)$$

where $\beta_1^2 = k^2(c^2/v_{S1}^2 - 1)$, $\beta_2^2 = k^2(1 - c^2/v_{S2}^2)$ $\qquad (44)$

The resulting solutions with $\omega = kc$ are:

$$u_{y1} = A_1 \exp\{i(kx-\beta_1 z-\omega t) + B_1 \exp\{i(kx+\beta_1 z-\omega t)\} \qquad (45a)$$

$$u_{y2} = A_2 \exp\{-\beta_2 z+i(kx-\omega t) + B_2 \exp\{\beta_2 z+i(kx-\omega t)\} \qquad (45b)$$

For $z \to \infty$ the amplitude u_{y2} must vanish. This leads to $B_2 = 0$. The solution Eq. (45) represents plane waves propagating back and forth within the layer.
A plane wave solution based on $\beta_2^2 < 0$ would not be capable of displaying the behaviour observed in seismology, since it would represent refracted waves carrying energy away from the layer. Such a wave system would quickly loose its energy and not be of significance a long distance away. By substitution of the solu-

tion Eq.(45a,b) in the boundary conditions Eq.(41a,b,c), we obtain the following homogeneous system of equations to determine the constants A_1, B_1, A_2 and the wave velocity $c=\omega/k$

$$\exp(i\beta_1 d)\ A_1 - \exp(-i\beta_1 d)\ B_1 = 0$$

$$-i\mu_1\beta_1(A_1-B_1) + \mu_2 B_2 = 0 \tag{46}$$

$$A_1 + B_1 - A_2 = 0$$

The resulting determinant of coefficients gives the frequency equation

$$\tan \beta_1 d = \frac{\mu_2\beta_2}{\mu_1\beta_1} \tag{47}$$

i.e.

$$\tan^{-1}\frac{\mu_2\beta_2}{\mu_1\beta_1} = \beta_1 d + n\pi$$

A surface wave in the layer can exist if Eq.(47) can be satisfied. This is possible if β_1 and β_2 are real, so that

$$v_{S1} < c < v_{S2}$$

Eq.(47) shows that the phase velocity of the Love wave, as a SH-wave in a layer is generally called, depends upon the frequency and therefore the propagation is dispersive. The appea-

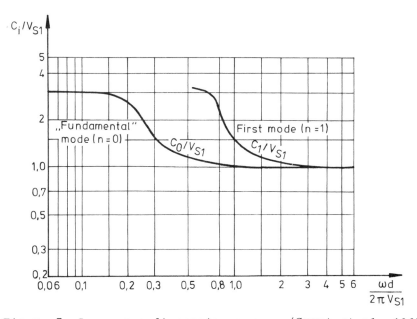

Figure 7. Love wave dispersion curves (Grant et al. 1965).

237

rance of the undetermined integer n in this expression suggests
that more than one wave may be propagated through the layer.
Different values of n correspond to different "normal modes"
of propagation. An important feature of Love wave dispersion
is that each mode, except the fundamental mode (n=0), has a
low-frequency cutoff, below which unattenuated progagation
through the layer cannot occur (Figure 7 after Grant, F.S.,
West, G.F. 1965). Only in the fundamental mode it is theoretically possible for all wave lenghts to propagate without horizontal attenuation.

The dispersion of SV-waves (including Rayleigh waves) within
a surface layer may be calculated by applying the same mathematical procedures to the displacement components u_x and u_z.
The essential features of the analysis are the same as those
for Love waves; only the algebra is more cumbersome. The frequency equation for the phase velocity is a rather complicated
transcendental equation which possesses several branches. Exemp-
les of SV-dispersion curves from Tolstoy and Usdin are shown in
Figure 8 after Grant, F.S., West, G.F. 1965. An interesting
point about these solutions, however, is that in the lowest
mode the high frequency limit for the phase velocity is the
solution of the Rayleigh equation for the upper layer, when
this layer is treated as a half-space. Thus the fundamental
mode at large distances corresponds to a Rayleigh wave, which
now appears to be dispersive.

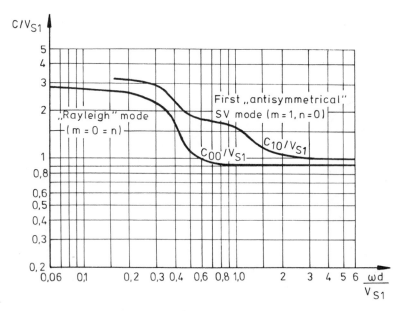

Figure 8. SV-wave dispersion curves (Grant et al. 1965).

4 FORCED VIBRATIONS

4.1 Integral transforms

We shall now consider some problems of forced vibrations of the
half-space. The vibration source lies on the surface of the
half-space and will be expressed in the form of stress-boundary
conditions. As mentioned above, whether a steady state or a
transient input exists or not, the governing equations are
Helmholtz's equations Eq.(33). A general method of finding
solutions for this homogeneous equation, satisfying the inhomogeneous boundary conditions, is that of integral transforms.
The integral transform $\overline{f}(\xi)$ of a function $f(x)$ is defined by
the integral equation:

$$\overline{f}(\xi) = \int_a^b f(x)K(\xi,x)dx \tag{48}$$

$K(\xi,x)$ is called the kernel of the transform. Different transforms differ by the appearance of the function K and the integral limits a und b. In elastodynamic problems the most usual
integral transforms are:
the Laplace transform

$$\overline{f}(\xi) = \int_0^\infty f(x)\exp(-\xi x)dx \tag{49a}$$

the Fourier transform

$$\overline{f}(\xi) = \int_{-\infty}^\infty f(x)\exp(i\xi x)dx \tag{50a}$$

and the Fourier-Bessel (or Hankel) transform

$$\overline{f}(\xi) \int_0^\infty xf(x)J_n(x\xi)dx \tag{51a}$$

All these formulae can be interpreted as linear operators trans-
forming f to $\overline{f}$. In the case of functions with several variables,
multiple integrals are used, concerning the transformation of
each variable.

By performing an integral transform on the governing equations
and boundary conditions of the problem, we obtain ordinary
differential equations instead of partial differential equations.
Thus it is easier to find solutions satisfying the boundary
conditions in the transformed space. Afterwards we have to invert the solutions, by inversion formulae, in the initial space.
Inversion formulae for the above transforms can be obtained from
the so-called Fourier's integral formula. These are:

for the Laplace transform

$$f(x) = \frac{1}{2\pi i} \int_{\gamma-i\infty}^{\gamma+i\infty} \overline{f}(\xi) \exp(\xi x) d\xi \qquad (49b)$$

for the Fourier transform

$$f(x) = \frac{1}{2\pi} \int_{-\infty}^{\infty} \overline{f}(\xi) \exp(-i\xi x) d\xi \qquad (50b)$$

for the Fourier-Bessel transform

$$f(x) = \int_{0}^{\infty} \xi \overline{f}(\xi) J_n(x\xi) d\xi \qquad (51b)$$

We shall use the method of integral transforms to find solutions for the wave field of the following problems.

4.2   Integral solutions for plane harmonic surface sources

The plane of the wave propagation is the x-z-plane, and z=o gives the surface of the half space (Figure 9). The governing equations are Eq.(33)

$$(\nabla^2 + k_1^2) \Phi = o$$

$$(\nabla^2 + k_2^2) \Psi = o$$

$$(33) \longleftrightarrow (52)$$

where

$$k_1 = \frac{\omega}{v_P}, \quad k_2 = \frac{\omega}{v_S}$$

Ψ is here a scalar function.

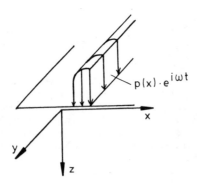

Figure 9. Plane problem.

The boundary conditions are:

$$z = 0 \quad \tau_{yz} = 0$$
$$z = 0 \quad \sigma_z = p(x)\exp(i\omega t)$$

(53)

The Fourier transform concerning x of the governing equations gives the following two ordinary differential equations

$$\frac{d^2\overline{\Phi}}{dz^2} - (k^2 - k_1^2)\overline{\Phi} = 0$$

$$\frac{d^2\overline{\Psi}}{dz^2} - (k^2 - k_2^2)\overline{\Psi} = 0$$

(54)

where $\overline{\Phi}(k,z)$ and $\overline{\Psi}(k,z)$ are the Fourier transforms of Φ and Ψ. The Fourier transform of the boundary conditions gives:

$$z = 0 \quad \overline{\tau}_{yz} = 0$$
$$z = 0 \quad \overline{\sigma}_z = \overline{p}(k)\exp(i\omega t)$$

(55)

The solution of Eq.(54), with the consideration that the wave amplitudes at infinity shall vanish, are:

$$\overline{\Phi} = A(k)\exp(-\nu_1 z)$$
$$\overline{\Psi} = B(k)\exp(-\nu_2 z)$$

(56)

where $\nu_1^2 = k^2 - k_1^2$, $\nu_2^2 = k^2 - k_2^2$ and the radiation conditions require

$$\mathrm{Re}\{\nu_1\} > 0 \quad , \quad \mathrm{Re}\{\nu_2\} > 0$$

The two coefficients $A(k)$ and $B(k)$ can be calculated from the two boundary conditions Eq.(55), and substituting in Eq.(56) we obtain

$$\overline{\Phi}(k,z) = \frac{2k^2 - k_2^2}{F(k)} \frac{\overline{p}(k)}{\mu}$$

$$\overline{\Psi}(k,z) = \frac{2ik\nu_1}{F(k)} \frac{\overline{p}(k)}{\mu}$$

where $F(k) = (2k^2 - k_2^2)^2 - 4k^2 \nu_1 \nu_2$ is the Rayleigh function.

From $\overline{\Phi}$ and $\overline{\Psi}$ we obtain the transformed displacements, and in particular for the surface z=o they take the form

$$\overline{u}_{xo} = \overline{u}_x(k,o) = \frac{1}{\mu} \frac{ik(-2k^2 + k_2^2 + 2\nu_1 \nu_2)}{F(k)} \overline{p}(k)\exp(i\omega t)$$

$$\overline{u}_{zo} = \overline{u}_z(k,o) = \frac{1}{\mu} \frac{\nu_1 k_2^2}{F(k)} \overline{p}(k)\exp(i\omega t)$$

(57)

By applying the inverse Fourier transform (Eq.(50b)) to Eq.(57) the surface displacements can be obtained

$$u_{xo} = u_x(x,o) = \frac{1}{2\pi\mu} \int_{-\infty}^{\infty} \frac{ik(-2k^2+k_2^2+2\nu_1\nu_2)}{F(k)} \bar{p}(k)\exp\{i(\omega t-kx)\}dk$$

(58)

$$u_{zo} = u_z(x,o) = \frac{1}{2\pi\mu} \int_{-\infty}^{\infty} \frac{\nu_1 k_2^2}{F(k)} \bar{p}(k)\exp\{i(\omega t-kx)\}dk$$

For an impulse line source (Figure 10a) we obtain $\sigma_z=P\delta(x)\exp(i\omega t)$ where $\delta(x)$ is the Dirac delta function.

$$\bar{\sigma}_z = \bar{p}(k)\exp(i\omega t) = \int_{-\infty}^{\infty} P\delta(x)\exp\{i(\omega t+kx)\}dx = P\exp(i\omega t)$$

A uniforme source (Figure 10b) gives:

$$\sigma_z = \frac{P}{2a}\{\mathcal{U}(x-x_1) - \mathcal{U}(x-x_2)\}$$

$\mathcal{U}(x-x_i)$ is the step function. The Fourier transform of σ_z is

$$\bar{\sigma}_z = \bar{p}(k)\exp(i\omega t) = \frac{P}{2aik}\{\exp(ikx_2)-\exp(ikx_1)\}\exp(i\omega t)$$

For $x_1 = -a$ and $x_2 = a$ we obtain:

$$\bar{\sigma}_z = \bar{p}(k)\exp(i\omega t) = \frac{P}{a}\frac{\sin(ka)}{k}\exp(i\omega t)$$

In the last expressions k=o is a new pole of the integrand in the inversion formula.

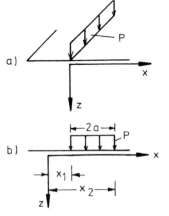

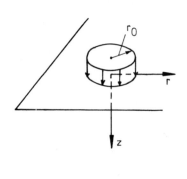

Figure 10. Line and uni-
forme source

Figure 11. Circular source.

242

4.3 Integral solutions for circular source

The case of a rotationally symmetric surface load, located at the surface of the half-space, has more applications than the two dimensional problems previously considered. It occurs in the case of a dynamically loaded circular plate with the radius r_o (Figure 11). We assume symmetry about the z-axis. The displacements u_z and u_r, parallel and perpendicular to the z-axis, presented in terms of potentials Φ and Ψ are:

$$U_r = \frac{\partial \Phi}{\partial r} + \frac{\partial^2 \Psi}{\partial r \partial z} \ , \qquad U_z = \frac{\partial \Phi}{\partial z} - \frac{\partial^2 \Psi}{\partial r^2} - \frac{1}{r}\frac{\partial \Psi}{\partial r} \tag{59}$$

where $u_r = U_r \exp(i\omega t)$ and $u_z = U_z \exp(i\omega t)$.

The Helmholtz's equations have the same form as Eg.(33).

$$(\nabla_r^2 + k_1^2)\Phi = o \ , \qquad (\nabla_r^2 + k_2^2)\Psi = o \tag{33} \leftrightarrow (60)$$

where

$$\nabla_r^2 = \frac{\partial^2}{\partial r^2} + \frac{1}{r}\frac{\partial}{\partial r} + \frac{\partial^2}{\partial z^2} \tag{60a}$$

The boundary conditions are:

$$z=o: \quad \tau_{zr} = o \qquad z=o: \ \sigma_z = \begin{cases} p(r) = \dfrac{P}{\pi r_o^2}\exp(i\omega t) & r \leqslant r_o \\[2mm] p(r) = o & r > r_o \end{cases} \tag{61}$$

We apply the Fourier-Bessel transform, concerning the variable r, to Helmholtz's equations Eq.(60) and the boundary conditions Eq.(61). The further operations are analogous to those mentioned above for the plane source. Once the displacements satisfying the boundary conditions in the transformed space have been calculated, the inverse transformation can be applied.

For the surface displacements we obtain:

$$u_{ro} = u_r(r,o) = \frac{P}{\mu\pi r_o} \int_o^\infty \frac{k(-2k^2+k_2^2+2\nu_1\nu_2)}{F(k)} J_1(kr_o)J_1(kr)dk \ \exp(i\omega t) \tag{62}$$

$$u_{zo} = u_z(r,o) = \frac{P}{\mu\pi r_o} \int_o^\infty \frac{\nu_1 k_2^2}{F(k)} J_1(kr_o)J_1(kr)dk \ \exp(i\omega t)$$

This solution for the displacement of a dynamically loaded circular surface area was first given by Reissner 1936, in the form

$$u_{zo} = \frac{P}{\mu\pi r_o}(f_1+if_2)\exp(i\omega t) \tag{63}$$

243

4.4 Evaluation of the integral solution. Contour integration

The integrals obtained as a solution in the problems treated above cannot be evaluated by direct integration. A useful approach is to replace the variable of integration k by the complex variable ζ and apply contour integration in the complex plain, by using Cauchy's theorem. A detailed description of the method is given by Lamb 1904, Ewing et al. 1957 and Bath 1968. Only a summary will be given here.

If $\zeta = k+i\tau$ is the complex variable, the integrals in Eq.(58) and Eq.(62) have poles, determined by the zeros of the Rayleigh equation

$$F(\zeta) = (2\zeta^2 - k_2^2)^2 - 4\zeta^2 \nu_1^2 \nu_2^2 \qquad (64)$$

and branch points introduced by the zeros of the radicals

$$\nu_1 = \sqrt{\zeta^2 - k_1^2} \quad \text{and} \quad \nu_2 = \sqrt{\zeta^2 - k_2^2} \qquad (65)$$

Before Cauchy's theorem is applied, the integrand must be made uniform by introducing cuts in the complex plane. The lines defined by $Re\{\nu_1\}=o$ and $Re\{\nu_2\}=o$ can be used as such cuts. According to the restrictions, that the displacements at infinity must be zero, the sheet of the Riemann surface, for which $Re\{\nu_1\}> o$ and $Re\{\nu_2\} > o$, must be selected.

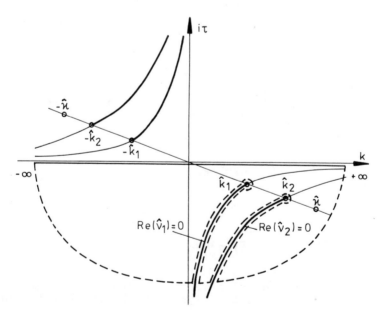

Figure 12. Integration path in the complex plane (Ewing et al. 1957).

By using this method, complex values of ω have also to be considered. We put:

$$\hat{\omega} = \omega - i\sigma \tag{66}$$

so that the real magnitudes k_1, k_2 and $\nu_{1,2}$ become complex

$$k_1 = \frac{\omega}{v_P} \quad \longrightarrow \quad \hat{k}_1 = k_1 - ik_1^{im}$$

$$k_2 = \frac{\omega}{v_S} \quad \longrightarrow \quad \hat{k}_2 = k_2 - ik_2^{im} \tag{67}$$

$$\nu_{1,2} \quad \longrightarrow \quad \hat{\nu}_{1,2} = \nu_{1,2} + i\nu_{1,2}^{im}$$

For the cut $\text{Re}\{\hat{\nu}_{1,2}\} = 0$ we obtain

$$\hat{\nu}_{1,2}^2 = \zeta^2 - \hat{k}_{1,2}^2 = k^2 - \tau^2 + 2ik\tau - (k_{1,2}^2 - k_{1,2}^{2\,im} - 2ik_{1,2}k_{1,2}^{im}) \tag{68}$$

$\text{Re}\{\nu_{1,2}\} = 0$ requires that the above expression must be real and negative i.e.

$$k\tau = k_{1,2}\,k_{1,2}^{im}$$

$$k^2 - \tau^2 = k_{1,2}^2 - k_{1,2}^{2\,im} \tag{69}$$

These equations define parts of hyperbolas to be used as a cut. These hyperbolas and the integration path is shown in Figure 12.

Of course, other branch cuts can be used, as was shown in Lamb (Lamb 1904) and described in detail by Bath (Bath 1968).

The integrals involved after the contour integration can be treated numerically, either by numerical integration or by series expansions. The results of these investigations are the displacement functions (for example f_1, f_2 in Eq.(63)) or the compliance of the foundations; they will be treated in the next chapter.

5 NUMERICAL RESULTS

Reissner (Reissner 1936) was the first to present the integral solution for a uniform loaded circular area (Eq.(63)). His main interest was in the calculation of the vertical vibration amplitude of the middle point. Lysmer (Lysmer 1965) expanded the numerical results to other points inside and outside the loaded area. He also varied the value of the dimensionless frequency $a_o = r_o\,\omega/v_S$ within a large range: $0 \leqslant a_o \leqslant 8.0$. The real and imaginary parts of the displacement functions are given in Figure 13.

The three dimensional case for a rectangular uniform loaded area (Figure 14), as a stress boundary value problem, was solved by

245

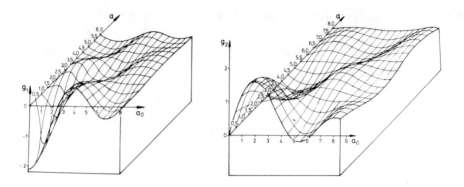

Figure 13. Real and imaginary parts of Reissner's displacement
functions (Lysmer 1965)

Thomson and Kobori (Thomson, W.T., Kobori, T. 1963) and Holzlöhner (Holzlöhner 1969). By normalising the displacement
functions

$$w = \frac{P}{C} (F_1 + iF_2) \exp(i\omega t) \tag{70}$$

where C is the static spring constant of the quadratic foundation, the influence of Poisson's ratio becomes negligable (Savidis 1975).

In the analytical formulations in the last chapter we used
stress boundary conditions. In rigid foundations, displacements
are prescribed beneath, and stresses outside the footing. The
analytical treatment of such a mixed boundary value problem is
very difficult. This difficulty can be avoided by sub-dividing
the base of the footing into small areas and supposing a constant pressure distribution in each element-area. Lysmer (Lysmer
1965) used this method to calculate the displacement-functions
of a rigid circular base (ring method). Elorduy et al. (Elorduy
J. et al. 1967) used a similar approach to calculate the displacement-functions of rigid rectangular footings. They divided

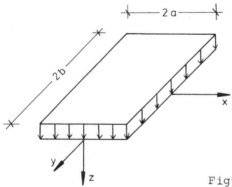

Figure 14. Rectangular uniform
loaded area.

246

the base area into rectangular elements and set point-forces
into the middle of each element, using Pekeris' solution (Pekeris, C.L. 1955) for the surface displacements due to a point
load. The calculated side ratios and frequency ranges are not
very large.

By modifying the solution of Holzlöhner to get the displacement
at the corner-point of a uniformly loaded rectangular area,
Savidis (Savidis, S.A. 1977) calculated the displacement-
functions of a rigid rectangular foundation after Figure 15 for
a wide range of side rations b_o (b_o = a/b = 1.0, 1.5, 2.0, 4.0
and 8.0) and dimensionless frequency a_o ($0 \leqslant a_o \leqslant 8.0$). The results are shown in Figure 16. From these displacement-functions
the spring and damping constants of a lumped parameter system
were calculated. These frequency independent constants are given
in table 1. They are valid for constant- and rotating mass-
excitation. Figure 17 gives the magnification factor of a rectangular foundation with the side ratio b_o = a/b = 2.0 for these
two kinds of excitations, and compares the results of the lumped
parameter system (dashed line) and the half-space theory.

The problem of the dynamic interaction of two rigid rectangular
foundations can also be solved with this method. A general view
of the problem after Savidis et al. (Savidis, S.A., Richter, T.
1977) is shown in Figure 18. The two rectangular foundations
are at a distance x_o, y_o from each other and are loaded with
harmonic, vertical forces and rocking moments of the same excitation frequency. When the shear stresses are neglected, unknown

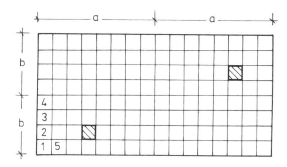

$$S_{ij} = S_i^{AEIG} - S_i^{CEIH} - S_i^{BFIG} + S_i^{DFIH}$$

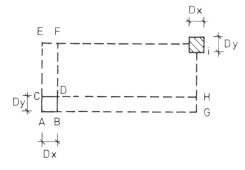

Figure 15. S_{ij}-Matrix for
the calculation of the displacement functions of a
rectangular footing.

247

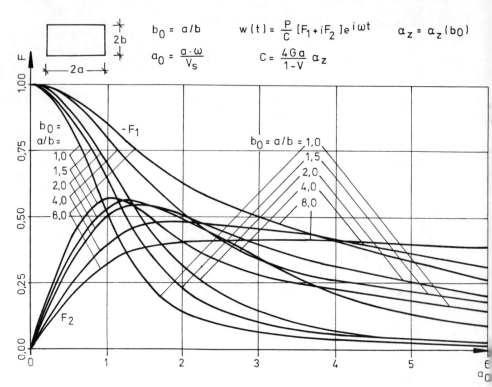

Figure 16. Displacement functions for a rigid rectangular
footing (after Savidis 1977).

Mode of vibration	Mass ratio	Damping coeff.	Spring coeff.
Vertical	$B_z = \dfrac{1-\nu}{4\alpha_z}\dfrac{m}{\rho_a{}^2 b}$	$c_z = \beta_z\dfrac{a^2}{1-\nu}\sqrt{\rho G}$	$k_z = \alpha_z\dfrac{4Ga}{1-\nu}$

$b_o = a/b$	1.0	1.5	2.0	4.0	8.0
α_z	1.065	0.870	0.776	0.605	0.479
β_z	3.843	2.593	2.030	1.195	0.594

Table 1. Spring and damping coefficients for rigid rectangular
foundation resting on elastic half-space (after Savidis 1977).

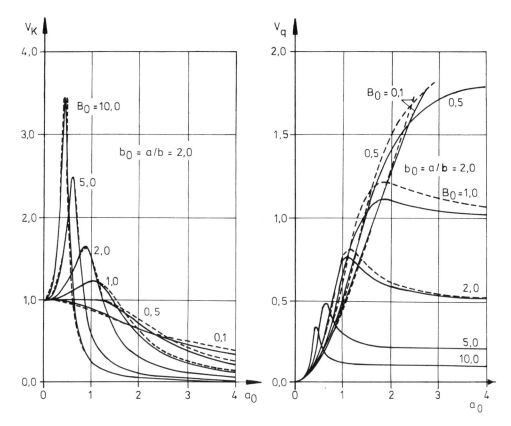

Figure 17. Magnification factor of a rectangular rigid foundation with the side ratio $b_o = a/b = 2.0$ (after Savidis 1977).

normal stresses alone are acting on the base of the foundation.
These stresses can be composed into the reaction forces P_i and
reaction moments T_i

$$\underline{p}^T = [T_{y1}, \; T_{y2}, \; T_{x1}, \; T_{x2}, \; P_1, \; P_2] \tag{71}$$

Each foundation possesses two rotations $\Psi_{xi}, \; \Psi_{yi}$ about the axes
x_i and y_i and one vertical displacement of the middle point
W_i:

$$\underline{\delta}^T = [\Psi_{y1}, \; \Psi_{y2}, \; \Psi_{x1}, \; \Psi_{x2}, \; W_1, \; W_2] \tag{72}$$

The displacement vector $\underline{\delta}$ and the reaction force vector $\underline{p}$ are
associated with the linear relation

$$\underline{\tilde{\delta}} = \underline{\Delta} \; \underline{p} \exp(i\omega t) \tag{73}$$

where $\underline{\Delta}$ is a displacement matrix which describes the dynamic

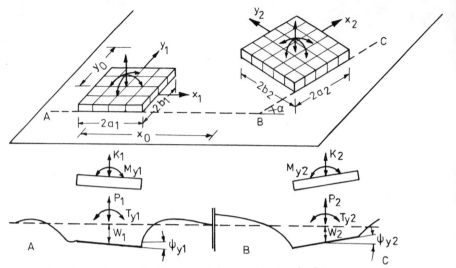

Figure 18. Geometry of the problem and symbols.

interaction of both rigid foundations, and which is dependent on the frequency ω, the material constants of the soil ρ, G and the geometrical parameters of the problem. To calculate the Δ-Matrix, the base areas are subdivided into a grid of rectangular elements for which the soil pressure q_i is assumed to be uniform. The vertical displacement at the centre of the grid element i, which results from the element soil pressures of both foundations, is:

$$w_i = S_{ij} q_j \tag{74}$$

The compatibility conditions of displacement require the settlements under each foundation to lie on a plane:

$$h_1 x_i + h_2 y_i + h_3 + w_i = 0 \tag{75a}$$

for the first footing, and

$$h_4 x_j + h_5 y_j + h_6 + w_j = 0 \tag{75b}$$

for the second footing.
The dynamic conditions of equilibrium for each foundation require for the vertical forces and for the rocking moments:

$$\sum_{i=1}^{m,n} q_i A_i = P_{1,2} \exp(i\omega t)$$

$$\sum_{i=1}^{m,n} q_i A_i = T_{y1,2} \exp(i\omega t) \tag{76}$$

$$\sum_{i=1}^{m,n} q_i A_i = T_{x1,2} \exp(i\omega t)$$

250

where m and n are the numbers of the elements in the first and second foundations and A_i is the area of a grid element. Equations (75) and (76) provide a linear algebraic system of $m + n + 6$ complex equations, for the calculation of the unknowns h_i ($i = 1\div6$) and the soil pressures q_j ($j = 1\div m+n$). The solution of this system gives the elements of the matrix $\underline{A}$. The vibration amplitudes of the foundations, which possess mass, are calculated from the equation

$$\underline{M} \ \ddot{\tilde{\delta}} + \underline{p} \ exp(i\omega t) = \underline{F} \ exp(i\omega t) \tag{77}$$

where $\underline{M}$ is a diagonal matrix which contains the masses M_i and the inertia moments Θ_i. $\underline{F}$ is the column matrix of the excitation forces. $\underline{p}$ is substituted in Eq.(77) from Eq.(73) and with the harmonic statement

$$\tilde{\underline{\delta}} = \underline{\delta} \ exp(i\omega t) \tag{78}$$

the following complex system of equations is obtained:

$$(\underline{A}^{-1} - \omega^2 \ \underline{M}) \ \underline{\delta} = \underline{F} \tag{79}$$

Magnification functions, e.g. V_{w1}, V_{w2} for the vertical amplitudes of the first and second foundation are formed by dividing the amplitudes $\underline{\delta}$ row by row, with the corresponding static values for $a_o = 0$.

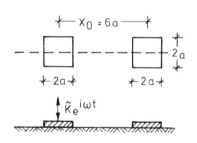

Numerical results for V_{w1} are given in Figure 19 for two quadratic footings, in the case of the first footing being subjected to a constant harmonic force with the frequency ω, while no excitation exists on the second foundation.

$a_o = \dfrac{a_1 \omega}{v_S}$ is the dimensionsless

frequency and $B_i^z = \dfrac{m}{\rho \, a_i^2 \, b_i} \ \dfrac{1-\nu}{4\alpha_z}$

is the mass ratio.

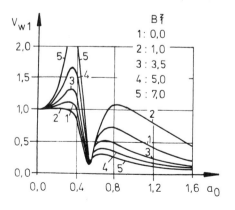

Figure 19. Magnification function V_{w1} for two square footings.

251

6 CONCLUSIONS

Analytical methods of calculating wave fields and the response of dynamically loaded foundations are presented here. In the analytical investigations and numerical results an elastic homogeneous half-space was assumed. It was demonstrated that analytical methods in the case of simple boundary conditions supply useful solutions for practical problems. In addition, such solutions can be used to check the results of numerical methods, such as the finite element method etc. Analytical solutions of the visco-elastic half-space (see Veletsos and Verbic 1973, Lee 1970 et al.) and analytical investigations which treated the soil as a porous medium (Bhattacharyya, R.K. 1965, Jones, J.P. 1962, Paul, S. 1976), were not dealt with in this paper.

7 REFERENCES

Bath, M. 1968, Mathematical aspects of Seismology, Amsterdam, London, New York, Elsevier Publ. Co.

Bhattacharyya, R.K. 1965, Rayleigh waves in granular medium, Pure and Appl. Geophys. 62,III.

Elorduy, J. et al. 1967, Dynamic response of bases of arbitrary shapes subjected to periodic vertical loading, Proc. Int. Sym. Wave Propagation, New Mexico.

Ewing, W.M. et al. 1957, Elastic waves in layered Media, New York, Mc Graw Hill Book Co.

Grant, F.S., West, G.F. 1965, Interpretation Theory in Applied Geophysics, New York, Mc Graw Hill Book Co.

Holzlöhner, U. 1969, Schwingungen des elastischen Halbraumes bei Erregung auf einer Rechteckfläche, Ing. Archiv, 38, 6.

Jones, J.P. 1962, Rayleigh waves in porous elastic saturated solid, J. Acoust. Soc. Am. 34,36.

Kolsky, H. 1963, Stress Waves in Solids, Dover Publication.

Lamb, H. 1904, On the Propagation of Tremors over the Surface of an Elastic Solid, Phil Trans. R. Soc. London A, Vol. 203.

Lee,T.M. 1970, Vibratory surface loadings on a viscoelastic half-space, Cold Reg. Res. Engng. Lab. Res. Rep. RR 286.

Lysmer, J. 1965, Vertical Motion of Rigid Footings, Contract Report No. 3-115, U.S. Army Engnr. WES.

Pao, Y.-H., Mow, C.-C. 1973, Diffraction of elastic waves and dynamic stress concentrations, New York, Crane Russak.

Paul, S. 1976, On the Displacements Produced in a Porous Elastic Half-space by an Impulsive Line Load, Pageoph., Vol. 114.

Pekeris, C.L. 1955, The Seismic Surface Pulse, Proc. of N.A.S., Vol. 41.

Reissner, E. 1936, Stationäre, axialsymmetrische, durch eine schüttelnde Masse erregte Schwingungen eines homogenen elastischen Halbraumes, Ing. Archiv, Vol. 7.

Richart, F.E., Hall, J.R., Woods, R.D. 1970, Vibrations of Soil and Foundations, Prentice Hall Inc.

Savidis, S.A. 1975, Wechselwirkung zweier starrer, dynamisch belasteter Rechteckfundamente, Festschrift zum 70. Geburtstag von Prof. H. Lorenz, TU Berlin.

Savidis, S.A. 1977, Einmassenanaloga für vertikale Schwingungen
  von starren Rechteckfundamenten. Veröffentl. des Grundbau-
  institutes der TU Berlin.
Savidis, S.A., Richter, T. 1977, Dynamic Interaction of Rigid
  Foundations, Proc. IX Int. Conf. SMFE, Tokyo.
Sternberg, Gintin 1962, Proc. U.S. Congress Appl. Mech.
Thomson, W.T., Kobori, T. 1963, Dynamical Compliance of Rectan-
  gular Foundations on an Elastic Half-space, Journal of Appl.
  Mech., ASME.
Veletsos, A.S., Verbic B. 1973, Vibration of Viscoelastic Foun-
  dations, Int. J. Earthqu. Engnr. Str. Dyn., Vol. 2

Proceedings of DMSR 77 / Karlsruhe / 5-16 September 1977 / Volume 1

Numerical methods for the computation of steady-state harmonic wave fields

WOLFGANG A. HAUPT
University of Karlsruhe, Karlsruhe, Germany

SYNOPSIS

For the analysis of wave propagation problems usually numerical methods must be applied, among which the analysis of steady-state harmonic wave fields by the Finite-Element-method is a suitable way. It allows the consideration of complicated geometries and the elimination of the time variable, thus yielding a formally static problem. The application of the dynamic conditions at the boundaries of the FE-grid, appropriate to the total system, is an integral equation problem, which can be solved analytically only for special cases. The influence-matrix boundary condition is a general solution, based on a numerical procedure. It provides a perfect boundary condition replacing the infinite homogeneous half-space. Furthermore, with the influence-matrix procedure any wave field within a continuum can be generated easily. Finally some results of a FE-analysis concerning a Rayleigh wave field are presented and discussed.

1. INTRODUCTION

The analysis of the propagation of waves in the ground is one of the. major topics in soil dynamics. In the design of an important building with respect to earthquake resistance or a vibration-sensitive installation the engineer usually must have some knowledge of the type, the direction of propagation and the magnitude of the waves which could occur in the ground and affect the structure. The better this wave propagation process is known the more accurate the prediction of the actual vibrations of the structure can be.

Only in very simple cases can the wave field be calculated analytically. Usually numerical methods have to be applied because the results depend highly on the geometry of the problem and simplifying assumptions may be of great influence. Basically there are two methods of finite approach to the continuum: the Finite-Difference-method (FD) and the Finite-Element-method (FE). The FD-method is a powerful tool and it seems, that there are types of problems, where it can be applied more economically

255

than the FE-method. It is based on the replacement of differen-
tials by finite differences. There are, however, cases, where
the boundary conditions cannot easily be defined or only defined by use of nonsymmetric finite differences. This latter
case would yield a nonsymmetric system of linear equations.

There are essentially three different ways of treating dynamic problems with the FE-method:

1) Direct integration
   This method consists of the integration of the differential
   equation of wave propagation with respect to time, where the
   time interval to be considered is divided into small finite
   "steps" and the integration is performed by a numerical
   finite approach. This method yields the displacements due
   to a given load function at any given instant of the time
   interval. However, the required computer time can be considerable and under certain circumstances the calculation
   can become numerically instable or the solution can include
   an artificial damping.

2) Modal response spectrum analysis
   In this method the eigenvalues and the corresponding eigenmodes of the total system are calculated. By the appropriate
   superposition of these eigenmodes the displacement function
   can be found corresponding to the respective load function.
   The disadvantages of this method are the restriction to relatively small systems and the impossibility of applying dynamic boundary conditions.

3) Steady-state harmonic wave field method
   In many cases the wave field is generated by vibrations which
   are steady-state and harmonic, for instance those of the
   footing of an unbalanced rotary machine. Non-harmonic or
   transient vibrations can be considered as the superposition
   of harmonic steady-state vibrations found from a Fourier-
   analysis. By this steady-state harmonic wave field method
   large FE-systems can be computed with a relatively small
   amount of computer time. Further, there are some dynamic
   boundary conditions available, which is of great importance,
   as will be seen later.

This paper deals with the third method of analysing dynamic
problems, which are considered in the following to be plane.
The material behaviour is assumed to follow a linear stress-
strain-law, which implies,for instance,a linear elastic or a
linear visco-elastic material, respectively. The linearity of
the stress-strain-law is necessary for the elimination of the
time variable but the approach to non-linear stress-strain-
behaviour, developed for static problems, can well be applied
here. However, in this case the half-space boundary conditions
are no longer exactly valid, and some errors in the results
have to be taken into account.

2. GENERAL CONSIDERATIONS

2.1 Elimination of Time Variable

In steady-state harmonic problems the time variable can be eliminated and the problem can formally be considered as a static
one, as shown by the following. The general equation of the
dynamic equilibrium in the x-direction at a point (x, y, z)
within an elastic body is:

$$\rho \; \frac{\partial^2 u_x(x,y,z,t)}{\partial t^2} \; = \; (\lambda+G) \; \frac{\partial \bar{e}}{\partial x} \; + \; G \; \nabla^2 \; u_x(x,y,z,t) \tag{1}$$

In this equation u_x is the displacement in x-direction, $\bar{e}$ the
volumetric strain and ∇^2 is the Laplace operator. The parameters of a linear elastic material are the Lamé's constants λ
and G and ρ is the density. As volume forces only the inertia
forces are taken into account. If a one-dimensional P-wave propagating in the x-direction is considered, this equation reduces to:

$$\frac{\partial^2 u_x(x,t)}{\partial t^2} \; = \; \frac{(\lambda + 2G)}{\rho} \; \frac{\partial^2 u_x(x,t)}{\partial x^2} \tag{2}$$

For a linear visco-elastic material, for which the stress-
strain-law is given by

$$\underset{\sim}{\sigma} \; = \; \underset{\sim}{E} \cdot \underset{\sim}{\varepsilon} \; + \; \underset{\sim}{E}' \cdot \frac{\partial \varepsilon}{\partial t} \tag{3}$$

($\sim$ means a tensor)eq.(2) becomes:

$$\frac{\partial^2 u_x(x,t)}{\partial t^2} \; = \; \frac{(\lambda+2G)}{\rho} \; \frac{\partial^2 u_x(x,t)}{\partial x^2} \; + \; \frac{(\lambda'+2G')}{\rho} \; \frac{\partial^3 u_x(x,t)}{\partial x^2 \partial t} \tag{4}$$

The solution of this equation is assumed to be harmonic with
respect to time:

$$u_x(x,t) \; = \; u(x) \; e^{i\omega t} \; , \tag{5}$$

where ω is the circular frequency. Inserting this into eq.(4)
one obtains:

$$-\omega^2 \; u(x) \; e^{i\omega t} \; = \; \frac{1}{\rho} \left[\; \lambda+i\omega\lambda'+2G+i\omega 2G' \; \right] \frac{\partial^2 u(x)}{\partial x^2} \; e^{i\omega t} \tag{6}$$

if the term in brackets is expressed by the complex modulus of
elasticity, E_b^*, and the terms $e^{i\omega t}$ at the left hand and the
right hand side, respectively, are eliminated, eq.(6) becomes:

$$k_p'^2 \cdot u(x) \; + \; \frac{\partial^2 u(x)}{\partial x^2} \; = \; 0, \tag{7}$$

where k_p' is the complex wave number, corresponding to the relation:

$$k_p'^2 = \frac{\omega^2}{E_b^*/\rho} = \frac{\omega^2}{v_p^2} \qquad (8)$$

Equation (7) describes formally a purely static problem. It can easily be shown, that the solution of this differential equation is a complex number, irrespective of whether the wave number is complex, as in the case of viscos damping, or not:

$$u(x) = Re(u) + i\ Im(u) = u_0\ e^{-ik_p'x} \qquad (9)$$

The total solution of eq.(4) then is the function:

$$u_x(x,t) = u_0\ e^{i(\omega t - k_p'x)} \qquad (10)$$

In this equation u_0 is the amplitude, which is determined by initial conditions and, in the case of damping, depends on the coordinate. Eq.(10) describes the propagation of a one-dimensional P-wave in the positive x-direction. It may be represented as in fig.1.

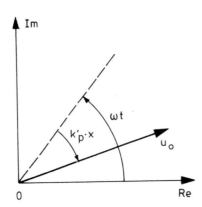

Figure 1. Representation of eq.(10) in the complex plane.

In a linear visco-elastic material, for which in the case of harmonic waves a complex Young's modulus has been obtained, the velocity of propagation of the wave is higher than in a purely linear elastic body. The reduction of a steady-state dynamic problem to one, including only the space variables, has been shown here at a one-dimensional P-wave, but it is of course valid for all types of harmonic waves in a two- or three-dimensional space.

2.2 FE-Method

In this method the continuum is divided into single finite elements, which are connected to each other by the nodal points on

258

their boundaries. The FE-method is basically the Ritz-method, where the displacement functions are defined to have non-zero values only within the respective element. These functions must be chosen in a way, that the continuity of the displacements is assured across the boundaries of the elements. The displacements of the nodal points in the x- and the y-direction, respectively, are the degrees of freedom of this system. By making use of the principle of virtual work the equation of the dynamic equilibrium for each of the degrees of freedom within the considered region can be established, thus yielding a system of linear equations:

$$[A] \{u\} = \{q\}. \tag{11}$$

[A] is the structural matrix, which is usually small banded, {q} is the vector of external forces and {u} the vector of the displacements in all degrees of freedom. The structural matrix contains the mass matrix [M], by which the inertia forces are introduced and the stiffness matrix [K]. In the case of material damping the imaginary damping matrix [D] is also included, which yields a complex structural matrix.

$$[A] = -\omega^2 [M] + i\omega[D] + [K] \tag{12}$$

In a steady-state harmonic problem the external forces as well as volume forces vary in a harmonic function with time, thus forcing all displacements to vary in the same time function. Furthermore, the displacements have to be complex, thus the displacement vector can be written as

$$\{u_t\} = \{u\} \, e^{i\omega t} = \{Re(u) + i \, Im(u)\} \, e^{i\omega t}. \tag{13}$$

Because the calculation can be performed at any instant of a period the external forces must be complex, too. Consequently the system of linear equations finally becomes:

$$[A] \{u\} \, e^{i\omega t} = \{q\} \, e^{i\omega t}, \tag{14}$$

where the term including the time variation can again be eliminated. In this equation the matrix and the vectors are all complex. There are special routines available for the operation of complex numbers enabling a fast computation and an easy programming. The dublication of the required storage space for the calculation due to the complex numbers can become very important, if large systems of equations are involved.
The amplitude of a dynamic displacement is calculated by

$$|u| = \sqrt{Re^2(u) + Im^2(u)} \tag{15}$$

and the phase-angle, relative to a reference vibration, by

$$\phi = -arc \, tan \, \frac{Im(u)}{Re(u)} \tag{16}$$

(see fig.1).

3. BOUNDARY CONDITIONS

The two major problems in the analysis of steady-state wave fields by the FE-method are:
  1) the application of the appropriate conditions at the boundaries of the FE-grid
  2) the handling of large systems of equations.

The importance of the boundary condition (in the following abreviated by BC) for the accuracy of the result of the analysis will be demonstrated by the following example.

3.1 Example

A long, straight, homogeneous bar is considered, which is free at one side and fixed at the other. At the open end normal stresses σ_0, varying harmonically with time, act in the direction of the axis of the bar, see fig.2a. Let the original length of the bar be 20 λ_p, when λ_p is the wave length of the P-wave travelling through the bar. The normal stress along its axis for this case is referred to as the "exact" solution.

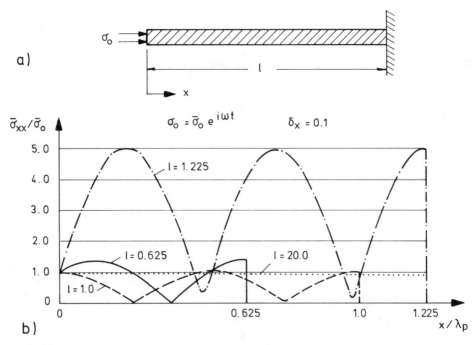

Figure 2. Normal stress in a bar of different length.
a) System   b) Amplitude functions.

Now, only a short part of the bar near to the open end is analysed and it is assumed that for lack of a BC, which could replace the rest of the bar, the fixed end is retained.

260

In fig.2b the amplitude-functions of the normal stress in the
bar, $\bar{\sigma}_{xx}$ normalized on $\bar{\sigma}_o$, for the cases $1 = 0.625 \lambda_p$, $1.0 \lambda_p$
and $1.225 \lambda_p$, respectively, are presented. These functions
are highly dependent on the length of the bar and do not at all
coincide with the exact solution, which is indicated by the
dotted line. The error may rise even up to several hundred per
cent. If the exact solution is unknown, no evaluation of the
error and its direction is possible.
    The reason for this absolutely unacceptable result is the
application of the BC of the fixed end, which is not adequate
to the total system and by which the incoming waves are reflected totally.

3.2 Integral Equation Problem

The principle difficulty concerning the dynamic BC will be outlined briefly in the following: The ground is represented by an
infinite half-space. A finite region G_I is imagined to be cut
out of this half-space and its boundary is denoted by C, see
fig.3.

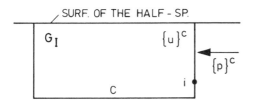

Figure 3. Finite region
in the half-space.

In order that no reflexion - not even partial - of the incoming
waves at C occurs the dynamic stresses of reaction must act on
this boundary. They are - as usual in the FE-method - represented by a vector of nodal forces, $\{p\}^C$. These forces depend
on the properties of the incoming waves, which are: type of the
wave, direction of propagation α, wave velocity v, frequency ω,
and a parameter of the material damping δ. The nodal force in
one degree of freedom on C, for example i, p_i^C, then is a function of these parameters and is also related directly to the
displacement in this degree of freedom. Hence, this force can be
written in the form:

$$p_i^C = f(\text{type}, \alpha, v, \omega, \delta) \ u_i^C \qquad (17)$$

    All elements of the list in brackets are determined by the
totality of the displacements on the boundary C, represented by
the vector $\{u\}^C$. Therefore the boundary forces $\{p\}^C$ depend on
the displacements in the following way:

$$\{p\}^C = [F]\{u\}^C \qquad (18)$$

[F] is an operator, which contains all the elements of the list
in brackets in some unknown form and which connects each

261

component of $\{u\}^C$ with each component of $\{p\}^C$ in a linear function. Usually [F] is unknown and cannot be introduced into the
calculation, because it depends on the wave field, which has to
be analysed, and which is only obtained exactly if [F] is known.
In general, [F] can be determined only by solving a problem of
integral equations, but it is uncertain that a solution can always be found analytically. There are, however, some special
cases, where the direct analytical calculation of the boundary
forces depending on the boundary displacements is possible.
    In the following the most important dynamic BC will be de-
scribed briefly.

3.3 Full-Space Boundary-Condition (FS-BC)

Lysmer and Kuhlemeyer (1969) assumed the dynamic reaction stresses, replacing an infinite, homogeneous isotropic body with
respect to the propagation of body waves, to be:

$$\sigma_{xx} = a \cdot \rho \cdot v_p \cdot \dot{u}_x (x_c, y, t)$$

$$\sigma_{xy} = b \cdot \rho \cdot v_s \cdot \dot{u}_y (x_c, y, t). \tag{19}$$

In this relations $\dot{u}_x$ and $\dot{u}_y$ are the particle velocities at the
boundary in the x- and y-direction, respectively,(see fig.4a)
and v_p and v_s are the respective wave velocities of the P- and
the S-wave. The authors found that a good absorption of the
incoming body waves can be obtained, if both a and b are chosen
to be 1.0 and the angle α is less than about 60°. This BC represents only an approximation to the exact one (exact for
$\alpha = 0°$), but no better FS-BC is available at present, which
can be easily applied. Valliapan and cooperators (1976) extended this BC to an anisotropic material.

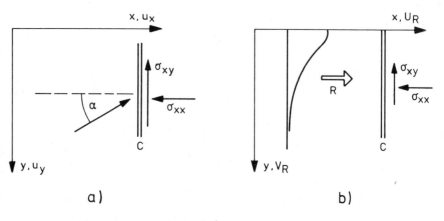

a)   b)

Figure 4. Notation for a) FS-BC   b) RW-BC.

3.4 Rayleigh Wave Boundary Condition (RW-BC)

Soil dynamic problems very often deal with a half-space and in this case the surface wave is of special interest. In an unlimited homogeneous half-space the solution of the homogeneous differential equation of the wave propagation is represented by the free Rayleigh wave (R-wave). To this type of wave the FS-BC mentioned above is not applicable. As in the case of a free R-wave all elements, listed within the brackets of eq.(17), are known, therefore the BC replacing the infinite, homogeneous, isotropic quarter-space can be calculated analytically. For the notation see fig.4b.

Lysmer and Kuhlemeyer (1969) started from relations analogue to eq.(19), introducing the horizontal and vertical displacement components of the R-wave, $U_R(x_c,y,t)$ and $V_R(x_c,y,t)$, respectively. a and b are now functions of the depth and can be calculated from:

$$\left.\begin{aligned}
a(y) &= \frac{1}{k_p}\left[k_R + (1 + 2\kappa^2)\frac{V'_{RO}(y)}{U_{RO}(y)} \right] \\[2ex]
b(y) &= \frac{1}{k_s}\left[k_R - \frac{U'_{RO}(y)}{V_{RO}(y)} \right].
\end{aligned}\right\} \quad (20)$$

In this equations k_p, k_s and k_R are the wave numbers of the P-, S- and R-wave, respectively, and $\kappa = k_p/k_s$. $U_{RO}(y)$ and $V_{RO}(y)$ represent the displacement functions of the R-wave components with depth, the prime indicates the derivation with respect to y. The functions a(y) and b(y) are presented for $\nu = 0.25$ in fig.5, where λ_R denotes the Rayleigh wave length.

As $U_{RO}(y)$ vanishes at a point near to the surface, the function a(y) has a pole at a depth of about $0.2\,\lambda_R$. Hence, in this upper layer of the half-space the FE-grid has to be chosen very dense to avoid an unadmissible inaccuracy from the finite approach. It should be noted, that the application of this RW-BC yields a non-symmetric structural matrix. It has been found, however, that by using the expression

$$\frac{1}{2}\,([A] + [A]^T) \qquad\qquad (21)$$

for the structural matrix, the error introduced into the numerical solution is negligible.

This BC is exact in the case of a free Rayleigh wave incoming to a vertical boundary. However, if a wave source or an obstacle within the half-space is located near to the boundary no free Rayleigh wave exists and a partial reflexion of the wave at the boundary will occur. As the Rayleigh wave develops from the body waves with distance the BC described above is applicable again if the boundary is removed several wave lengths away from the source. Thus, usually the FE-grid has to be chosen much larger than the region of interest to be analysed only to fulfill the requirement for the applicability of the RW-BC. This implies a considerable waste of computer-time.

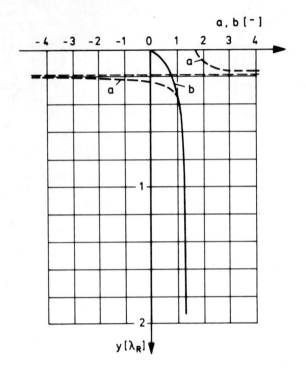

Figure 5. Functions
a(y) and b) for RW-BC.

3.5 Boundary Condition for Layered Half-Space (LHS-BC)

Often the subsurface ground consists of different layers with
more or less different dynamic material parameters, or the water
table is not far below the surface. In these cases a(y) and b(y)
usually cannot be calculated, because an analytical solution of
the differential equation of wave propagation in a multiple
layered half-space is established only for special cases (Ewing,
Jardetsky and Press 1957). The problem has been solved first
for plane condition by Lysmer and Waas (1972), who developed a
boundary condition for a layered half-space, based on a numerical solution. Kausel, Roesset and Waas (1975) extended it to
the axisymmetric case.
    The relation between the boundary displacements and the
boundary forces is derived at that part of the half-space, which
has been cut off from the region to be analysed. This part is a
semi-infinite region which is divided into a number of horizontal layers. The lower boundary is represented by an infinitely stiff layer, i.e. the displacements at this boundary are
forced to be zero. The displacement within one layer is assumed
to vary linearly with depth and harmonically in the horizontal
direction. Following the normal FE-procedure, which is applied
in this case to an assemblage of semi-infinite elements, a homogeneous system of linear equations is obtained. The eigenvalues
of this system are the wave numbers of the corresponding eigenmodes. The boundary forces $\{p\}^c$ now can be expressed by the

stiffness matrix [A] of the semi-infinite layered region, the matrix of the eigenmodes [V] and the boundary displacements:

$$\{p\}^C = i \; [A][V][K][V]^T \; [A]^T \; \{u\}^C \qquad (22)$$

[K] is a diagonal matrix of the eigenvalues. Equation (22) represents the desired relationship, expressed by eq.(18), because all matrices can be calculated without considering the actual boundary displacements. Because by this equation each component of $\{p\}^C$ is related to each component of $\{u\}^C$, this relationship represents in fact a numerical solution of the integral equation problem mentioned above.

This LHS-BC is a powerful tool in the analysis of steady-state dynamic problems by the FE-method. It has been proven to yield very good results, even if applied very close to a wave source. There is, however, one restriction to the applicability of this BC. Because the lower boundary of the FE-grid is represented by an infinitely stiff layer, this condition can be used only, if in reality such a very stiff layer is underlaying a soft layer, which certainly will often be the case. However, if the half-space is rather homogeneous the assumption of an infinitely stiff layer will yield erroneous results. There is, of course, the possibility to shift the stiff layer far down into the half-space but here again the FE-grid has to be chosen much larger than that to be analysed originally. Furthermore, as a total reflection of the incoming waves at the stiff layer definitely takes place, the wave field will always be affected and the error can be in the order of that of the example presented earlier.

Therefore, a boundary condition for the homogeneous half-space has been developed, which is applicable at any distance of a wave source within the half-space. As this so-called "influence-matrix boundary condition" in addition provides a considerable advantage concerning the size of the system of equations to be solved, it will be outlined here in more detail.

4. INFLUENCE-MATRIX BOUNDARY CONDITION (EM-BC)

4.1 Principle

Besides the FE-grid G_I (see fig.3) a second one, G_{II}, is considered, which is imagined to have been cut off from the first one along the vertical boundary C. Hence, these boundaries in the two partial regions are identical and therefore the respective vectors of the nodal displacements on these boundaries are identical as well, see fig.6.

The forces of reaction between the two regions act on the boundaries in opposite directions. The structural matrix of the region G_{II} is denoted by $[A]_{II}$.

Now the case of a force of magnitude "1", acting in one degree of freedom on the boundary C in G_{II}, for example i, is considered. This load case is analysed following the normal FE-procedure, yielding the total displacement vector within G_{II}, $\{u\}_{IIi}$, but especially the vector of the displacements on the

265

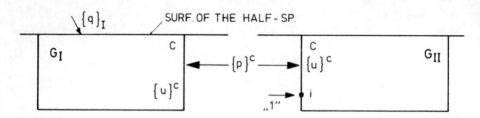

Figure 6. Two finite regions in the half-space

boundary C, $\{u\}^C_{IIi}$. In the following an analogic computation is
made for all degrees of freedom on this boundary. All the displacement vectors of the region G_{II} obtained by this procedure
are collected to be the columns of a matrix $[U]_{II}$. The vectors
of the boundary displacements are collected in the same way in
the matrix $[EM]$. One line of this matrix, for example i, is
nothing more than the influence-function - as it is known from
statics - for the displacement in the degree of freedom i, in a
discretized form as usual in the FE-method. The matrix $[EM]$,
which collects all these influence functions, consequently is
called influence-matrix (in german "Einfluß-Matrix").

The columns of the matrix $[EM]$ contain the displacements on
the boundary C, due to the forces of value "1". Hence, if this
matrix is multiplied by the vector of actual external nodal
forces, $\{p\}^C$, the vector of the actual nodal displacements on
C, $\{u\}^C$, is obtained:

$$\{u\}^C = [EM] \{p\}^C \tag{23}$$

By inverting this equation one gets:

$$\{p\}^C = [R] \{u\}^C, \tag{24}$$

where $[R]$ is the so-called reaction matrix, which is calculated
from:

$$[R] = [EM]^{-1}. \tag{25}$$

Both, the influence matrix as well as the reaction matrix are
square, symmetrical and complex.

Provided the region G_{II} is not subjected to external forces,
the problem of the boundary condition replacing that part, which
has been cut off, is solved by eq.(24), because the operator $[F]$
is represented now by the reaction matrix $[R]$.

For the analysis of the total region ($G_I + G_{II}$), in the case
of an external load vector $\{q\}_I$ acting on G_I, only the system
of linear equations for the region G_I has to be established,

$$[A]_I \{u\}_I = \{q\}_I - \{p\}^C. \tag{26}$$

In this equation $[A]_I$ represents the structural matrix of G_I
and $\{q\}_I$ is the vector of external forces acting on this region.

The boundary forces $\{p\}^C$ are considered as external forces with negative sign. Introducing equation (24), one obtains:

$$[A]_I \{u\}_I = \{q\}_I - [R] \{u\}^C \qquad (27)$$

As the displacements on the boundary, $\{u\}^C$, are included within the displacement vector $\{u\}_I$ of the total region G_I, eq.(27) can be written:

$$([A]_I + [\bar{R}]) \{u\}_I = \{q\}_I . \qquad (28)$$

In this equation $[\bar{R}]$ represents the reaction-matrix, adapted to the dimensions of $[A]_I$. Solving this system of equations, the displacements within the region G_I are such, as if the total region $(G_I + G_{II})$ has been analysed, because $[\bar{R}]$ includes all properties of G_{II} with respect to wave propagation, like geometry, boundary conditions and material properties.

The displacement vector within region G_{II} now can be found from the actual boundary displacements by simple matrix multiplication: Since the matrix $[U]_{II}$ contains in its columns the displacement vectors of G_{II}, due to the forces of value "1" on the boundary C, it must be multiplied by the vector of the actual boundary forces to yield the actual displacement vector of G_{II}:

$$\{u\}_{II} = [U]_{II} \{p\}^C \qquad (29)$$

Introducing eq.(24) one gets:

$$\{u\}_{II} = [U]_{II} [R] \{u\}^C , \qquad (30)$$

where $\{u\}^C$ is known, because it is part of $\{u\}_I$.

Thus, the influence-matrix procedure provides two possibilities:

1) The part G_{II} of the total region $(G_I + G_{II})$ can be replaced perfectly with respect to G_I by the appropriate boundary condition represented by the reaction matrix $[\bar{R}]$.

2) Once $[R]$ has been established, the displacement vector of the region G_{II} can be calculated by simple matrix multiplication making use of the displacement vector on the boundary C in region G_I.

In case the total region has to be analysed several times - for instance different load vectors are acting on G_I - the advantage is obvious: only the system of equations for the region G_I has to be solved each time, thus providing a saving of computer time and required storage space.

The principle of the influence-matrix is not only valid for dynamic problems, but also for static ones in soil mechanics or structural engineering. The only condition to be accomplished is, that different load cases can be superimposed. This means a linear stress-strain behaviour of the material and small deformations. Since with this BC each component of $\{p\}^C$ is

267

related to each component of $\{u\}^C$, it also represents the numerical solution of an integral equation problem with respect to the vertical boundary C.

4.2 Technique of Calculation

The economic application of the EM-procedure depends essentially on the method of calculating [EM], from which [R] is easily obtained by eq.(25). Therefore, the technique of establishing [EM] will be described in more detail in the following.

Let the two FE-grids G_I and G_{II} have a different number of nodal points, say k and n, respectively. Due to other requirements the global numbering of these grids will usually be such, that the corresponding nodal points on the common boundary C will not coincide in their numbers. For these boundary nodal points a local numbering j = 1,m is defined, which may be again different in the two grids, and which yields 2m degrees of freedom on C, if a plane problem is considered.

The application of the force of magnitude "1" in the degree of freedom i (local numbering) on C is a load case, which is represented by the system of equations

$$[A]_{II} \{u\}_{IIi} = \{b\}_{IIi} . \tag{31}$$

In this equation $[A]_{II}$ is a symmetric (2n x 2n)-matrix and the vectors are 2n-dimensional. The load vector $\{b\}_{IIi}$ contains only zeros except one single "1", which is located according to the global number of that respective degree of freedom. The load vectors corresponding to the 2m load cases are assembled to be the columns of the load matrix $[B]_{II}$. They are arranged according to the local numbering of the degrees of freedom on C in G_{II}, see for example fig.7a. The displacement vectors, each belonging to one of the load vectors, are assembled in the matrix $[U]_{II}$, corresponding to the same arrangement. Thus the matrix equations

$$[A]_{II} [U]_{II} = [B]_{II} \tag{32}$$

is obtained.

In order to solve eq.(32) a FE-program has been extended by inserting more than one vector on the right hand side of the system of equations. By this the expensive triangularization of the coefficient matrix by use of the Gauss' algorithm has to be performed only once. The program MLV (multi load vector), which has been developed for this purpose, requires for the analysis of 48 different load cases only twice the time necessary for one single case. Hence, if the total grid (G_I + G_{II}) has to be analysed more than two times, the application of the EM-procedure already provides an economical advantage.

Once eq.(32) is solved the influence matrix is obtained from:

$$[EM] = [B]_{II}^T [U]_{II} , \tag{33}$$

and [R] from eq.(25), both being symmetric (2m x 2m)-matrices.

268

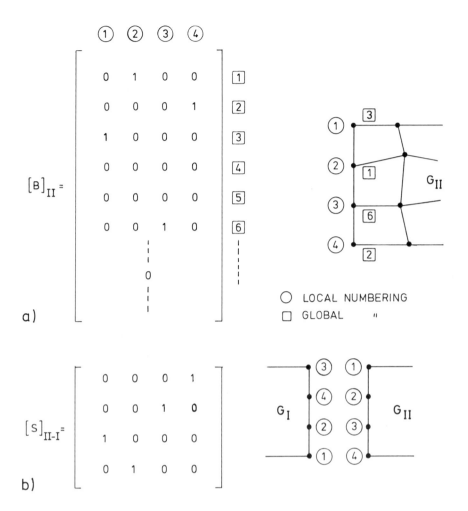

a)

b)

Figure 7. a) Example for $[B]_{II}$ with difference in local and global numbering   b) Example for $[S]_{II-I}$

As $[R]$ ist built up corresponding to the lokal numbering on C in G_{II} a rearrangement of the matrix elements has to be performed for its adaption to $[A]_I$. For this purpose a $(2m \times 2m)$-matrix $[S]_{II-I}$ is defined. Its columns are arranged according to the lokal numbering on C in G_{II} whereas the arrangement of the rows follows the local numbering on C in G_I. This matrix contains only zeros except the numbers "1" at those locations which connect the identical degrees of freedom of the two regions. For an example see fig.7b. The adapted reaction matrix then is obtained from:

$$[\bar{R}] = [B]_I [S]_{II-I} [R] [S]_{II-I}^T [B]_I^T , \qquad (34)$$

where $[B]_I$ is built analogical to $[B]_{II}$.

The fast computation of $[EM]$ by the program MLV, which solves eq.(31) for all the 2m load cases at the same time, is the condition necessary for the economic application of the EM-procedure.

4.3 Boundary Condition for Homogeneous Half-Space

Homogeneous Quarter-Space

If the conditions at the vertical boundary C of G_I are bound to replace the adjacent homogeneous quarter-space, extending infinitely to the right, the region G_{II} has to be homogeneous as well and must be chosen in a special way: At the lower boundary of this FE-grid the FS-BC is used, whereby it is assumed, that the waves incoming to this boundary are perfectly absorbed. This is justified because some small reflection at this lower boundary does not affect the displacements on C.

At the vertical boundary opposite to C the RW-BC must be applied, which has been mentioned earlier and which can be established analytically. The region G_{II} now must be chosen long enough, so that the body waves, radiating from a point source located on boundary C, have developed to a Rayleigh wave until they reach the opposite vertical boundary.

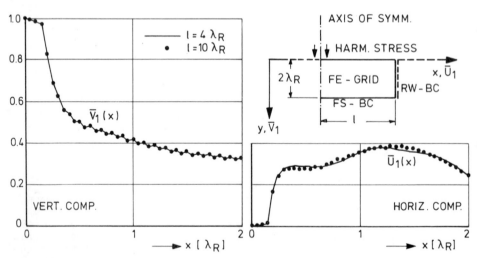

Figure 8. Amplitudes of the vertical and the horizontal displacements at the surface for different length of G_{II}

In fig.8 the vertical and horizontal displacement amplitudes at the surface, due to a harmonically varying, vertical load

270

at the surface, are plotted for a length of the FE-grid of $4\lambda_R$
and of $10\lambda_R$, respectively, where λ_R represents the Rayleigh
wave length. At the point of load the coincidence of the corresponding amplitudes is very good. This proves that the length
of $4\lambda_R$ of the region G_{II} is sufficient for the reflection of
the waves at the opposite boundary being negligible at C. Therefore region G_{II} can be considered as behaving like the upper
left corner of the infinite, homogeneous right quarter-space
with respect to the propagation of waves. Thus, in this case
the reaction matrix [R] represents a boundary condition, which
simulates this right homogeneous quarter-space.

Homogeneous Half-Space

If the total half-space has to be included in the analysis, the
influence and the reaction matrix, respectively, for the left
vertical boundary of G_I have to be established. In the following $[EM]^l$ and $[R]^l$ refer to these matrices and G_{II}^l means region G_{II} on the left hand side of G_I. The index r will refer
to the right hand side. If G_{II}^l is different from G_{II}^r with respect to the material properties or the geometry, $[EM]^l$ has to
be calculated following the procedure described in chapter 4.2.
    However, if G_{II}^l includes the same material as G_{II}^r and if
furthermore these two FE-grids are mirror-symmetric concerning
the global geometry as well as the division into finite elements (at least in the vicinity of C^l), then $[EM]^l$ can be derived easily from $[EM]^r$. It is necessary only to convert the
sign of a part of the matrix elements. The basic relations
are demonstrated in fig.9. If one element of [EM] is denoted
by em_{ij}, the procedure to be followed for the conversion of
the signs, based on these relations, is:

$$
\left.
\begin{aligned}
em_{ij}^l &= em_{ij}^r \qquad && \text{for } i+j = \text{even number} \\[2mm]
em_{ij}^l &= -em_{ij}^r \qquad && \text{for } i+j = \text{odd number.}
\end{aligned}
\right\} \qquad (35)
$$

    With both matrices $[R]^l$ and $[R]^r$ applied at the left and the
right vertical boundary of region G_I and the full-space condition used at its lower boundary the total infinite, homogeneous half-space is included in the dynamic analysis of region G_I, subjected to harmonically varying external forces.
If $[EM]^l$ is to be derived from $[EM]^r$ the two regions G_{II}^l and
G_{II}^r have to be mirror-symmetric, whereas no restriction exists
for G_I with respect to the material properties and the global
geometry. The only condition is that the left and right
boundaries coincide with the corresponding ones of the two
regions G_{II}, whereby these boundaries have not necessarily to
be straight and vertical, as they have been chosen here.

271

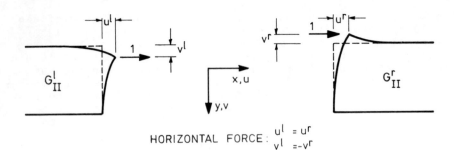

HORIZONTAL FORCE : $\begin{matrix} u^l = u^r \\ v^l = -v^r \end{matrix}$

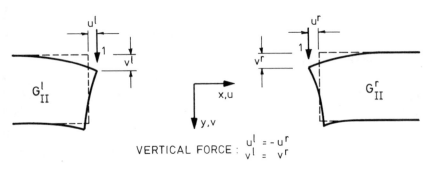

VERTICAL FORCE : $\begin{matrix} u^l = -u^r \\ v^l = v^r \end{matrix}$

Figure 9. Sign conversion at $[EM]^r \rightarrow [EM]^l$

4.4 Generation of Rayleigh Wave Field in the Half-Space

If the source of the wave field to be analysed - for instance a vibrating machine foundation - is not located within region G_I, the wave field has to be defined along the boundaries of G_I by external forces or by specified displacements. For this purpose the influence-matrix procedure can be applied very usefully as is demonstrated in the following. The case of a free Rayleigh wave at the surface of the half-space is considered, which is an important problem in soil dynamics, see fig.10a.

The left and the right vertical boundaries of G_I will be denoted by C^l and C^r, respectively, in the following and on the right hand side G_{II}^r, representing the right quarter-space (see chapter 4.3), is replaced by $[R]^r$.

Let us assume that G_I contains a local inhomogeneity at whic the free Rayleigh wave, penetrating in its theoretical mode fro the left into G_I, is partially reflected. Then C^l must not be a free boundary at which the waves, travelling back from the inhomogeneity, are again reflected. Furthermore, no displacements must be specified at this boundary, either, in order to avoid a partial reflection of these waves. Therefore, the adjacent left, homogeneous quarter-space, which is represented by G_{II}^l, has to border on the region G_I at this side.

In this total system ($G_{II}^l + G_I + G_{II}^r$) the former boundary C^l

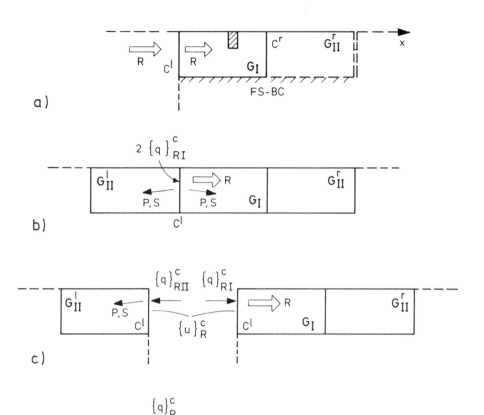

Figure 10. a) System for free R-wave analysis
b) $2\{q\}^c_{RI}$ applied to line C^1 c) G^1_{II} and $(G_I+G^r_{II})$ separated
d) Generation of R-wave at line C^1.

now represents a vertical line in the half-space (plane problem),
along which external nodal forces must be acting such, that a
free R-wave is generated, propagating to the right.

   The stresses within the half-space, due to a free R-wave
travelling in the positive x-direction are represented by the
vector of nodal forces $\{q\}^c_{RI}$. If these forces act in the degrees
of freedom on C^1 the displacements corresponding to the free
R-wave, $\{u\}^c_R$, will be produced only, if C^1 represents the free

boundary of the region $(G_I+G_{II}^r)$, see right hand side of fig.10c. If, however, the total system $(G_{II}^l+G_I+G_{II}^r)$ is considered, where C^l represents a line, the vector $2 \cdot \{q\}_{RI}^C$ will generate, besides the Rayleigh wave, body waves, spreading out from C^l to the left as well as to the right (fig.10b). The reason for this is the difference in the phase-angle of $\pi/2$ between the vertical and the horizontal displacement component of the R-wave. This phase shift is coupled to the direction of propagation.

In the following the regions G_{II}^l and $(G_I+G_{II}^r)$ are considered to be seperated. If eq.(24) is applied to G_{II}^l and the displacements on C^l are assumed to be those of the free R-wave, $\{u\}_R^C$, one obtains::

$$\{q\}_{RII}^C = [R]^l \{u\}_R^C . \tag{36}$$

Hence, the vector of forces $\{q\}_{RII}^C$ generates the same displacements on C^l in G_{II}^l as does the vector $\{q\}_{RI}^C$ on C^l in $(G_I+G_{II}^r)$ (see fig.10c).

The first of Castigliano's relations is given by

$$\frac{\partial E}{\partial u_i} = k_i , \tag{37}$$

where E represents the total potential energy of an elastic body, due to the external, single forces k_i. The displacements u_i are taken at the location and in the direction of action of the forces (Becker and Bürger 1975). It can be shown, that eq.(37) is valid, too, for steady-state harmonic problems in a linear visco-elastic body (Christensen 1971). This equation, applied to the two FE-regions under consideration, yields:

$$\left. \begin{aligned} \frac{\partial E_{II}}{\partial \{u\}_R^C} &= \{q\}_{RII}^C \\[2em] \frac{\partial E_I}{\partial \{u\}_R^C} &= \{q\}_{RI}^C , \end{aligned} \right\} \tag{38}$$

where E_I refers to region $(G_I+G_{II}^r)$.

If these two regions are connected again and no other external forces than those on C^l are acting in either region, the potential energy of both regions must be added. Therefore, by use of eq.(38) one obtains:

$$\frac{\partial(E_I+E_{II})}{\partial\{u\}_R^C} = \frac{\partial E_I}{\partial\{u\}_R^C} + \frac{\partial E_{II}}{\partial\{u\}_R^C} = \{q\}_{RI}^C + \{q\}_{RII}^C =: \{q\}_R^C \tag{39}$$

Hence, the vector $\{q\}_R^c$, acting along the line C^1 in the total Region $(G_{II}^1 + G_I + G_{II}^r)$, will generate the displacements $\{u\}_R^c$, which belong to a Rayleigh wave, propagating to the right of C^1 through G_I (see Fig.10d). If this wave is partially reflected at the in-homogeneity within G_I, the returning waves will pass the line C^1 without being reflected again, becauce G_{II}^1 (i.e. the homo-geneous left quarter-space) is simulated perfectly by $[R]^1$.

The wave field within G_{II}^r, $\{u\}_{II}^r$, can now easily be calcula-ted from eq.(30) by introducing the appropriate reaction matrix $[R]_{II}^r$ and the displacement vector on C^r in G_I, $\{u\}_I^{cr}$, obtained from the FE-analysis of that region:

$$\{u\}_{II}^r = [U]_{II} [R]_{II}^r \{u\}_I^{cr} . \tag{40}$$

For the calculation of the wave field within C_{II}^1 more compli-cated considerations are necessary (Haupt 1978a).

3. FREE RAYLEIGH WAVE FIELD ANALYSIS

An investigation has been performed on the influence of an in-homogeneity within the half-space to the propagation of sur-face-waves (Haupt 1978a, 1978b). Here only the procedure of the FE-analysis and the results concerning the free, undisturbed Rayleigh wave field are presented.

The FE-grid was chosen to be rectangular with the dimensions of 10 by 2 λ_R, see fig.11a.

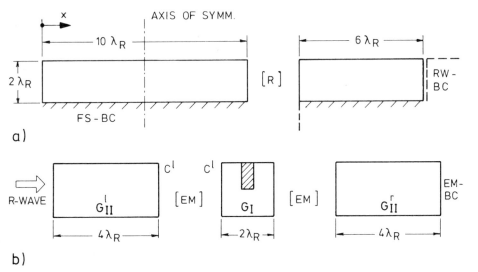

Figure 11.   a) FE grid with BC.   b) Division of FE-grid.

The density of the grid was 20 nodal points within one wave length in the upper part of the half-space and decreased to about 10 within one λ_R in the lower part, by this yielding a system of linear equations of about 7500 komplex unknowns. At the boundaries at both sides of the grid the EM-BC, replacing the homogeneous half-space, was applied. It had been calculated using a grid of the length of 6 λ_R following the procedure described in chapter 4.2 and 4.3. The vertical boundaries of these FE-grids included 24 nodal points. Hence, both, [EM] and [R] were (48 x 48)-matrices. Since the total geometrical system was chosen to be mirror-symmetrical (see fig.11a) the matrices at the left hand side could be derived easily from those at the right hand side (see chapter 4.3). The Rayleigh wave was generated at the left boundary following the procedure described in chapter 4.4.

The total FE-grid was subdivided into two outer grids, G_{II}^l and G_{II}^r, of 4 λ_R length and an inner one, G_I, of 2 λ_R length, which contained the inhomogeneity, see fig.11b. At the interfaces between G_{II}^l and G_I and between G_I and G_{II}^r, respectively, the influence-matrices were established. At the left boundary C^l of region G_I again the free Rayleigh wave was generated (chapter 4.4). In this case, however, the displacement vector on C^l was not calculated analytically but obtained from the preceding FE-analysis of region G_{II}^l. Thus, G_I only was subjected to external forces (on C^l) and had therefore only to be calculated following the normal FE-method. The displacements within G_{II}^l and G_{II}^r were then obtained by the EM-procedure.

By subdividing the FE-grid with application of the EM-procedure a reduction of the system of equations to one, containing only 1500 komplex unknowns, was obtained. Only by this considerable saving of computer-time a very large number of different cases concerning the geometrical and material parameters of the inhomogeneity could be analysed (Haupt 1978b).

5.1 Results

In fig.12 the amplitude functions $\bar{U}_{RO}(y)$ and $\bar{V}_{RO}(y)$, and the phase-functions, $\bar{\psi}_{RO}(y)$ and $\bar{\varphi}_{RO}(y)$, respectively, of the free R-wave with depth are presented. The respective curves obtained from the analytical and from the numerical analysis are compared. The solid lines indicate the analytical solution without material damping and the dotted lines that with viscos damping included. The difference between these two cases is too small to be plotted for the amplitudes. The solid circles represent the amplitude and the phase-angle, respectively, of the nodal points obtained from the FE-analysis. As can be seen from this figure the amplitudes do agree very well, even at the point of zero displacement of the horizontal component.

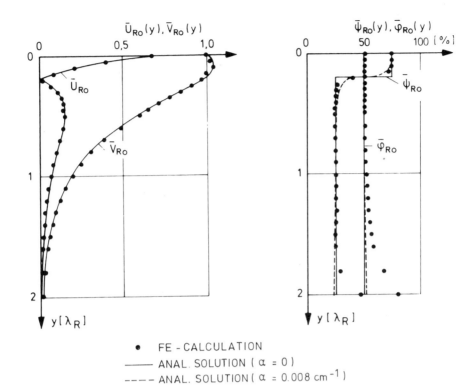

Figure 12. Comparison of free Rayleigh wave displacements from
num. and from analyt. analysis.

The functions of the phase-angles agree also very well in
the upper part of the half-space. Even the phase shift of 50 %
of one period of the horizontal component is realized very well.
This indicates the grid in this part of the half-space being
chosen dense enough. Near to the lower boundary, however, the
agreement between the analytical solution and the numerical one
is not good. This is due to the FS-condition at this boundary,
which is a very bad approximation to the exact one for the extreme case of a wave travelling parallel to the boundary. This
unfavorable behaviour of the FS-BC, however, does not at all
affect the amplitudes.

Fig.13 shows the amplitudes and the phase-angles of the horizontal and the vertical component, respectively, at the surface over the total length of the FE-grid. Due to the material
damping the amplitudes attenuate as an exponential function.
Hence, they appear as straight lines on a semi-logarithmic
scale, see the left ordinate. The phase-angles are expressed in
percent of one period. No influence from the boundaries or from
the subdivision of the FE-grid can be observed in any of these
curves.

The displacement field and the stress field respectively obtained from the FE-analysis are always a more or less good
approximation to the true ones. From the principle of the minimum of the potential energy follows, that each approximate

277

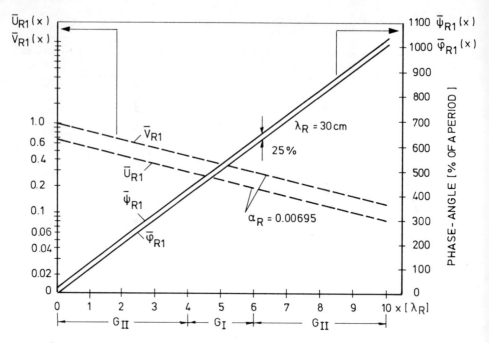

Figure 13. Free Rayleigh wave displ. funct. at the surface of the half-space.

solution has a higher level of potential energy than the true one. This overestimation of the potential energy includes an underestimation of the dissipated energy. Hence, in the FE-analysis the material damping is always smaller than it has to be corresponding to the material properties. In this case from the material constants an attenuation factor of the Rayleigh wave of $\alpha = 8.0 \cdot 10^{-3}$ cm^{-1} ($\delta_x = 0.24$) should have been obtained. However, from the amplitude data this factor was found to be $\alpha = 6.95 \cdot 10^{-3}$ cm^{-1}.

From the overestimation of the potential energy also follows an overestimation of the stiffness of the system. This is obvious if the reduction of the infinite number of degrees of freedom of the continuum to a finite number in the FE-approach is considered. From this increase of stiffness follows an increase of the wave velocity, in this case of 1.4 %. Systematic investigations concerning this point are not known, but it is possible that these considerations could yield a criterion for the upper limit of the element size.

6. SUMMARY AND CONCLUSION

Vibrations or shocks travelling through the ground can be considered as a superposition of steady-state harmonic waves. Therefore the analysis of these waves by the FE-method is an appropriate way to investigate the dynamic processes in the ground.
Since the time variable can be eliminated, the problem can be
treated formally as a static one. The essential problem in the
steady-state harmonic wave field analysis is the application of
the conditions, appropriate to the total geometrical system,
at the boundaries of the limited FE-grid. There are analytical
solutions available for the full-space and for the half-space,
if dealing with a free Rayleigh wave. By a numerical procedure
the boundary condition, simulating a layered half-space, can be
obtained.

In this paper the so-called "influence-matrix" boundary condition is presented. It is based on a numerical solution of the
integral equation problem, which represents essentially the
boundary condition problem. By making use of the Rayleigh wave
boundary condition the infinite, homogeneous quarter-space adjacent to a limited FE-grid is simulated by this condition. The
generation of any wave field within the considered region can
easily be performed.

The application of the influence-matrix procedure allows any
subdivision of the FE-grid to be analysed without losing any
information about the sub-grids. Therefore, if only a small part
of the FE-grid is subject to the variation of the geometry or
material properties, a large number of calculations can be performed with a minor amount of computer time.

Thus, by use of the influence-matrix procedure the FE-method,
contrary to its usual application for the solution of single,
complicated boundary-value-problems, can be applied as an investigation method, which is able to yield general relations.
By the economic method of subdividing the FE-grid the influence-
matrix procedure may possibly be a way of treating truly three-
dimensional problems by the FE-method.

REFERENCES

Ewing, W.M., W.S. Jardetzky, F. Press 1957, Elastic Waves in
    Layered Media. New York, Intern.Series on Earth Science,
    McGraw-Hill
Haupt, W.A. 1976, Influence-Matrix Boundary Condition for the
    Analysis of Dynamic Problems by FE-Method,Proc.Int.Symp.Num.
    Meth. in Soil and Rock Mech. (NMSR 75), Karlsruhe
Haupt, W.A. 1978a, Verhalten von Oberflächenwellen im inhomo-
    genen Halbraum mit besonderer Berücksichtigung der Wellenab-
    schirmung, Veröffentlichung des Instituts für Bodenmechanik
    und Felsmechanik (IBF), Universität Karlsruhe, Vol.74, 1978
Haupt, W.A. 1978b , Surface-Waves in Non-Homogeneous Half-
    Space, Proc.Dyn.Methods in Soil and Rock Mech. (DMSR 77),
    Vol. I, Rotterdam, A.A. Balkema Publ.
Kausel, E., J.M. Roesset, G. Waas 1975, Dynamic Analysis of
    Footings on Layered Media, Proc. ASCE, No. EM5, Oct.1975

Lysmer, J., R.L. Kuhlemeyer 1969, Finite Dynamic Model for In-
    finite Media, Proc. ASCE, No. EM4, Aug.1969
Lysmer, J., G. Waas 1972, Shear Waves in Plane Infinite Struc-
    tures, Proc. ASCE, No. EM1, Feb.1972
Valliapan, S., W. White, I.K. Lee 1976, Energy Absorbing
    Boundary for Anisotropic Material, Num.Meth.in Geom.,
    Publ. by ASCE, Vol. II

Primary and secondary interferences in wavefields

B. PRANGE
University of Karlsruhe, Karlsruhe, Germany

SYNOPSIS

Interferences in the amplitude- and phase-curves of surface wavefields are related to the superposition of Rayleigh-, S- and P-
waves in the nearfield and to anomalies in the subsoil in the
farfield. Additional interferences will be introduced by noncompensated vibration transducers due to inertia effects.

INTRODUCTION

The experimental investigation of surface waves in the field is
aimed at the determination of the dynamic parameters of the subsoil as well as the disclosure of local anomalies. In either
case, wavefields are produced artificially to be able to control
amplitude and frequency of the wave source in order to vary both
the dynamic strain amplitude and the wave length.

The different methods of excitation of wavefields are not subject of this paper, and their discussion will therefore be omit-
ted here. We will restrict ourselves to wavefields with stationary, periodic and in particular monochromatic excitation. The
time-dependency of the vibrations at all points of the wavefield will therefore be $e^{i\omega t}$, ω being the angular frequency of
the wave source. Amplitude and phase angle are then the only
variables at the surface controlled by the dynamic parameters
of the subsoil and their anomalies.

The restriction to stationary periodic excitation leads us to
a phenomenon many wavefields have in common, interference.

In the case of the determination of the dynamic subsoil parameters, the data of the wavefield, namely the complex vibration vectors at different distances from the wave source enable
us to calculate the attenuation due to geometric damping and due
to material damping as well as the elastic constants G and ν,
provided we know the specific density of the subsoil. Stationary
periodic excitation of wavefields does not allow us to distinguish between the measurement of surface wave components (R-
waves) and space wave components (P- and S-waves). Component
separation can only be achieved by correlation methods (taking
advantage of the different wave velocities). Since at least in
the nearfield around the wave source the different components

are present at the same time with comparable amplitudes, interferences between the different wavetypes will take place. These
interferences, caused by no other reason than the presence of
different types of waves will be called primary interferences.

Having gained sufficient knowledge about wavefields in the
absence of local anomalies in the subsoil, we may next consider
experiments where from the interpretation of anomalies in the
wavefields conclusions may be drawn on the location of anomalies in the subsoil like cavities, enclosures, zones of low
density etc. Such anomalies do reflect and refract the different
types of waves and may hence be considered as secondary wave
sources. Consequently, they give rise to another type of interferences of waves at the surface which will be called secondary
interferences.

There exists yet another type of disturbance of the wavefield at the surface. In most cases, the complex vibration vectors of the wavefield are determined by contact-transducers
which introduce an additional local inertia effect due to the
dynamic soil-transducer interaction. This STI is very much
affected by the local soil conditions because of the usually
small dimensions of the transducer in relation to the wavelength. Such small-scale local conditions would never show up
to the same extend in the free wavefield because of the inte-
grating effect within the wavelength. Due to the very fact of
measuring, however, the freefield will be distorted. The resulting interferences which we find in the experimental data
will also be called secondary interferences because they do not
result from the superposition of different types of waves.

PRIMARY INTERFERENCES

In order to establish a simple physical model and in agreement
with experimental data at low strains we assume the soil to be
linearly elastic, neglecting a small amount of material damping.
We also assume homogeneity, knowing of course that due to the
well known stress-dependency of G real soils are not behaving
exactly homogeneous. This shall be considered a secondary
effect, adding, however, to the interferences we will find.

The wavefield investigated should be circular-symmetric
because in the experiment we will generate circular-symmetric
fields by exciters vibrating vertically.

Furthermore, we will not introduce a large error if we
assume the vibrator being a point-source, provided the radius
of the exciter area is small in comparison with the wavelength
at the surface.

It is well known, e.g. Bath (1968), that the solutions for
the dynamic displacements in the circular-symmetric case can
be gained from the application of the following differentical
operators to the similar solutions in the two-dimensional case:

$$q = - \frac{1}{\pi} \int_0^\infty \cosh u \, \frac{\partial}{\partial x}(u_0) du; \quad w = - \frac{1}{\pi} \int_0^\infty \frac{\partial}{\partial x}(v_0) du \qquad (1)$$

$$x \rightarrow r \cosh u$$

where u_0 = horizontal displacement component
               2-dimensional case
        v_0 = vertical displacement component
               2-dimensional case
        q   = horizontal displacement component
               circular-symmetric case
        w   = vertical displacement component
               circular-symmetric case
        x   = x-coordinate,2-dimensional case
        r   = r-coordinate, circular-symmetric case

If we know the complex displacement components of the two-
dimensional wavefield due to a line source we immediately find
the components of the circular-symmetric wavefield due to a
vertical point source by applying the above operators.
    In order to establish the two-dimensional wavefield we
follow the basic investigation of Lamb (1904) and use the notations of Bath (1968).
    Ommitting the time dependency $e^{i\omega t}$ we find

$$u_0 = -\frac{iQ}{2\pi G} \int_{-\infty}^{+\infty} \frac{\xi(2\xi^2-k^2-2a\beta)e^{i\xi x}}{F(\xi)} \, d\xi \tag{2}$$

$$v_0 = -\frac{Q}{2\pi Q} \int_{-\infty}^{+\infty} \frac{k^2_\alpha \, e^{i\xi x}}{F(\xi)} \, d\xi \tag{3}$$

By some transformation which shall be ommitted here the inte-
grales of equ. 2 and 3, by putting $\xi = \zeta + i\eta$, can be written
as follows:

$$\int_{-\infty}^{+\infty} \frac{\xi(2\xi^2-k^2-2a\beta)e^{i\xi x}d\xi}{F(\xi)} = -2\pi i \, H \, e^{-i\kappa x} +$$

$$+ e^{-ikx} \int_0^\infty [\frac{2\zeta^2-k^2-2a''\beta''}{(2\zeta^2-k^2)^2-4\zeta^2a''\beta''} - \frac{2\zeta^2-k^2+2a''\beta''}{(2\zeta^2-k^2)^2+4\zeta^2a''\beta''}]\zeta e^{-\eta x}i d\eta$$

$$+ e^{-ihx} \int_0^\infty [\frac{2\zeta^2-k^2-2a'\beta'}{(2\zeta^2-k^2)^2-4\zeta^2a'\beta'} - \frac{2\zeta^2-k^2+2a'\beta'}{(2\zeta^2-k^2)^2+4\zeta^2a'\beta'}]\zeta e^{-\eta x}i d\eta$$

$$= -2\pi i \, H \, e^{-i\kappa x} + \qquad \qquad \text{①}$$

$$+ 4ie^{-ikx} \int_0^\infty \frac{k^2(2\zeta^2-k^2)a''\beta'' \, \zeta \, e^{-\eta x}}{(2\zeta^2-k^2)^4+16\zeta^4(\zeta^2-h^2)(k^2-\zeta^2)} \, d\eta + \qquad \zeta = -k+i\eta$$

$$\qquad \qquad \qquad \text{②}$$

$$+ 4ie^{-ihx} \int_0^\infty \frac{k^2(2\zeta^2-k^2)a'\beta' \, \zeta \, e^{-\eta x}}{(2\zeta^2-k^2)^4+16\zeta^4(\zeta^2-h^2)(k^2-\zeta^2)} \, d\eta \qquad \zeta = -h+i\eta \tag{4}$$

$$\qquad \qquad \qquad \text{③}$$

$$\int_{-\infty}^{+\infty} \frac{k^2}{F(\xi)} \, \alpha \, e^{i\xi x} \, d\xi = + \, 2\pi i \, K \, e^{-i\kappa x} \, + \quad \textcircled{4}$$

$$+ \, 8i \, e^{-ikx} \int_0^{\infty} \frac{k^2 \zeta^2 (\zeta^2 - h^2) \beta'' \, e^{-x\eta} d\eta}{(2\zeta^2 - k^2)^4 + 16\zeta^4 (\zeta^2 - h^2)(k^2 - \zeta^2)} \, + \qquad \zeta = -k+ih$$

$$\textcircled{5}$$

$$+ \, 2i \, e^{-ihx} \int_0^{\infty} \frac{k^2 (2\zeta^2 - k^2)^2 a' \, e^{-\eta x} \, d\eta}{(2\zeta^2 - k^2)^4 + 16\zeta^4 (\zeta^2 - h^2)(k^2 - \zeta^2)} \qquad (5)$$

$$\zeta = -h+ih$$

$$\textcircled{6}$$

indicating the contributions of the Rayleigh-Wave, the S-Wave and the P-Wave by κ, k and h, respectively.

If the respective values for ζ in equ. 4 and 5 are inserted, the integrals reduce to the more general form

$$\int_0^{\infty} \eta^{1/2} f(\eta) \, e^{-\eta x} \, d\eta \qquad (6)$$

Putting $\eta x = \mu$ and therefore $x d\eta = d\mu$,
Equ. 6 can be expanded into a Taylor-series

$$f(\eta) = f(\mu/x) = f(0) + \frac{\mu}{x} f'(0) + \frac{\mu^2}{2!x^2} f''(0) + \ldots \qquad (7)$$

Using only the first three terms of this series, the integral equ. 6 then reads

$$\int_0^{\infty} \eta^{1/2} f(\eta) e^{-\eta x} d\eta = \int_0^{\infty} \frac{\mu^{1/2}}{x^{1/2}} f(\mu/x) e^{-\mu} \frac{d\mu}{x} =$$

$$= \frac{1}{x^{3/2}} \int_0^{\infty} \mu^{1/2} e^{-\mu} d\mu [f(0) + \frac{\mu}{x} f'(0) + \frac{\mu^2}{2!x^2} f''(0) + \ldots] =$$

$$= \frac{f(0)}{x^{3/2}} \underbrace{\int_0^{\infty} \mu^{1/2} e^{-\mu} d\mu}_{\Gamma(3/2)} + \frac{f'(0)}{x^{5/2}} \underbrace{\int_0^{\infty} \mu^{3/2} e^{-\mu} d\mu}_{\Gamma(5/2)} + \frac{f''(0)}{2!x^{7/2}} \underbrace{\int_0^{\infty} \mu^{5/2} e^{-\mu} d\mu}_{\Gamma(7/2)} + \ldots$$

with the definition of the Γ-function:

$$\Gamma(n) = \int_0^\infty e^{-x} x^{n-1} dx; \quad \text{convergent } n > 0 \tag{9}$$

Lamb (1904), Bath (1968) and similarly Ewing, Jardetzky and Press (1957) used only the first term of equ. 7 to calculate the displacement components of the circular-symmetric wavefield. This obviously is correct only for the farfield, where the contributions of the second, third and further terms may be neglected.

Since we include in our experiments in many cases the nearfield at distances as close as one wavelength to the wavesource, we must consider the contributions of these higher terms as well.

To find out the order of these contributions, the influence upon the wavefield components of the sequential consideration of the second and the third term of equ. 8 is studied.

Calculating the first and second derivative of $f(\eta)$ at $\eta = 0$ and putting the results of equ. 8 gained such into equ. 4 and 5 (for the intermediate steps see Prange 1978) we find the following equations for the horizontal and vertical displacement respectively:

$$u_o \frac{G}{Q} = \quad - e^{-i\kappa x} \cdot H \qquad \qquad \Big|\, R$$

$$+ \sqrt{\frac{2}{\pi}} \frac{1}{(kx)^{3/2}} e^{-i(kx+\pi/4)} H_1 ; \quad H_1 = \sqrt{1-m}$$

$$- \frac{3}{2} \sqrt{\frac{2}{\pi}} \frac{1}{(kx)^{5/2}} i e^{-i(kx+\pi/4)} H_1 \cdot H_2 ; \quad H_2 = 21 - 32\,m \qquad \Big|\, S$$

$$- \frac{60}{32} \sqrt{\frac{2}{\pi}} \frac{1}{(kx)^{7/2}} e^{-i(kx+\pi/4)} H_1 \cdot H_3 ; \quad H_3 = 556 - 1472m + 1024m^2$$

$$+ \sqrt{\frac{2}{\pi}} \frac{1}{(hx)^{3/2}} i e^{-i(hx+\pi/4)} H_4 ; \quad H_4 = \frac{\sqrt{m^3 - m^4}}{(2m-1)^3}$$

$$+ \frac{3}{2} \sqrt{\frac{2}{\pi}} \frac{1}{(hx)^{5/2}} e^{-i(hx+\pi/4)} H_4 \cdot H_5 ; \quad H_5 = \frac{-48m^4 + 80m^3 - 48m^2 + 4m + 1}{(2m-1)^4} \qquad \Big|\, P$$

$$- \frac{60}{32} \sqrt{\frac{2}{\pi}} \frac{1}{(hx)^{7/2}} i e^{-i(hx+\pi/4)} H_4 \cdot H_6 ;$$

$$H_6 = \frac{992m^4 - 1104m^3 + 360m^2 - 28m - (576m^3 - 320m^4)H_5}{(2m-1)^4}$$

$$m = (h/k)^2 = (v_s - v_p)^2 = \frac{1 - 2v}{2(1-v)}$$

$$\bar{h} = (v_r/v_p)^2; \qquad \bar{k} = (v_r/v_s)^2$$

$$H = \frac{-(2-\bar{k}-2\sqrt{(1-\bar{k})(1-\bar{k})}\,)}{8[(2-\bar{k})-\sqrt{(1-\bar{h})(1-\bar{k})}-\frac{1}{2}\sqrt{\frac{1-\bar{k}}{1-\bar{h}}}-\frac{1}{2}\sqrt{\frac{1-\bar{h}}{1-\bar{k}}}]}$$

$$\bar{k}^3 - 8\bar{k}^2 + \bar{k}(24-16m) + 16m - 16 = 0$$

(Rayleigh-Equation)

$$v_0 \frac{G}{Q} = \qquad\qquad i\, e^{-i\kappa x}\, .\, k \qquad\qquad\qquad\qquad \big| R$$

$$+\; 2\sqrt{\frac{2}{\pi}}\,\frac{1}{(kx)^{3/2}}\, i\, e^{-i(kx+\pi/4)}\, K_1; \qquad K_1 = 1-m$$

$$-\; 3\sqrt{\frac{2}{\pi}}\,\frac{1}{(kx)^{5/2}}\, e^{-i(kx+\pi/4)}\, K_1 . K_2; \qquad K_2 = -20+30m \qquad \big| S$$

$$-\; \frac{15}{4}\sqrt{\frac{2}{\pi}}\,\frac{1}{(kx)^{7/2}}\, i\, e^{-i(kx+\pi/4)}\, K_1 . K_3; \qquad K_3 = 860-2518m+1920m^2$$

$$+\; \frac{1}{2}\sqrt{\frac{2}{\pi}}\,\frac{1}{(hx)^{3/2}}\, i\, e^{-i(hx+\pi/4)}\, K_4; \qquad K_4 = \frac{m}{(2m-1)^2}$$

$$+\; \frac{3}{4}\sqrt{\frac{2}{\pi}}\,\frac{1}{(hx)^{5/2}}\, e^{-i(hx+\pi/4)}\, K_4 . K_5; \qquad\qquad\qquad \big| P$$

$$K_5 = \frac{-32m^4+64m^3-48m^2+8m}{(2m-1)^4}$$

$$-\; \frac{30}{32}\sqrt{\frac{2}{\pi}}\,\frac{1}{(hx)^{7/2}}\, i\, e^{-i(hx+\pi/4)}\, K_4 . K_6; \qquad\qquad\qquad (11)$$

$$K_6 = \frac{416m^4-512m^3+176m^2+8m+(192m^4+320m^3+192m^2-32m)K_5}{(2m-1)^4}$$

$$K = \frac{-\bar{k}\sqrt{1-\bar{h}}}{8\,(2-\bar{k})-\sqrt{(1-\bar{h})(1-\bar{k})}-\frac{1}{2}\sqrt{\frac{1-\bar{k}}{1-\bar{h}}}-\frac{1}{2}\sqrt{\frac{1-\bar{h}}{1-\bar{k}}}}$$

Applying the differential operator equ. 1 to the solutions
equ. 10 and 11, we have to differentiate with respect to x,
put rcoshu instead of x and integrate over u between 0 and ∞.
This can be done separately with the terms of the Rayleigh-
wave (R), the S-wave (S) and the P-wave (P). Each of the latter

ones consists of 3 terms (due to the consideration of 3 terms in the Taylor expansion equ. 7). Each term is the product of two factors, one describing the geometrical damping and the other one describing the phase relation.

Differentiation, therefore, must lead to a twinset of solutions (differentiating a product), alltogether 12 for the P- and the S-Waves. To formalize the rather lengthy calculation, for details see Prange (1978), we introduce the following terms

ik	X_{ik}	Y_{ik}
0	$1\ H\ 1$	$1\ K\ 1$
11	$J_1 H_1 1\ P_1\ /\ \hat{k}$	$L_1 K_1 1\ P_1\ /\ \hat{k}$
12	$J_1 H_1 1\ P_2\ /\ \hat{k}^2$	$L_1 K_1 1\ P_2\ /\ \hat{k}^2$
21	$J_2 H_1 H_2 P_3\ /\ \hat{k}^2$	$L_2 K_1 K_2 P_3\ /\ \hat{k}^2$
22	$J_2 H_1 H_2 P_4\ /\ \hat{k}^3$	$L_2 K_1 K_2 P_4\ /\ \hat{k}^3$
31	$J_3 H_1 H_3 P_5\ /\ \hat{k}^3$	$L_3 K_1 K_3 P_5\ /\ \hat{k}^3$
32	$J_3 H_1 H_3 P_6\ /\ \hat{k}^4$	$L_3 K_1 K_3 P_6\ /\ \hat{k}^4$
41	$J_4 H_4 1\ P_1\ /\ \hat{h}$	$L_4 K_4 1\ P_1\ /\ \hat{h}$
42	$J_4 H_4 1\ P_2\ /\ \hat{h}^2$	$L_4 K_4 1\ P_2\ /\ \hat{h}^2$
51	$J_5 H_4 H_5 P_3\ /\ \hat{h}^2$	$L_5 K_4 K_5 P_3\ /\ \hat{h}^2$
52	$J_5 H_4 H_5 P_4\ /\ \hat{h}^3$	$L_5 K_4 K_5 P_4\ /\ \hat{h}^3$
61	$J_6 H_4 H_6 P_5\ /\ \hat{h}^3$	$L_6 K_4 K_6 P_5\ /\ \hat{h}^3$
62	$J_6 H_4 H_6 P_6\ /\ \hat{h}^4$	$L_6 K_4 K_6 P_6\ /\ \hat{h}^4$

$$(12)$$

i	J_i	L_i	P_i
1	1	2	$1/(2\pi^2)$
2	3/2	3	$3/(8\pi^3)$
3	15/8	15/4	$1/(4\pi^3)$
4	1	1/2	$5/(16\pi^4)$
5	3/2	3/4	$1/(8\pi^4)$
6	15/8	15/16	$7/(32\pi^5)$

$$(13)$$

with the values of H_i and K_i taken from equ. 10 and 11 respectively and

$$\hat{k} = v_r/v_2; \qquad \hat{h} = v_r/v_p$$

With the terms of equ. 12 and 13, the real (cos) and imaginary (sin) components of the horizontal and vertical displacement vectors can be written as

$$q \frac{G}{Q} \lambda_r \genfrac{}{}{0pt}{}{RE}{IM} = \qquad\qquad w \frac{G}{Q} \lambda_r \genfrac{}{}{0pt}{}{RE}{IM} =$$

$$X_0 / R^{1/2} \genfrac{}{}{0pt}{}{\cos}{\sin} (-2\pi R - 3\pi/4) \qquad Y_0 / R^{1/2} \genfrac{}{}{0pt}{}{\cos}{\sin} (-2\pi R - \pi/4)$$

$$+ X_{11}/R^2 \genfrac{}{}{0pt}{}{\cos}{\sin} (-2\pi \hat{k} R) \qquad + Y_{11}/R^2 \genfrac{}{}{0pt}{}{\cos}{\sin} (-2\pi \hat{k} R + \pi/2)$$

$$+ X_{12}/R^3 \genfrac{}{}{0pt}{}{\cos}{\sin} (-2\pi \hat{k} R - \pi/2) \qquad + Y_{12}/R^3 \genfrac{}{}{0pt}{}{\cos}{\sin} (-2\pi \hat{k} R)$$

$$+ X_{21}/R^3 \genfrac{}{}{0pt}{}{\cos}{\sin} (-2\pi \hat{k} R - \pi/2) \qquad + Y_{21}/R^3 \genfrac{}{}{0pt}{}{\cos}{\sin} (-2\pi \hat{k} R + \pi)$$

$$+ X_{22}/R^4 \genfrac{}{}{0pt}{}{\cos}{\sin} (-2\pi \hat{k} R - \pi) \qquad + Y_{22}/R^4 \genfrac{}{}{0pt}{}{\cos}{\sin} (-2\pi \hat{k} R + \pi/2)$$

$$+ X_{31}/R^4 \genfrac{}{}{0pt}{}{\cos}{\sin} (-2\pi \hat{k} R + \pi) \qquad + Y_{31}/R^4 \genfrac{}{}{0pt}{}{\cos}{\sin} (-2\pi \hat{k} R - \pi/2)$$

$$+ X_{32}/R^5 \genfrac{}{}{0pt}{}{\cos}{\sin} (-2\pi \hat{k} R + \pi/2) \qquad + Y_{32}/R^5 \genfrac{}{}{0pt}{}{\cos}{\sin} (-2\pi \hat{k} R - \pi)$$

$$+ X_{41}/R^2 \genfrac{}{}{0pt}{}{\cos}{\sin} (-2\pi \hat{h} R + \pi/2) \qquad + Y_{41}/R^2 \genfrac{}{}{0pt}{}{\cos}{\sin} (-2\pi \hat{h} R + \pi/2)$$

$$+ X_{42}/R^3 \genfrac{}{}{0pt}{}{\cos}{\sin} (-2\pi \hat{h} R) \qquad + Y_{42}/R^3 \genfrac{}{}{0pt}{}{\cos}{\sin} (-2\pi \hat{h} R)$$

$$+ X_{51}/R^3 \genfrac{}{}{0pt}{}{\cos}{\sin} (-2\pi \hat{h} R) \qquad + Y_{51}/R^3 \genfrac{}{}{0pt}{}{\cos}{\sin} (-2\pi \hat{h} R)$$

$$+ X_{52}/R^4 \genfrac{}{}{0pt}{}{\cos}{\sin} (-2\pi \hat{h} R - \pi/2) \qquad + Y_{52}/R^4 \genfrac{}{}{0pt}{}{\cos}{\sin} (-2\pi \hat{h} R - \pi/2)$$

$$+ X_{61}/R^4 \genfrac{}{}{0pt}{}{\cos}{\sin} (-2\pi \hat{h} R - \pi/2) \qquad + Y_{61}/R^4 \genfrac{}{}{0pt}{}{\cos}{\sin} (-2\pi \hat{h} R - \pi/2)$$

$$+ X_{62}/R^5 \genfrac{}{}{0pt}{}{\cos}{\sin} (-2\pi \hat{h} R - \pi) \qquad + Y_{62}/R^5 \genfrac{}{}{0pt}{}{\cos}{\sin} (-2\pi \hat{h} R - \pi)$$

(14)   (15)

Where Q is the intensity of the exciter point source, λ_r the wavelength of the Rayleigh-Wave and R the normalized distance r/λ_r.

Recalling the well-known figure for the attenuation due to geometrical damping (Richart, Hall, Woods 1970) for the three wave-types, fig. 1,

288

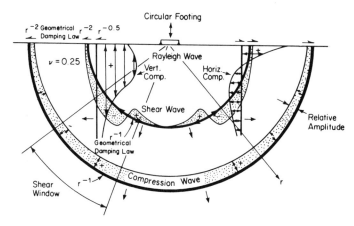

Fig.1   Wave propagation from circular footing
        (from Richart, Hall and Woods, 1970)

we identify the attenuation in figure 1 with that of the respective first terms for the R-, S- and P-Waves in equ. 14 and
15. It is obvious, that in figure 1 only the first term of the
Taylor-expansion equ. 7 was considered, even ommitting the
second part of the twinset dying out with $1/R^3$. The consideration of the second term of equ. 7 leads to additional contributions, dying out with $1/R^3$ and $1/R^4$, respectively. The consideration of the third term of equ. 7 leads to further contributions, dying out with $1/R^4$ and $1/R^5$, respectively. Of course,
all contributions have different phase relations, as seen from
the cos- or sin-arguments.
    The resulting vector-amplitude follows from equ. 14 and 15:

$$\left| q \, \frac{G}{Q} \, \lambda_r \right| = (RE_q^2 + IM_q^2)^{1/2} = q^*$$

$$\left| w \, \frac{G}{Q} \, \lambda_r \right| = (RE_w^2 + IM_w^2)^{1/2} = w^*$$

(16)

and the phase angle between the exciting vertical point source
and the displacement vector

$$\varepsilon_q = atan(IM_q/RE_q)$$

$$\varepsilon_w = atan(IM_w/RE_w)$$

(17)

Equ. 16 and 17 yield the horizontal and vertical vibrations of
the wavefield at the surface:

$$q(t) = \frac{Q}{G} \frac{1}{\lambda_r} q^* \, e^{i(\omega t + \varepsilon_q)}$$

$$w(t) = \frac{Q}{G} \frac{1}{\lambda_r} w^* \, e^{i(\omega t + \varepsilon_w)}$$

289

Equations 16 and 17 were computed at the UNIVAC of the University of Karlsruhe, considering the contributions of the first three terms of the Taylor-expansion equ. 7 sequentially and varying the Poisson's-ratio ν between 0,30 and 0,45.

In the following figure, amplitudes and phase-angles of equ. 16 and 17 resp. are plotted against normalized radial distance $R = r/\lambda_r$. The numbers indicate:

1   first term equ. 7 only
2   first and second term of equ. 7
3   first three terms of equ. 7
4   Rayleigh-term only

In all the amplitude plots we find the above mentioned interferences, resulting in a sequence of minima and maxima of equal radial distance, called the interference-wavelength λ_i in the following evaluation. The phase curve is considerably less affected by interferences, since, at least for Poisson's ratios $\nu \geqslant 0,35$, the Rayleigh-wave prodominates even in the near-field.

There is some justification to limit the farfield to normalized radii $R > 5$, since in this region at least for Poisson's ratios $\nu \geqslant 0,35$ the surface wave pattern consists of almost the Rayleigh-waves alone.

Another feature of the nearfield is the steeper rise of the amplitude curve towards the wave source with increasing order of considered terms. This, of course, is due to the higher attenuation factors of the higher terms.

If we calculate the interference-wavelenghts λ_i (i.e. radial distance between minima or maxima resp.) for the three different wave types R, S and P without attenuation due to geometrical damping, we find the results plotted in figure 10. The letters on the curves indicate the wave-types interfering with each other. By assuming no attenuation we introduce only a small error for $R \geqslant 5$, since the amplitudes do not change drastically.

In the same figure 10 the normalized wavelengths taken graphically from figures 2 through 9 are presented. They almost exactly coincide with the R/P interference pattern. This leads to the conclusion that the main source of interference is obviously the superposition of the surface - contributions of the Rayleigh-wave and the P-wave. The pattern of interference between the R- and the S-waves cannot be seen clearly in fig. 2 - 9, firstly because the S-contribution on the surface due to a vertical point-source is obviously small and secondly the interference wavelengths are very large ($\lambda^* > 14$). However, it is astonishing that the P-wave plays such an important role in the surface interference pattern taking into account that the distribution of radiated energy of a circular plate vibrating vertically equals
67,4 % - 25,8 % - 6,9 % for R-S-P waves
resp., see Miller, Pursey (1955).

The interference between R- and S-waves can more easily be seen in the case of a circular plate vibrating horizontally. Cherry (1962) calculated the displacement vectors following the same procedure as indicated above and again considering

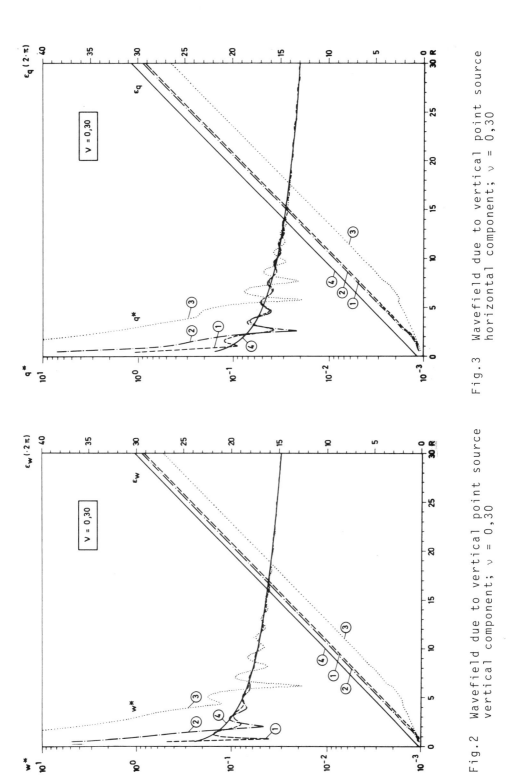

Fig.3   Wavefield due to vertical point source
horizontal component; $\nu = 0,30$

Fig.2   Wavefield due to vertical point source
vertical component; $\nu = 0,30$

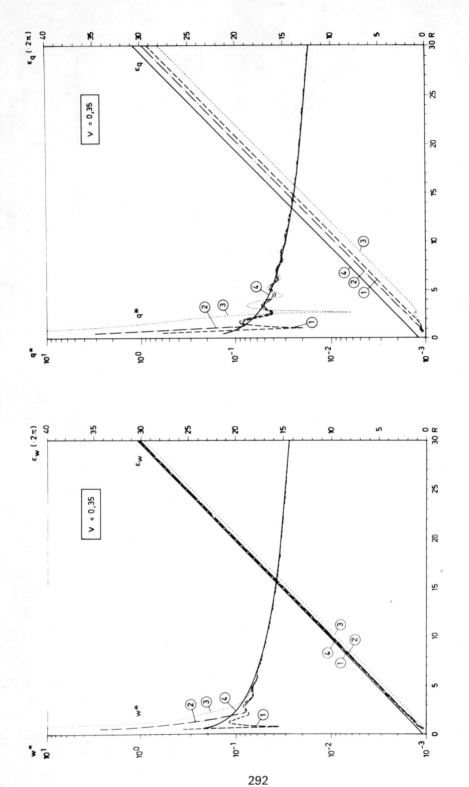

Fig.4   Wavefield due to vertical point source
vertical component; $\nu = 0{,}35$

Fig.5   Wavefield due to vertical point source
horizontal component; $\nu = 0{,}35$

292

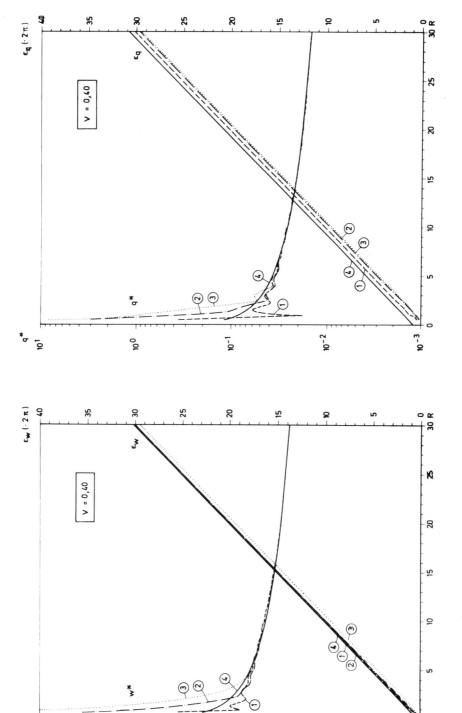

Fig.7   Wavefield due to vertical point source
         horizontal component; $\nu = 0,40$

Fig.6   Wavefield due to vertical point source
         vertical component; $\nu = 0,40$

293

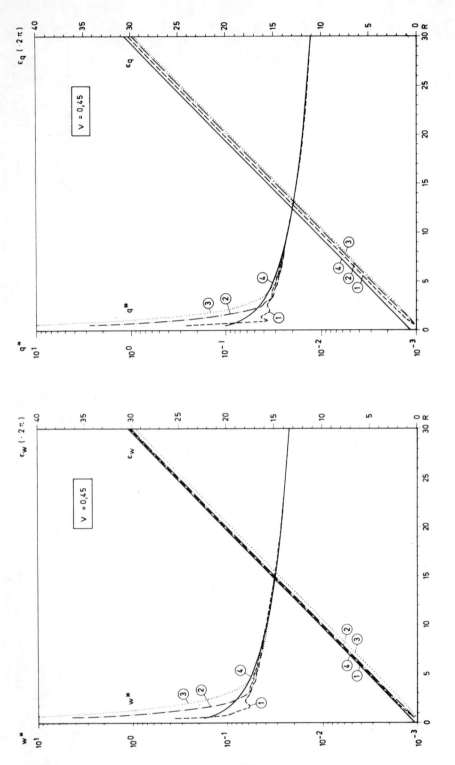

Fig.8 Wavefield due to vertical point source
vertical component; $\nu = 0,45$

Fig.9 Wavefield due to vertical point source
horizontal component; $\nu = 0,45$

294

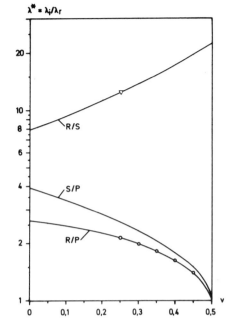

Fig.10 Normalized interference-
wavelengths λ^*
plane wavefield
no attenuation

only the first term in the Taylor expansion. Since horizontally
vibrating sources do not radiate any P-waves at the surface,
the interference pattern R/S is more pronounced since it is not
overlain by P/R-interferences.

Figure 11 shows amplitudes and phase angles of the three
displacement components calculated after Cherry (1962), the
subscripts indicating the coordinates (z vertical, r radial,
t tangential). For q_t the normalized interference wavelength
λ_i/λ_r is found to be 12. It is included in figure 10 for $\nu=0.25$
(triangle), coinciding well with the R/S interference pattern.

TEST RESULTS

We shall now focus our attention to experimental investigations
to find out how well test results do agree with the interference patterns calculated theoretically.

Two effects hitherto assumed to be neglegible will influence
the experiments:

Inhomogeneity

The shear modulus G depends on stress and consequently on depth.
This will not only effect the surface waves which penetrate
approximately one wavelength into the subsoil (for exact values
see e.g. Richart, Hall, Woods 1970), but also the space wave
contributions of the S- and P-waves. The space waves will travel
on continously refracted wave paths resulting in an additional
transport of dynamic energy to the surface.

Material Damping

The attenuation exponents due to material damping differ for

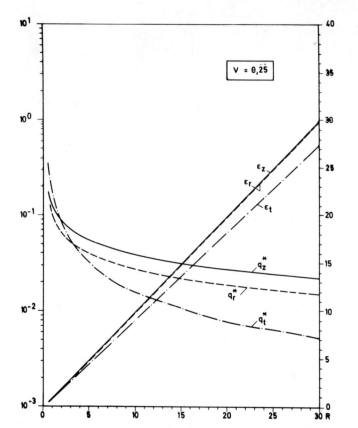

Fig.11   Wavefield due to horizontal point source

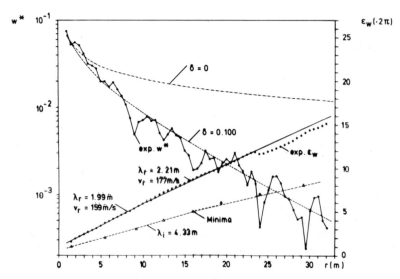

Fig.12 Experimental wavefield, vertical component axis 5; 80 Hz

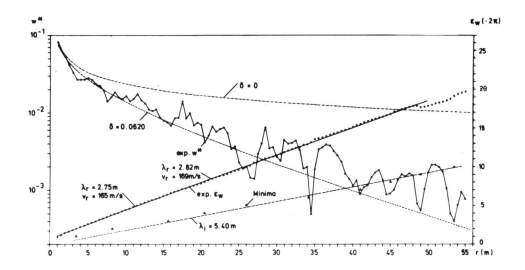

Fig.13 Experimental wavefield, vertical component axis 5; 60 Hz

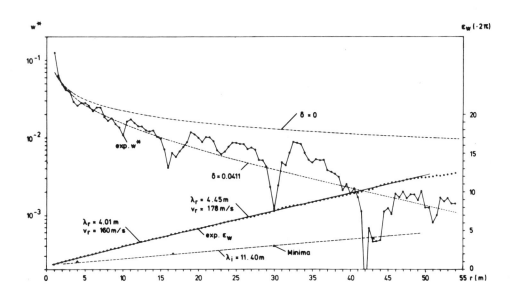

Fig.14 Experimental wavefield, vertical component axis 5; 40 Hz

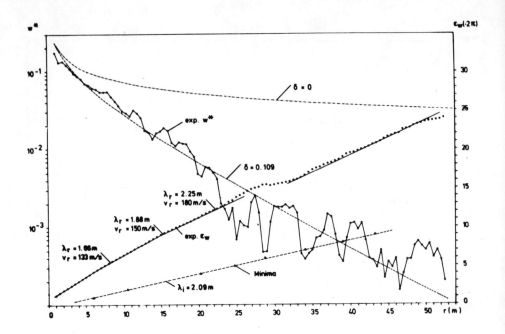

Fig.15 Experimental wavefield, vertical component axis 4; 80 Hz

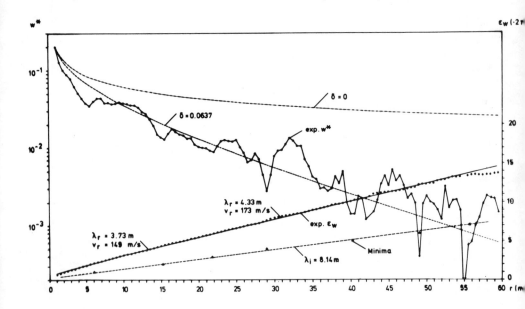

Fig.16 Experimental wavefield, vertical component axis 4; 40 Hz

the three types of waves. The amplitude curves will therefore not exactly match those given in equ. 16. Similarly, material damping will increase the wave velocity. However, the errors introduced are only small for sands with small material damping.

The site investigated was located in the Campus of the University of Karlsruhe, the subsoil consisting of medium sand to fine gravel. An electro-dynamic exciter vibrating vertically and capable of delivering up to 2 kW mechanical power was placed at one end of a test axis. Amplitude and phase angle of the generated surface wave field was measured by means of a compensated vibration transducer, consequently producing no field distortion due to inertia effects (see below and Prange 1977).

The experimental results of two different axes are plotted in figures 12 - 16 in a semi-logarithmic plot.

The dashed-dotted line indicates the theoretical geometrical damping of the pure Rayleigh-wave (attenuation $r^{-0.5}$), the dashed line averages the measured data (averaging the maxima and minima of the interferences), circles represent the free-field data measured.

Plotting the phase angle and measuring the wavelength from this curve the resultant surface-wave velocity is gained.

Similarly, measuring the radial distance between the minima of the interference pattern, the interference wavelength λ_i is obtained (minima indicated by triangles).

In all cases, only the vertical component of the surface vibration was measured and plotted as w^* without dimension (a scale factor resulting in a parallel shift of no interest here).

Two main features become immediately evident from the experiments:

The soil exhibits hysteretic damping (see also Prange 1978a) with $\tan\phi$ beeing of the order of 0.07 (equalling D = 0.04), which justifies the neglection of the material damping in the theoretical calculations.

Pronounced interference patterns are found in the amplitude curves, the interference wavelength depending on frequency. The phase curves are very much less affected by interferences.

To find out about the Poisson's ratio ν, tests were performed with a stationary sequence of stochastic impulses. From the cross-correlation function between the generated signal sequence at the wave source and the measured signals, the wave velocities of the R-wave and the P-wave can be easily obtained (see Roesler, 1978). Fig.17 shows the resultant correlograms.

From velocity ratio of $v_p/v_r = 2$ we find $\nu = 0.3$ which would result in a normalized interference wavelengths $\lambda^* = 2$ (see fig. 10).

To prove this value by the experiments, the Rayleigh-wave-lengths λ_r and the interference wavelengths λ_i from a greater number of experiments were plotted against frequency in a double-logarithmic plot in figure 18. Theoretically, they must lay on a straight line of -45°, the two lines beeing apart by a vertical shift (factor) of 2.

They fall, in fact, into two bands with a reasonable band-width, which are apart by a vertical shift 2.

299

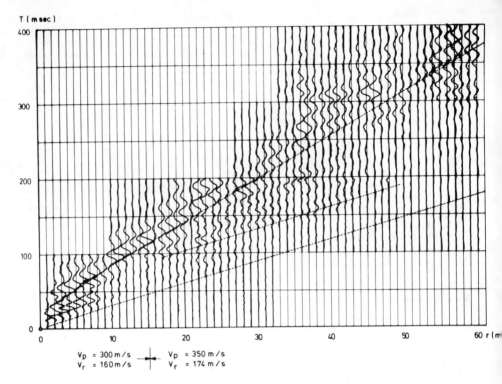

T (m sec)

V_p = 300 m/s V_p = 350 m/s
V_r = 160 m/s V_r = 174 m/s

Fig.17   Travel times of P- and R-Waves;
         stochastic excitation and measurement
         of cross-correlation function at axis 5

Yet another prove of $\lambda^* = 2$ is the distribution of the measured values λ^*. Figure 19 shows a Gaussian distribution, however crude, with a mean value of $\lambda^* = 2.02$.
   The excellent agreement between theoretical and experimental
results encourages us to relate the interference pattern, even
in the case of natural soils (not homogeneous), in the near-
field to the superposition of the three types of waves R, S
and P.

SECONDARY INTERFERENCES

From the theoretical and experimental findings mentioned above
we must likewise deduce that for the purpose of disclosing
local anomalies only the farfield should be applied. Only in
the farfield we find minor primary interferences in the amplitude curve and approximately no interferences in the phase
curve. Interferences showing up in the farfield would therefore merely be secondary interferences due to suboil anomalies.
We will see in the following, that secondary interferences will
effect the phase curve much more pronounced than in the case
of primary interferences.

300

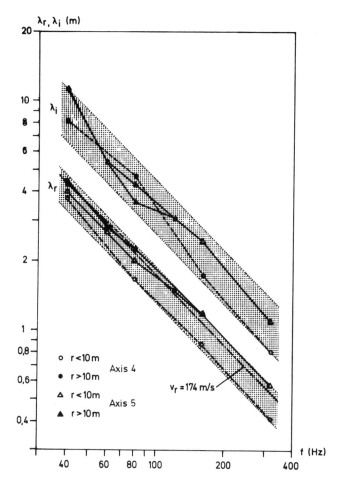

Fig.18   Rayleigh-wavelengths and interference-wavelengths
         vs. frequency; axis 4 and 5

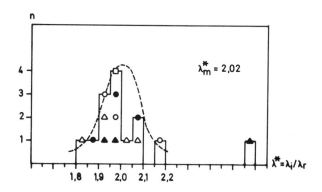

Fig.19   Distribution of normalized interference wavelengths λ^*

Secondary interferences may be caused by cavities or obstacles in the subsoil. Figure 20 shows the heavy distortion of the phase curve at 115 Hz due to a sewer running perpendicular to axis 41 in the ground, Prange (1971).

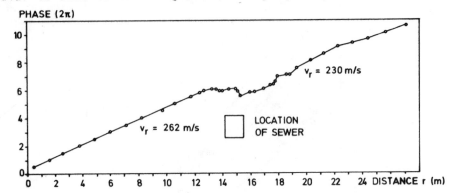

Fig.20   Distortion of phase-curve due to underground sewer (from Prange 1971)

The experiments were performed in connexion with locating medieval foundations of the Salzburg-cathedral by means of dynamical methods. Figure 21 shows the archeological situation of the medieval apse of 1181, together with the isophases (lines of equal phase angle) in the same field. The exciter was placed at S 73. We quite clearly see the coincidence between the location of the foundations and the heavy distortion of the isophase-field.

Figure 22 shows the results of experiments in a sand bin in the laboratory. A concrete block was buried in the sand giving rise to interferences in the amplitude and phase curves. The dashed lines indicate the data of the freefield (no concrete block), Prange and Haupt (1974). Figure 23 shows the same situation for a frequency of 420 Hz in the form of the isophase-field. The obstacle can clearly be localized by the neavy distortion of the isophases.

SECONDARY INTERFERENCES DUE TO STI

As mentioned above, the soil-transducer-interaction (STI) gives rise to secondary interferences as well. The inert mass of the transducer upon the soil-surface together with the effective spring- and damping constants of the subsoil in the contact area control the resonance behavior of the system soil-transducer. Therefore, the local freefield amplitude and phase is changed depending on frequency. Details of the STI are given in Prange (1977) and Prange (1978b). Only one example of STI interference shall be mentioned here. To check the efficiency of the inertia-compensation method, the surface wave field was measured along an axis which led partly over sandy and grassy soil surface. In the grassy region, the coupling of the noncompensated transducer to the soil vibrations was more or less random. This results in the enormous scatter of the noncompensated experimental data, figure 24.

302

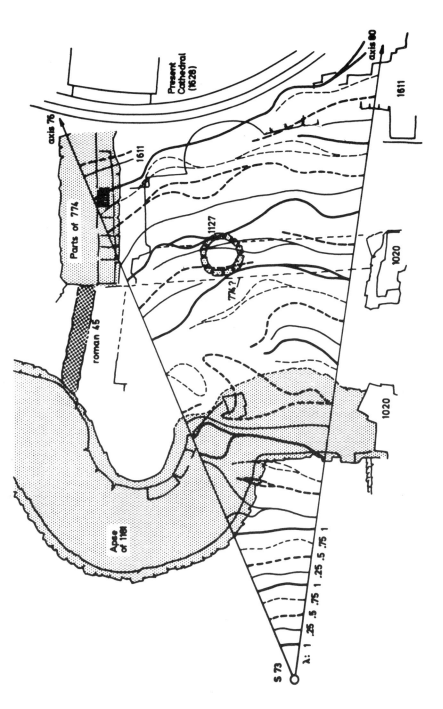

Fig.21 Distortion of isophase-field due to medieval foundations (1181) of the Salzburg cathedral; isophase-lines of 1/4 wavelength; source at S 73

303

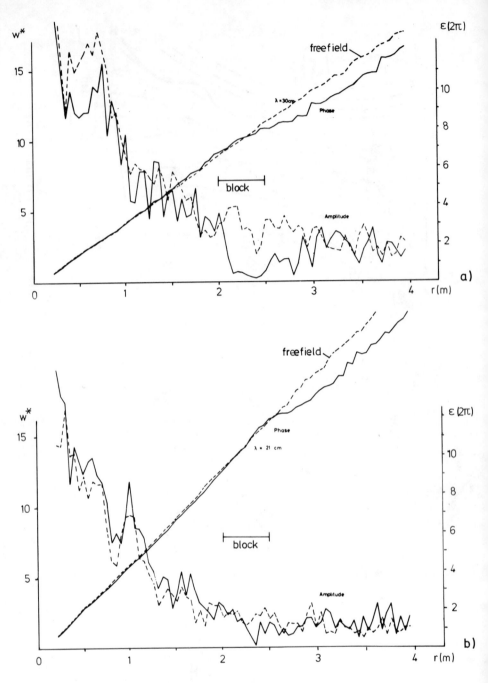

Fig.22   Distortion of wavefield (amplitude and phase)
due to concrete block 50 x 50 x 41 cm,
buried 6 cm deep in experimental sand bin
a) 320 Hz   b) 420 Hz

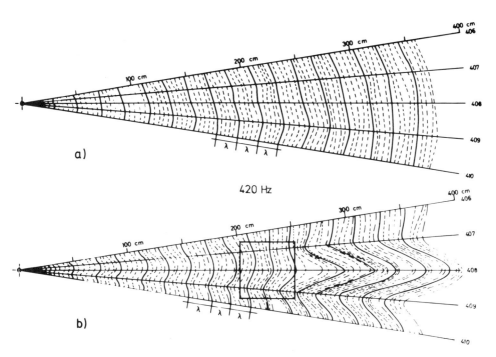

420 Hz

Fig.23 Distortion of isophase-field due to buried
concrete block in experimental sand bin
a) freefield b) distorted field (square = block)

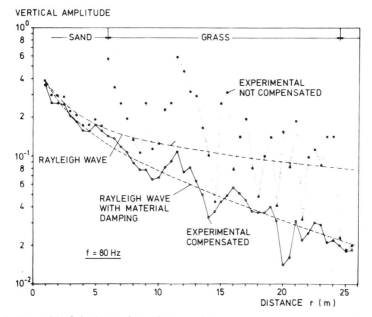

VERTICAL AMPLITUDE

Fig.24 Wavefield data (amplitude) from compensated and non-compensated vibration transducer on sandy and grassy surface

305

The interference pattern can hardly be recognized and averaging the amplitudes to determine material damping is almost impossible.

In the sandy region, coupling conditions are more constant, resulting in a slight amplitude amplification at 80 Hz (the resonance frequency beeing near 200 Hz).

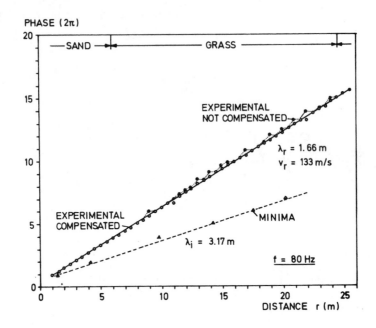

Fig.25 Wavefield data (phase) from compensated and non-compensated vibration transducer on sandy and grassy surface

In the respective phase plots, figure 25, the compensated data, even in the grassy region, lay on a straight line. The noncompensated data, however, show phase differences of almost 180°. This means, that an upward motion of the freefield vibration is indicated by the noncompensated transducer as a downward motion, the largest possible error.

In either case of primary interference (superposition of R-, S- and P-waves) and secondary interferences (anomalies beeing secondary wave sources), the wavefields should be measured by compensated transducers to overcome the problem of secondary interferences introduced by inertia effects.

Of course, the above mentioned example of partly grassy soil surface involves a most disadvantageous situation which will probably never be the common case in soil dynamics. It shows, however, the necessity and efficiency of inertia compensation methods in a very drastic manner.

CONCLUSIONS

All wavefields we measure in the field have in common the phenomenon interference. Considering the freefield, we find two
reasons for these interferences:

In the nearfield, space wave contributions of the S- and
P-wave at the surface are of comparable amplitude to those of
the Rayleigh-wave, resulting in characteristical interference
patterns due to the superposition of the three wave types. The
interference-wavelengths are controlled by frequency and
Poisson's ratio.

In the farfield, interferences are only due to subsoil anomalies. Their interpretation is therefore a means of localizing
such anomalies, in analogy to nondestructive testing methods by
ultrasonics.

Additional interferences will be introduced into any wavefield by measuring the surface vibrations with noncompensated
transducers due to inertia effects. Not considering the different local coupling conditions of the soil-transducer-interaction may lead to serious misinterpretation of wavefield data.

REFERENCES

Bath, M. 1968 Mathematical Aspects of Seismology
    Amsterdam, London, New York Elsevier Publ.
Cherry, I.T. 1962 The Azimuthal and Polar Radiation Patterns
    Obtained from a Horizontal Stress Applied at the Surface of
    an Elastic Half-Space. Bull.Seism.Soc.America Vol.52 No.1
Ewing, W.M., Jardetzky, W.S. and Press, F. 1957 Elastic Waves
    in Layered Media New York, McGraw-Hill
Lamb, H. 1904 On the Propagation of Tremors over the Surface
    of an Elastic Solid Phil.Trans.Royal Soc. London Ser. A
    Vol. 203
Miller, G.F. and Pursey, H. 1955 On the Partition of Energy
    between Elastic Waves in a Semi-Infinite Solid Proc.Royal
    Soc. London Ser. A Vol. 233
Prange, B. 1971 Soil-Dynamical Investigations of Medieval
    Foundations of the Cathedral of Salzburg. Report for the
    Gesellschaft für Salzburger Landeskunde. Institute of Soil
    and Rock Mechanics, University of Karlsruhe
Prange, B. 1977 Inertia Compensation of Vibration Transducers
    Proc. I.C.S.M.F.E. Tokyo
Prange, B. 1978 On the Interaction between a Body and the
    Wavefield. Publ. Inst. Soil and Rock Mech., University of
    Karlsruhe
Prange, B. 1978a Parameters Affecting Damping Properties
    Proceedings Dynamical Methods in Soil and Rock Mechanics,
    DMSR 77, Balkema, Rotterdam

Prange, B. 1978b Measurement of Vibrations, Proceedings
    Dynamical Methods in Soil and Rock Mechanics DMSR 77
    Balkema, Rotterdam
Prange, B. and Haupt, W. 1974 Anomalies of a Stationary Wave-
    field on the Surface of the Halfspace due to Subsoil Enclo-
    sures. Report for the German Research Council, Institute of
    Soil and Rock Mechanics, University of Karlsruhe
Richart, F.E., Hall, I.R. and Woods, R.D. 1970 Vibrations of
    Soils and Foundations Englewood Cliffs, N.J. Prentice Hall
Roesler, S.K. 1978, Correlation Methods in Soil Dynamics,
    Proc. Dynamical Methods in Soil and Rock Mechanics, DMSR 77,
    Balkema, Rotterdam

Proceedings of DMSR 77 / Karlsruhe / 5-16 September 1977 / Volume 1

Correlation methods in soil dynamics

S. ROESLER
University of Karlsruhe, Karlsruhe, Germany

SYNOPSIS

Correlation methods, already in use in fields like medicine, electronics and engineering, are introduced to soil dynamics, as a new tool in measurement technology. The immediate advantage of the new technique becomes obvious in details such as noise immunity, determination of phase- and group velocity of waves, detection of anisotropy in soils and nondestructive detection of subsoil anomalies. The principles of correlation are presented together with the results of laboratory experiments and field tests.

INTRODUCTION

Soil dynamics deal primarily with the exploration of soil parameters, which are related to wave propagation. We can apply it firstly to the measurement of shear moduli and Poisson's ratio on the basis of wave velocity, secondly to identify subsoil anomalies, like cavities, enclosures and layers and thirdly to judge the influence of vibrations and possibilities of protection.
It was tried to make use of this comparatively wellknown method in soil dynamics. It was possible to prove in various series of experiments that it is not only suitable, but that it even provides possibilities which were not available before.
So far we have only used in situ dynamic measurement techniques, i.e. periodical stimulation of systems or seismic measurement techniques, i.e. impulse stimulation. In both cases of traditional techniques of measurement we are faced with difficulties which can be avoided by applying the correlation method.
Since there are two types of waves which propagate in homogeneous subsoil, the primary wave and the shear wave and at the surface the Rayleigh wave as well, it is only possible to measure wave velocities with periodic stimulation if either a single wave type predominates or if it is possible to determine a second wave type by precise measurements of amplitude through interferences. Exact measurements of amplitude, however, are

rather difficult, and patterns of interference next to the
vibration exciter do not necessarily tell you exactly the interfering wave lengths. These conditions, i.e. the necessary
predominance of one single wave type and the precise measurement of amplitude, do not prevail in the correlation method.
It enables one to measure even the group velocity apart from
the phase velocity.

The second reason that the correlation method can be of
interest for soil dynamics concerns the properties of the soil
itself. Sand is by its very nature nonlinear, as far as its
stress-strain behaviour is concerned. If the behaviour of sand
is assumed to be linear with respect to wave propagation with
moderate precision, we have to meet an essential condition.
There has to be an inner state of stress which is not going to
be changed by additional vibrations, i.e. the strain which is
caused by vibration must only cause elastic deformations, which
have to be very small. At this stage we encounter the difficulties for seismic measurement techniques. In order to get a
large enough signal level at the location of the transducer,
one is nearly always forced to choose a rather strong excitation signal for the experiments, introducing an additional error
regarding the superposition principle. If one uses the correlation method one can choose such small excitation amplitudes
that the conditions for linearity are well met. Consequently,
one does not depend on signal amplitudes at the location of
the transducer which are distinctly different from all other
noise signals. Experiments were performed in which a broad band
test signal interfered with a broad band noise signal; in this
case the test signal was less than 3 % of the noise signal and
the result of the measurement was satisfactory.

PRINCIPLES OF CORRELATION

What is exactly understood by correlation method? It denotes
the correlation between a time dependent input signal y(t) and
an output signal x(t) of the same system. The correlation method establishes the relation if any between these two signals

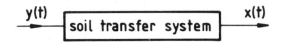

Fig. 1 Block diagram of soil transfer system

in order to draw conclusions about the system. If the properties of the system were known, one could determine the output-
signal for any input signal. In soil dynamics, these properties
are usually not known, therefore we proceed the other way
round. We try to establish the properties of the system by
analysing input and output signals. For this reason we introduce the cross correlation method between these two signals:

310

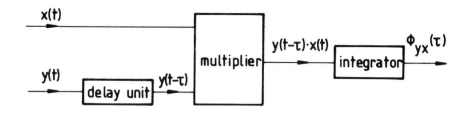

Fig. 2 Correlator block diagram

The input signal y(t) is delayed for a short period τ and multiplied with the output signal x(t). This process is repeated for
a number of delay times τ and all results are added up so that
we get as initial function the cross correlation function ϕ_{xy}
in the variable τ. The cross correlation function represents
the structure of relationship of both signals. The function indicates for which delay time τ both signals show the greatest
similarity. The following example indicates the procedure.
     Suppose a signal y(t) beeing the input signal of a transfer
system. The transfer system will only consist of a deadtime
system, i.e. the input signal is delayed by the deadtime without being changed in its structure and appears otherwise unchanged at the output. The deadtime system may be thought of as
a tape recorder with separate record head and playback head.
The deadtime equals the time which the tape requires to move
from the record head to the playback head. If one calculates
the cross correlation function between y(t) and x(t) the result will be as indicated in Fig. 3. The procedure of arriving
at these results is illustrated by Fig. 4.
     As I mentioned before, the input signal is delayed in the
correlator, in the upper case in Fig. 4 by $\tau = 1$. It is re-
cognizable, that we get as sum of this single product -1 times
+1 equalling -1. This value appears in the cross correlation
function at $\tau = 1$ in Fig. 3. For $\tau = 2$ we get two summands from
the products -1 times +1 and -1 times -1. The result is zero.
For $\tau = 3$ we have the products +1 x +1, -1 x -1, and -1 x -1.
The sum is 3. For $\tau = 4$, we only get two products +1 x -1 and
-1 x -1, the sum is zero. For $\tau = 5$, the sum is -1, and for all
higher values of τ, the sum is zero. The absolute maximum as
seen from the example in Fig. 3, is firstly a measure for the
amplitudes of both signals, and secondly, it appears at that
delay time, for which both signals have the greatest similarity.
The cross correlation function indicates the travel time and
the amplitude of the signal.

311

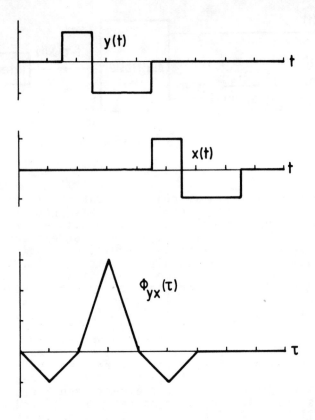

Fig. 3   Input-, output signal and CC-function of a deadtime
system

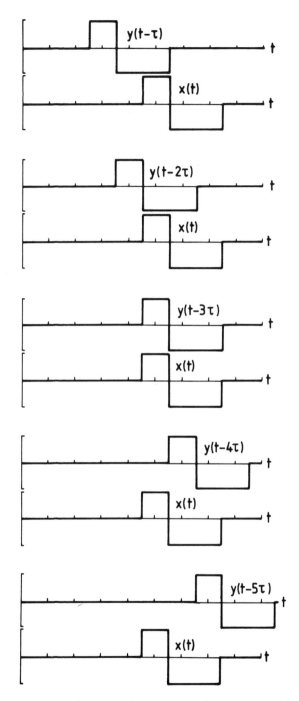

Fig. 4   Output signal and stepwise delayed input signals of a
deadtime system

TEST PROCEDURE

How is the cross correlation function determined in the experiments?

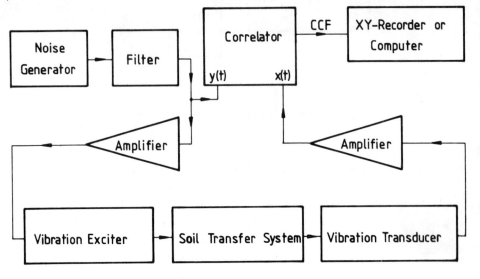

Fig. 5   Block diagram to determinate CC-functions

Fig. 5 shows the set up of the experiment which was followed
in most cases. A noise generator produces a stationary stochastic test signal which is connected as input signal to the
transfer system. The signal must not be periodic because the
output signal of the system would then also be periodic and
therefore we would get a correlation maximum for each period.
Therefore, in such only measurements of phase differences are
possible. Hence, the signal has to be stochastic and has to
cover a frequency range that is at least as wide as the range
which is relevant for the transfer behaviour of the soil. The
signal is filtered if necessary if we want to examine frequency
dependent transfer behaviour. It is then connected to the correlator as well as to the soil transfer system by means of an
amplifier and a vibration exciter. A vibration transducer re-
ceives the system output signal and transforms it into an
electric signal which is then connected to the correlator. The
correlator computes the cross correlation function and conveys
the data either onto an xy-recorder or to a computer for further
treatment.

TEST SIGNALS

In the following the advantages of various test signals in relation to their suitability for soil dynamic correlation measurements are described. White noise is a test signal which is

314

relatively easy to treat theoretically and which usually is also the reference of all error calculations even when other test signals are considered. White noise is a stochastic signal which has a continuous frequency range and whose power density is frequency independent. If we apply white noise to the transfer system, the cross correlation function will represent the weight function of the transfer system if it was measured in the manner described above. The weight function is the output function of a system excited by a Dirac impulse. The knowledge of the weight function implies the knowledge of the properties of the system. As white noise with infinite bandwidth cannot be realized experimentally, we have to prove which kind of test signals can be applied. One of them is the band limited noise, which is also called pink noise, in analogy to the colours of light. It has the same spectrum as white noise up to a cut-off frequency ω_g, from which point the power density decreases. Fig. 6 demonstrates which influence this noise yields upon the cross correlation function.

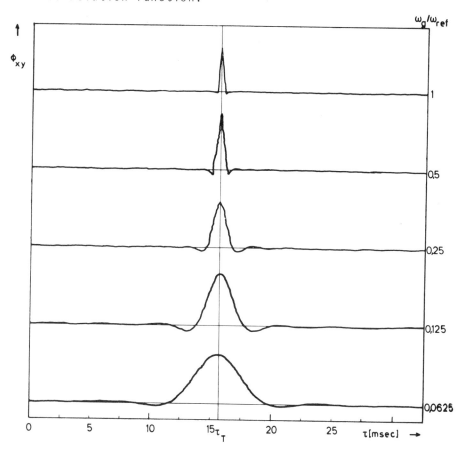

Fig. 6   Influence of ω_g to the CC-function

It is obvious that the representation of the cross correlation function of a deadtime system becomes wider with decreasing cut-off frequency ω_g, that the absolute maximum, however, remains at the correct deadtime. In other words: the exact measurement of the travel time is also possible with pink noise as test signal. Very often a binary noise signal is used as a test signal for reasons of simple production and considerable less complicated treatment with simple correlators. This signal has only two levels z_1 and z_2 and can change these only after the clockpulse. Graphically we can show the time behaviour of this signal by means of a Markoff diagram in Fig. 7.

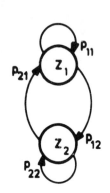

z_1 and z_2 signify the two levels and P_{ij} the transfer probabilities. The level z_1 therefore remains unchanged for the following clockpulse period with the probability P_{11} or it changes with the probability P_{12}. The same holds true for the level z_2. For the experiments we introduced no loss of generality if all probabilities were chosen 0,5. It can be shown that the clockpulse period of the binary signal sequence yields the same influence upon the cross correlation function as the upper cut-off frequency in the case of pink noise. Therefore we can measure travel times by using binary noise as well as by using pink noise. There exists a rather simple technology to generate binary noise.

Fig. 7 Markoff diagram of binary noise

It consists of a feedback shift register of the kind indicated in Fig. 8. If we put into this register any sequence of 0 and 1 and apply the clockpulse, we can take off at every output of

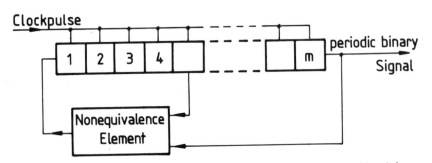

Fig. 8 Feedback shift register to produce periodic binary noise

the register a periodically binary noise signal. Apart of clockpulse frequency the period time T depends on the registerlength only. T must be chosen greater than all system travel times. This can be realized easily since the period time doubles with each further register position. What has been

said about the binary noise applies to the periodically binary
noise as well, if the period time is long enough.

All the kinds of noise which were mentioned, i.e. pink noise,
binary noise and periodically binary noise, are suitable for
the investigation of the frequency independent properties of
systems. As soon as the system shows dispersion, it is advisible
to make use of narrow band noise. This is the case e.g. in experiments with Rayleigh waves. Narrow band noise is produced by
filtering a narrow frequency band out of one of the above mentioned types of noise.

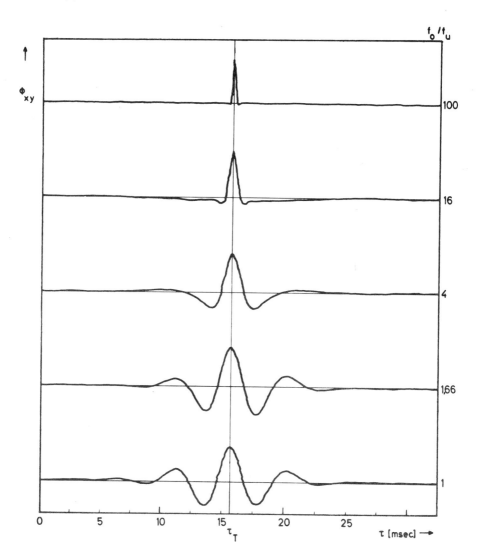

Fig. 9   Influence of small band noise to the CC-function

317

We can see in Fig. 9 the influence on the cross correlation
function of a deadtime system. The narrower we choose the band
limits the more distinctive secondary maxima are noticible.
Their amplitudes - in comparison to those of the main maximum
are also dependent on the filter characteristics, i.e. the
Bessel- or Tchebyscheff-type for example, and they are also dependent on the filter order. To prevent ambivalent measurement
results in natural soil transfer systems, which contain more
than just one deadtime, we have to compromize on filter characteristics, filter order and band limits in relation to the
measurement problem. In most cases this can well be achieved so
that it is easily possible to analyse slightly dispersive systems with narrow band noise.

NOISE IMMUNITY

The following example shows the vibration amplitudes the correlation method is capable of dealing with. An electrodynamic
exciter generated vibrations in the laboratory sand pit. We
measured the vibrations at a distance of two meters and
established the cross correlation function. The vibration
power was reduced by a ratio of $1:10^6$ and the scale of the
amplitudes (Fig. 10) increased by an equivalent ratio. The
vibration amplitudes at the surface in the vertical direction
amounted to between $2,5 \times 10^{-4}$ and 3×10^{-7} mm. We can see in
Fig. 10 that the cross correlation functions hardly differ
from each other with regards to the representation of the

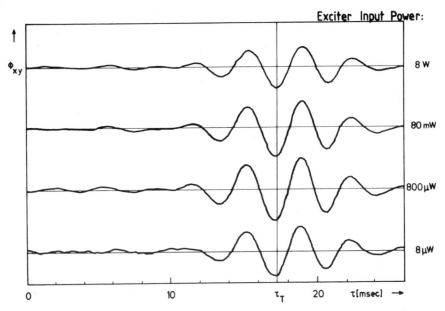

Fig. 10   Influence of different vibration power to the
          CC-function

318

system travel time τ_T. In this experiment the signal to noise
ratio (r.m.s.) due to amplifier noise amounted already to 0 dB
at a generator power of 800 μW, and to -20 dB at 8 μW. Generally speaking we can say that the influence of noise on the measurement result decreases with increasing dynamic range of the
correlator, with decreasing overlap of the frequency range of
the test signal and the noise signal, and - in cases of certain
noises - with increasing measurement time. We can give the
following resumée:

The correlation method is relevant for soil dynamics, because
firstly it allows an easy possibility to separate different
waves and even reflexions, which is not the case in periodic
measurement techniques; and secondly because it is susceptable
to noise to such a small extend that a linear interpretation
is justified in the case of small vibration amplitudes.

PHASE AND GROUP VELOCITY

A few examples may be mentioned which are supposed to show the
suitability of the correlation method when applied in practical
problems of soil dynamics. The first one deals with the behaviour of the shear modulus in inhomogeneous soil. It is well-
known that the shear modulus increases with the height of the
overlaying soil. Consequently the Rayleigh wave confined to the
surface shows a frequency dependent velocity(dispersion), because it's penetration is frequency dependent. Since it is
possible to separate the Rayleigh wave from the primary and
the shear wave by means of the correlation method and because
it is easy to measure its phase and group velocity, this
method is well suited for the analysis of the shear modulus
behaviour. Narrow band noise was chosen as a exciter signal
for this experiment.

As we have to distinguish between phase and group velocity
in dispersive systems we shall briefly discuss how the travel
times can be deduced from the cross correlation function. For
this purpose a cascade of lumped constant LC-units was set up
as demonstrated in Fig. 11, which clearly shows dispersion near
its cut-off frequency.

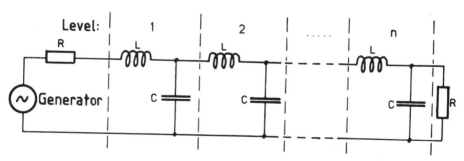

Fig. 11 Lumped LC-unit

319

This electrical analogon was chosen because only one single
wave exists in this case and the effect looked for in the cross
correlation function becomes especially visible and can be calculated easily. We can see in Fig. 12 the cross correlation
functions, which were determined between the system input and
the system output signals; the output signal was taken in each
case at the level of the LC-unit indicated in the illustration.

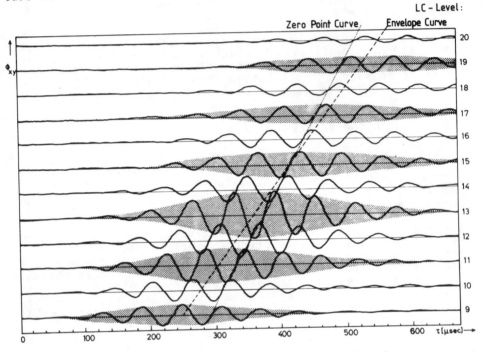

Fig. 12   CC-functions of lumped LC-unit

The input signal was narrow band noise. We can see that the enveloping function including the extremes - every second one is
underlaid with a dotted area - is shifted by other intervalls
than the equivalent zero crossing. The shift of the enveloping
function relates to the group velocity, whereas the shift of
the zero crossing relates to the phase velocity. Thus both
velocities can be determined from several cross correlation
functions.

RAYLEIGH WAVES

Fig. 13 shows the basic behaviour of the Rayleigh waves at
different frequences. We can see that the enveloping function
including the extremes of the cross correlation functions for
higher frequencies of narrow band noise shiftes to the right,
i.e. the group velocity decreases. We determined in extensive
series of measurements the phase and group velocities in laboratory sand pit over a distance of 6 meters for 8 frequency

320

ranges. Fig. 13 should only give an impression of the basic behaviour of the Rayleigh wave. Both velocities decrease with increasing frequency by practically the same measure and differ in the chosen frequency range by the factor 1.3.

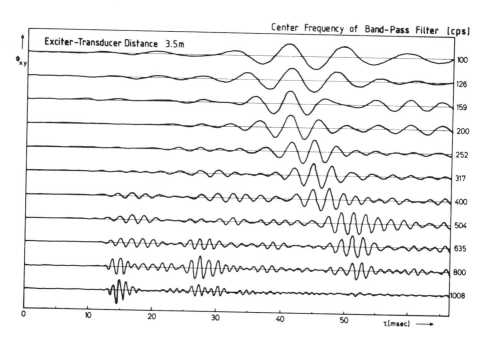

Fig. 13   CC-functions to show frequency dependence of Rayleigh waves

If the measurements of phase and group velocity are correct, it ought to be possible to infer from both values separately to one and the same phase response which shall be valid for the soil. There was good agreement in this case. We calculated a shear modulus increase from both velocities with increasing height of overlaying soil, increasing with the exponents 0,43 to 0,47. These values were more or less confirmed by other approaches, however, only by taking recourse to phase velocity.
Another example shall demonstrate how we can even recognize reflexions in the cross correlation function. For this purpose the following test was performed:
A wall, 100 cm wide, 6 cm thick and 30 cm high was installed in the laboratory sand pit, even with the sand surface. The wall had a mass density of about 2,4 g/cm^3 compared to that of the sand in the pit of 1,75 g/cm^3. The sand pit was filled with an almost homogenously compacted fine-to-medium sand with a lime content of less than 2 %. The pit dimensions are 9.3 x 9.3 x 3 m. We placed an electro-dynamic vibration exciter, vibrating vertically on the sand surface at a distance of 1.6 m perpendicular to the wall. Starting next to the exciter,

an acceleration transducer was placed onto the sand surface in steps of 10 cm along an axis to the wall and beyond it. For each individual step we determined the cross correlation functions between narrow band noise as input signal and the vertical component of the soil vibration as well as its horizontal component in the direction of the wave propagation.

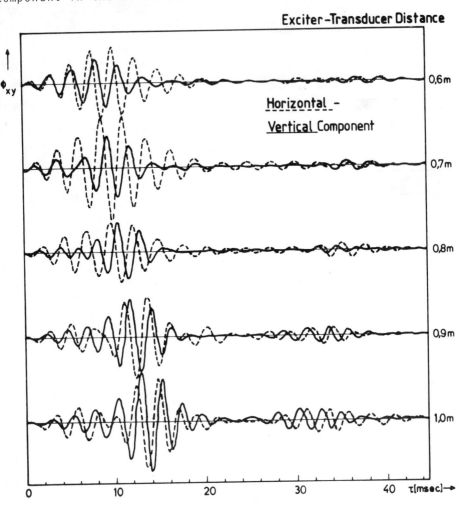

Fig. 14   CC-functions to show wave reflexions

Fig. 14 shows a part of the cross correlation functions for exciter-transducer distances of 0.6 to 1.0 m. Two properties of the cross correlation functions are immediately obvious. With decreasing distance to the wall the correlation maxima of the primary and the reflected wave approach each other. Apart from that the time delay between the horizontal and the verti-

cal component appears with reverse sign in the case of the reflected wave. We can distinguish the primary wave from the reflected wave by the sign of the phase delay between the horizontal and the vertical component of the surface wave. The
analysis of this sequence of experiments leads to the phase
travel time diagram for the vertical component, which we can
see in Fig. 15.

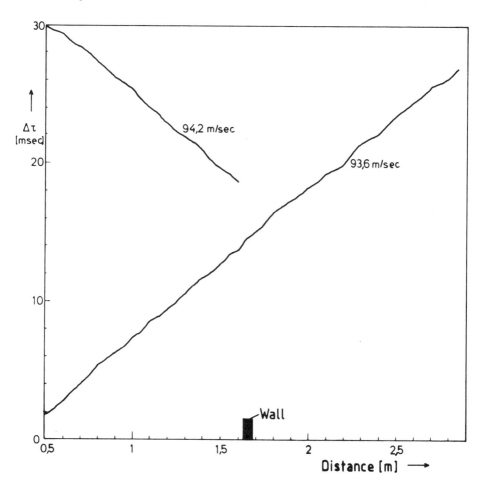

Fig. 15   Phase travel time diagram with reflexions

From the velocity data of an average of 93.6 m/sec and 94.3 m/sec
which we derived from the inclination we conclude that we must
be dealing with surface waves in both cases, because the p- and
s-wave velocities are much higher in the sand pit. In this diagram (Fig. 15), the travel time curves for primary and reflected waves do not necessarily have to meet at the location
of the wall, because we analysed different maxima of both waves.

323

EFFECTS OF ANISOTROPY

Another example of applying the correlation method which shall
be explained, has to our knowledge not been published so far.
It concerns measurements which prove the anisotropy in the case
of shear wave propagation in cohesionless soils. We have chosen
this example because it was not possible so far to measure this
particular soil behaviour with any of the other procedures used
today. For this purpose it was necessary to have a shear wave
generator which firstly generates only shear waves as far as
possible and which secondly allows to change the direction of
polarisation. Apart from this the generator must be applicable
in the correlation method. Without discussing preliminary experiments we arrived at a very simple and moderately inexpen-
sive solution: a D.C. motor with a permanent magnet in the
stator. If we run the motor with an A.C. signal, in this case
with the stochastic signal, the rotor, trying to start to ro-
tate in both directions, transmits the reaction moment to the
case. If we put the motor into the soil with good mechanical
contact we get an almost ideal shear wave generator. According
to the position of the motor axis it is easy to change the
polarisation direction of the shear wave. We undertook different measurements with this generator in the sand pit, arriving
at the following results:
    The generator was placed in the soil at a depth of 30 cm as
well as 3 transducers at different distances in the same depth.

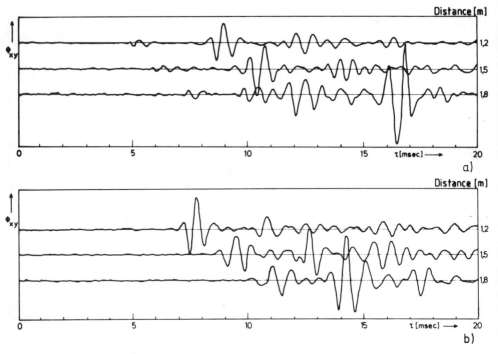

Fig. 16 CC-functions of shear waves
        a) vertical polarisation   b) horizontal polarisation

We can see in Fig. 16 the cross correlation functions which we arrived at by the procedure explained above. The upper 3 functions show the relations in the case of vertical polarisation, the lower 3 in the case of horizontal polarisation. It is obvious that the shear waves with horizontal polarisation are faster than the other ones. In this case the difference is about 14 %, for other depths of generator and transducers up to 17 % were measured. This means a shear modulus difference of about 37 %. Till now it was assumed that - apart from other factors - only the first invariant of stress determines the shear modulus of sand. These experiments proved that this cannot be true, since $\bar{\sigma}_0$ was not changed during the experiments.

In order to confirm the measurement results with an approved method and especially the conclusions which we derived we tested sand samples in the Resonant-column test. In this instrument, a cylinder shaped sample, being fixed on one side, is subjected to torsional vibrations. We can determine the axial shear wave velocity from the resonant frequency of the sample. The sample could only be put under isotropic pressure in our instrument. In order to study the stress anisotropy, beeing the main issue we modified the instrument to permit the application of a separate vertical pressure on the top of the sample. The stress in the direction of the shear wave propagation could thereby be chosen larger or smaller than the one perpendicular to it. The result was as follows: For those cases, in which the pressure applied to the sample in the direction of the shear wave propagation is larger than or equal to the pressure applied to the other plane, it seems to be true that the velocity is only dependent of the first invariant of stress. However, as soon as the stress in the direction of the shear wave propagation becomes smaller than the one perpendicular to it, this dependence is no more valid. In this case the stresses perpendicular to the direction are more important then the stress in direction of the propagation, in our experiment approx. 65 %. All publications, which deal with the Resonant-column tests, examine just the first case, i.e. stress increase only in the direction of the propagation. Their results were generalized for the other case without proof. Unfortunately it is not possible to find the dependence of influences of all separate pressures in the Resonant-column test. It was possible, however, to proof experimentally that an anisotropy of stress can cause anisotropy of velocity. The results of the correlation measurements in the R.c. test coincide with field tests for similar conditions for horizontal shear wave propagation in soils compared to those in the Resonant-column instrument with dimished stress in direction of propagation. This was the purpose of this excursion into already known measurement techniques.

We should like to indicate briefly the practical possibilities of applying this behaviour of the soil: It would be possible for example to control the stress variation in problematic slopes, if we observe the behaviour of the shear wave velocity. Generally it is possible to infer with this measurement technique the orientation of inner stress in soil - even a long time before possible failure.

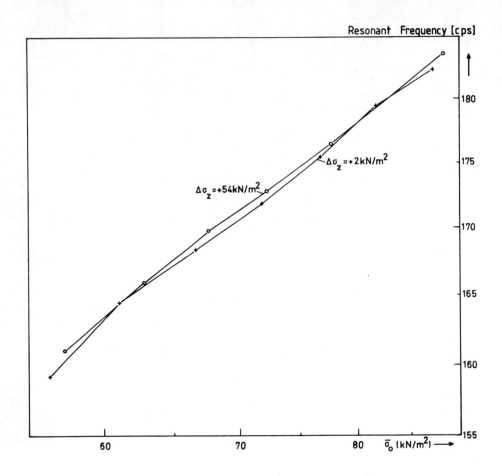

Fig. 17   Stress influence to the shear wave velocity

WAVE PATHS

The cross correlation functions show still another phenomenon:
The fact that the shear modulus in soil increases with greater
depts results in the body waves not moving by the shortest but
by the fastest path from the generator to the transducer. The
fastest path for this shear modulus configuration is a concave
curve. This longer path shows up in the presented measurement
results in such a way that the shear wave travel time does not
increase linear to the measured distance but less than that.
This proves that shear waves do not spread in linear direction.

326

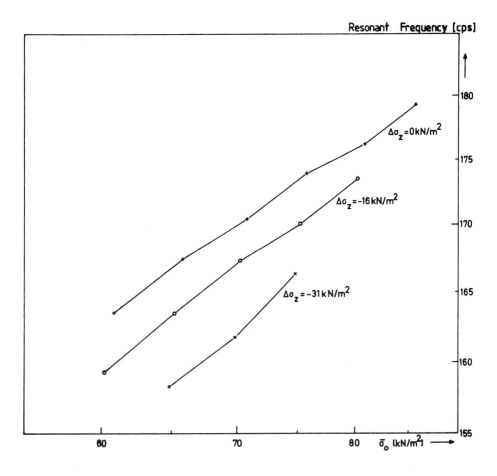

Fig. 18   Stress influence to the shear wave velocity

The following series of experiments can give another indication for the concave propagation path of the shear wave. We
assumed that the various degrees of moisture of the investigated sand should have an influence, however small, on the
phase travel times of all three types of waves. The important
point was not the absolute amount of the moisture influence but
that it should be possible to measure it. If the shear waves
were to move in bent paths below the propagation way of the
Rayleigh waves, the influence of a sudden watering of the sand
surface on the wave propagating in a lower plane would have to
start later, after the moisture had reached those deeper regions.

327

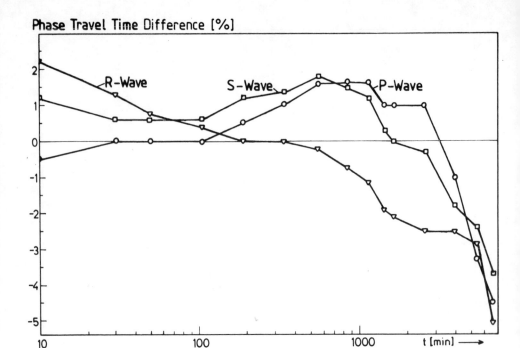

Phase Travel Time Difference [%]

Fig. 19   Influence of moisture to the wave velocity

From the cross correlation functions we found the phase travel
times of all three waves over a distance of 3.5 m for a noise
signal band-limited at about 400 cps. Then we sprinkled the
area between the generator and the transducer with about 6.6 mm
of water. We compared the phase travel times which were measured afterwards with those before sprinkling. It was observed
that the Rayleigh wave reacted very strong to the sprinkling,
and its travel time kept decreasing till the end of the ob-
servation after five days. The primary and the shear waves
changed their travel times considerably less as a result of
the sprinkling. The first obvious changes in their travel times
which could be measured occured after about 100 minutes and
continued in the subsequent time in a relative synchronous way.
Both observations, the delayed reaction and the constant behaviour in contrast to that of the surface wave, allowed the
conclusion that the body waves were mainly propagating along
paths below the Rayleigh wave.

DETECTION OF ANOMALIES

Another field of applying correlation measurements is offered
by the nondestructive exploration of anomalies in soils. In the
course of research in the correlation method, two different experiments have been performed. In the first case, the distortion

328

of the field of Rayleigh waves by a concrete block was taken as
a criterion. In the second case a tunnel cover of 18 to 27 meters was investigated and the changes of travel time of the
primary wave were analysed. The details of the experiment in
the first case is shown in Fig. 20.

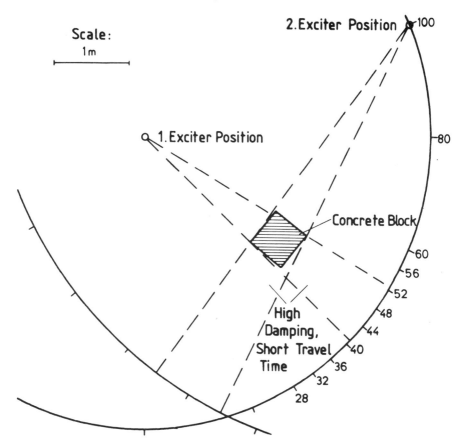

Fig. 20   Test structure to detect soil irregularities

The vibration exciter was placed at a distance of 2 meters from
the concrete block for the first series of measurements. The
concrete block was 41 cm thick buried 6 cm below the sand surface. On a circle, the transducer was placed successively in
the indicated positions (Fig. 20).
   Fig. 21 shows a part of the cross correlations functions of
this series of tests. It is easy to perceive that faster waves
are noticeable in the influence area of the concrete block and
that the other ones are damped. The faster waves may be explained by the higher velocity in the concrete block, and the
damping is caused among other factors by reflexions. Analogous
cross correlation functions can be obtained if we perform measurements in the same way with a second exciter position.

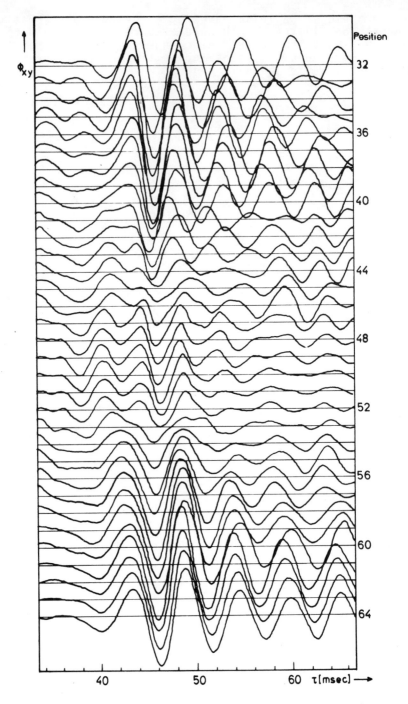

Fig. 21 CC-functions to detect soil irregularities

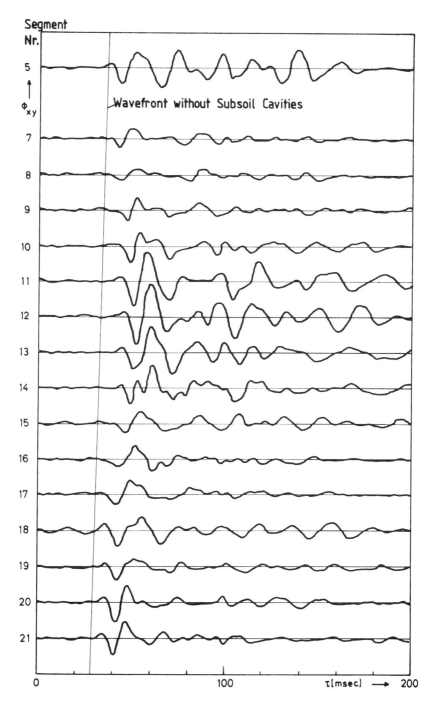

Fig. 22   CC-functions to detect cavities between a tunnel tube
          and the surface

331

Thus, the concrete block can be located from the surface.

Another series of locating anomalies were performed at the Hamburg motor-way-tunnel under the river Elbe. It was supposed that cavities had formed over the tunnel tube in the over-laying soil due to construction work.

The part of the tunnel in question was built by excavation with closed face shield. The object of the measurement was to locate these anomalies without drilling. For this purpose an electromagnetic vibration exciter was placed over the tunnel tube at the soil surface and the transducer vertically beneath in the interior of the cast iron tunnel lining segment. The construction work was still in process, concrete drills in particular caused a high noise level, which, however, had no influence on this noise immune measurement technique. Since the regular travel times in the individual soil layers were not known we considered those travel times as normal, which did not vary more than 10 % about an average value. Exciter and transducer were moved stepwise in such a way that we covered 110 meters with 95 measurements along the tunnel axis. The travel times in one part were up to 42 % higher than the average, as seen from the cross correlation functions in Fig. 22. The explanation was that the primary wave did not propagate directly from the generator to the transducer but that it had to travel around a cavity or to propagate through an area of low density. Standard Penetration Tests, which were performed later confirmed the findings of the correlation measurements. These measurements would not have been possible with dynamic measurement methods, since only the surface end and the tunnel end of the vertical measurement axis were accessible without drilling. Seismic measurements would have been possible theoretically, however, it is certain, that the inhabitants of this residental area would have objected to the noise and the vibrations in their lawns. The vibration exciter for the correlation measurements didn't cause any damage and made no more noise than a gas stove.

CONCLUSIONS

The application of the correlation method combined with stochastic stationary signals provides reliable and relevant information about the vibration transfer system, especially in the following conditions:

a) small amplitudes have to be used,
b) presence of high noise level,
c) large measurement distances to be bridged,
d) measurements near the exciter,
e) poor coupling of the transducer to the soil
f) unknown or artificial vibrations to be identified and analysed.

The following considerations may clarify the conditions:

ad a)
The application of small vibration amplitudes is necessary when we are supposed to examine the linear elastic properties of a system. The same condition applies to objects endangered by

332

large vibration amplitudes. Last not least the inconvenience of heavy vibrations and noise, e.g. detonations, may prevent the use of high amplitudes.

ad b)
All measurement techniques which do not operate with rejection of noise need test signals which are clearly distinguished from the noise level. The correlation method which operates with system dependent automatic noise suppression is able to elimina- te noise levels according to the type of test and noise signals, which may be up to a hundred times higher than the level of the test signal without incuring considerably loss of measurement precision.

ad c)
In the case of wave transmission over long distances we are as a rule confronted with high damping due to geometric and ma- terial damping. If we operate with high exciter amplitudes for conventional measurement methods, we are infringing all the points mentioned under a). With small amplitudes, we face the difficulties mentioned under b). The correlation method allows us to operate with small amplitudes and at the same time with high noise levels in order to bridge large measurement distan- ces.

ad d)
For soil analysis in areas close to the surface, we primarily use the propagation properties of the Rayleigh wave. At a sufficient distance from the exciter it predominates because of its low geometric damping in comparison to that of body waves. Near the exciter, however, the individual waves form an interference pattern if harmonically generated. Velocity measurements with dynamic methods may then lead to considerable misinterpretation.

ad e)
The coupling of a transducer to the system has to be regarded poor if no reproducible conditions can be achieved by install- ing the transducer various times. Under these conditions dy- namic measurement results may depend to an extraordinary degree on the special frequency and will be changed as far as ampli- tude and phase is concerned. The cross correlation function, however, averages contionuously over the chosen frequency band in the case of narrow band noise and thus eliminates accidental variations. The same aspects apply to the coupling of the vi- bration exciter.

ad f)
The disclosure of the paths of vibration transmission in a com- plex system is the domaine of the correlation technique. Since the cross correlation function evaluates the extent of simi- larity between two signals, we can locate the individual sources in a vibration mixture, if the different signals are statisti- cally independent from each other.

Specific examples to apply the correlation technique in soil dynamic are:
    the location of subsoil enclosures, cavities and layers,
    in situ measurements of the shear modulus
      from the surface for surface close areas,

333

in the soil, together with its dependence of direction,
measurements to evaluate stability,
measurements of the orientation of in situ states of stress,
control of vibration isolation measures, like concrete core
   walls, etc.,
measurements for the design of vibrating foundations on the
   subsoil,
analysis of systems with numerous vibration exciters.

This enumeration does not claim to list all relevant areas
of application. The importance of the areas mentioned justifies,
however, the future consideration of this new measurement technique as a powerfull tool in soil dynamics.

REFERENCES

Buchwald, V.T. 1959, Elastic Waves in Anisotropic Media,
   Proceedings of the Royal Society London, Vol. 253
Hardin, B.D., Black, W.L. 1966, Sand Stiffnes under Various
   Triaxial Stresses, Proceedings of the ASCE
Haupt, W. 1973, In Situ Shear Wave Velocity by Cross Hole
   Method, Discussion, Proceedings of the ASCE
Isermann, R. 1971, Experimentelle Analyse der Dynamik von Re-
   gelsystemen, Identifikation I, Bibliographisches Institut
   Mannheim, Wien, Zürich
Nyquist, H., Brand, S. 1930, Measurement of Phase Distortion,
   Bell Syst. Technical Journal Bd. 7
Prange, B., Haupt, W. 1974, Anomalien eines stationären Wellen-
   feldes an der Halbraumoberfläche infolge von Untergrundein-
   schlüssen, DFG Bericht Karlsruhe
Richard, F., Hall, J., Woods, R. 1970, Vibrations of Soils and
   Foundations, Prentice-Hall, Inc., Englewood Cliffs, N.J.
Roesler, S. 1975, Statistisches Meßverfahren zur Ortung von
   Anomalien im Baugrund, Straße Brücke Tunnel, Heft 12, Berlin
Wehrmann, W. 1972, Technische Anwendungen der Korrelations-
   analyse in der Signal- und Systemtheorie, Techn. Inf. 9
   Heft 2, Norma Wien

Surface-waves in non-homogeneous half-space

WOLFGANG A. HAUPT
University of Karlsruhe, Karlsruhe, Germany

SYNOPSIS

The isolating effect of solid obstacles near to the surface
against surface-waves is investigated by use of the FE-method
and by experiments. By application of the Influence-Matrix procedure a large number of calculations has been performed,
yielding systematic relations between the screening effect of
the obstacles and their geometrical and material parameters.
The detailed study of the subsurface wave propagation processes
provides the interpretation of the isolating effect by physical
phenomena.

1. INTRODUCTION

The propagation of surface-waves is an important problem in
soil dynamics, because all buildings or vibration-sensitive
installations at the surface are affected more or less by these
waves. In a homogeneous half-space the surface-wave is represented by the so-called Rayleigh wave (Rayleigh 1885). In a
non-homogeneous half-space, however, the displacement functions
of the waves are usually totally different from those of the
Rayleigh wave, according to the type of inhomogeneity. This
paper deals with the propagation of surface-waves in a half-
space, containing a local inhomogeneity, which consists of a
solid obstacle near to the surface of the half-space.
  This obstacle can be expected to have a more or less large
screening effect to the incoming waves. Thus, the practical
purpose of the investigation, which will be reported here, was
to study the screening effect of such obstacles. The results
to be presented are coefficients indicating the screening
efficiency of this inhomogeneity depending on its geometrical
and material parameters. But this does not provide an actual
understanding of the phenomena causing this effect. Therefore,
the wave propagation processes in the vicinity of these obstacles have been studied in detail and will be explained. By use of
a computer film these processes can be observed directly.
  The problem of the isolation of buildings or installations
against shocks and waves in the ground is becoming more and

more important. The reason, on the one hand, is the increase of intensity of the vibrations generated by oscillating machine foundations or by highway or railway traffic. On the other hand, an increasing sensitivity of the objects affected by vibrations can be observed, as for example computers or instrumentations for mechanical or optical measurement of high accuracy. Furthermore, today people, who are staying in the buildings are no longer prepared to easily accept the disturbance or inconvenience caused by vibrations.

Hence, the intensity of the vibrations due to travelling waves in the ground will often exceed the limit of tolerance. In those cases, the installations or buildings affected must be isolated through appropriate measures in the ground.

A wave source at the surface of a homogeneous half-space - for instance a vibrating machine foundation - generates both body waves(P- and S-wave)which radiate into the inner of the half-space,and mainly the Rayleigh-wave, see fig.1a. This wave type is coupled to the surface of the half-space and it transmits the major part of the energy emitted into the half-space by the wave source (Miller and Pursey 1954). It consists of a

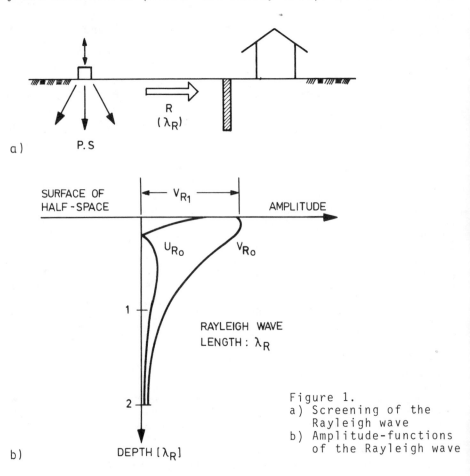

Figure 1.
a) Screening of the Rayleigh wave
b) Amplitude-functions of the Rayleigh wave

vertical and a horizontal displacement component the functions
of which with depth are presented in fig. 1b. The Rayleigh
wave length is slightly smaller than that of the S-wave. Due
to the higher geometrical damping of the body waves, the wave
field within some distance of the vibrating foundation, and
near to the surface, is determined almost by this Rayleigh wave
alone. Therefore, if a building or any shock-sensitive installation has to be isolated from ground vibrations, it is necessary
mainly to reduce the amplitudes of this wave type.

Prior Investigations

Experimental investigations in a model scale have been performed on open trenches by Woods (1968) in the near-field of
a wave source, where the body waves are dominant, as well as
in the far-field. The latter represents the Rayleigh wave isolation. He observed an average amplitude reduction of the vertical Rayleigh wave component down to 25 % at a trench depth
of slightly more than one Rayleigh wave length λ_R. Woods and
cooperators reported on first experiments on rows of boreholes as isolating measures using the holographic interfero-
metry-technique (Woods, Barnett and Sagesser 1974). In this
field, however, much research work remains to be done. Dolling
(1970) conducted measurements on the screening effect of plane
slurry filled trenches having a maximum depth of 6 m.
    However, open trenches of dimensions, necessary for practical
isolating purpose, are not feasible, and slurry filled trenches
are not stable over a long period without continuous main-
tenance, and are therefore inappropriate as permanent isolating
measures in practice. Thus, the investigation presented here
deals mainly with stiff solid constructions in the ground like,
for instance, concrete core walls. Those structures are long
stable and may be expected to have a vibration-isolating effect,
which is, however, not well known until now.
    The investigation has been performed theoretically by use of
the Finite-Element-method and, parallel to this, experimentally
in a model scale. The aim of these measurements was to control
the results of the calculations.

2. PERFORMANCE OF INVESTIGATION

2.1 FE-Calculations

The problem has been treated as a steady-state wave field
analysis, with the displacements varying harmonically with
time. This includes also non-harmonic processes, which can be
transformed to harmonic ones by a Fourier-analysis. In steady-
state harmonic problems, the time-dependency of the displacements can be eliminated and the problem to be solved becomes
formally a static one. Then all matrices and vectors in the
FE-analysis are complex.
    The material was chosen to be linear visco-elastic. Within
the range of strain, which is typical for the problems considered here, this stress-strain-law describes quite well the

337

true behaviour of the soil under dynamic load. The analysis has been performed as a two-dimensional, i.e. a plane strain problem Furthermore, a homogeneous isotropic half-space has been assumed. In a real soil the wave velocity usually increases with depth, but the results reported here, will, however, be applicable, if there is not a significant change of wave velocity in the ground. This has been confirmed by the measurements, as will be shown later. The FE-grid has been chosen to be rectangular with the surface of the half-space being the upper boundary. For the far-field analysis the dimensions were 10 λ_R in the horizontal direction and 2 λ_R in vertical direction (see fig. 2a), whereas for the near-field analysis the length was 14 λ_R. From this results a system of linear equations with about 7500 complex unknowns in the first case and 10400 in the second case.

Boundary Conditions

The two major difficulties in the analysis of steady-state harmonic wave fields is the application of the appropriate dynamic conditions at the boundaries of the FE-grid and - due to the complex numbers - the handling of very large systems of equations.
    In this analysis, at the lower boundary of the FE-grid, the simple "dashpot" condition, described by Lysmer and Kuhlemeyer (1969), had been applied. It replaces approximately the adjacent full-space (FS-BC). At the right and the left vertical boundaries, the so-called "influence-matrix" boundary condition (EM-BC) was used, which has been developed especially for this investigation (Haupt 1976, 1978a). By making use of the Rayleigh wave boundary condition, also described by Lysmer and Kuhlemeyer (1969), it replaces the adjacent homogeneous quarter-spaces at both sides of the FE-grid (see fig. 2a).

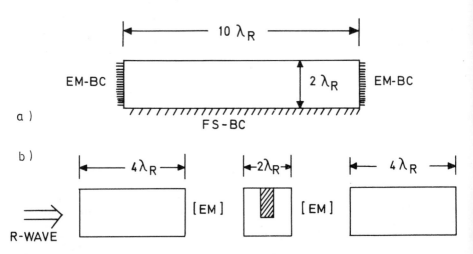

Figure 2.   a) Boundary conditions; b) Division of FE-grid, far-field analysis

But the influence-matrix procedure provides an additional advantage: The actual FE-grid has been divided into two outer grids of 4 λ_R length and a small inner grid of a length of $2\lambda_R$. This inner grid contained the obstacle and was only subjected to external forces. At the interfaces of the sub-grids, the influence matrix procedure has been applied. Now, for the analysis of one case, only the inner grid had to be calculated following the normal FE-procedure. The displacements within the outer grids were then obtained by simple matrix addition and multiplication. In this way, the system of equations to be solved could be reduced from one with 7500 complex unknowns to a system containing only 1500 unknowns. Only by this considerable reduction of computer time it was possible to perform an extensive parameter-study by analysing more than 370 different cases.

2.2 Experiments

Some results of the FE-analysis have been controlled by measurements in a model scale (Haupt 1978b). The experiments were conducted in a sand bin with the dimensions of 9.5 by 9.5 meters in plan and a depth of 3 meters, which was located in the laboratory of the Institute for Soil and Rock Mechanics (IBF) of the University of Karlsruhe. The soil was a dense medium sand with a moisture content of about 3 %, brought in place as homogeneously as possible. In the center of this sand bin, an electrodynamic vibrator has been located, generating a concentric harmonic Rayleigh wave field. The amplitude and the phase-angle of the vertical displacement component of this wave field was measured by placing an acceleration transducer on the surface at constant intervals along a radius, see fig. 3. The Rayleigh wave length varied between 24 and 40 cm.

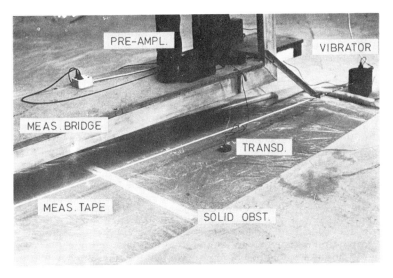

Figure 3.   Measurement of surface-wave field with obstacle

The measurements were performed first without any obstacle and then again with the obstacle burried into the ground. The obstacle material was especially prepared to yield the same ratios of wave velocities and densities as have been assumed in the FE-analysis.

2.3 Variation of Parameters

Different parameters can be expected to have an influence on the disturbing effect of an obstacle to surface-waves. The geometrical parameters of the obstacles, normalized on the Rayleigh wave length, are (see fig. 4):
- thickness β
- depth τ
- position below the surface
- inclination of the interfaces
- distance to the wave source ζ.

The normalized coordinates are ξ and ϑ. In this investigation, all obstacles are positioned so that the upper boundary coincides with the surface of the half-space. The thickness and the depth as well as the distance to the wave source have been varied independently within a wide range. The far-field isolation was investigated by assuming an infinitely large distance between the wave source and the obstacle.

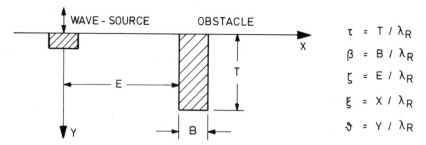

$$\tau = T / \lambda_R$$
$$\beta = B / \lambda_R$$
$$\zeta = E / \lambda_R$$
$$\xi = X / \lambda_R$$
$$\vartheta = Y / \lambda_R$$

Figure 4.   Geometrical parameters of the obstacle

The following material parameters may likely be of influence:
ratio of Poisson's ratios ν_k/ν
ratio of Young's moduli E_k/E
ratio of densities ρ_k/ρ
ratio of the coefficients of attenuation α_k/α

The subscript k indicates the obstacle material.
In this investigation, all values of Poisson's ratio have been chosen to be 0.25, yielding a ratio $\nu_k/\nu = 1.0$. In addition, the damping parameters have been assumed to be equal, because the internal damping of the obstacle material is expected to have small influence on the distorting effect, compared to that of other parameters. The variation of the moduli of elasticity and of the densities may be seen from the table 1, where nine series with different obstacle materials are listed. Due to the

340

equality of Poisson's ratios, the ratio of the wave velocities v_k/v is valid for all wave types in each series.

If a natural soil is assumed to have a specific weight of 18 kN/m³ (1.8 Mp/m³) and a wave velocity of 300 m/sec, the obstacle material of the series A corresponds to a concrete having a specific weight of 24 kN/m³ (2.4 Mp/m³) and a wave velocity of 1500 m/sec. As in practice this material will especially be of interest, the screening effect of these concrete obstacles has been studied in more detail. The procedure of the further analysis of the calculations and the basic physical processes will be described for this type of obstacle in the following.

Table 1.   Variation of the material-parameters

Series	E_k/E	ρ_k/ρ	v_k/v	$\sqrt{\dfrac{E_k \rho_k}{E\rho}}$
A	34.3	1.37	5.0	6.85
B	25.0	1	5.0	5.0
C	400.0	1	20.0	20
D	4.0	1	2.0	2
E	0.25	1	0.5	0.5
F	1	1.37	0.85	1.17
G	1	4	0.5	2.0
H	1	25	0.2	5.0
J	1	0.2	2.24	0.447

3. FAR-FIELD ISOLATION

In this study, the far-field was represented by a free Rayleigh wave, which was generated by the appropriate stresses acting at the left vertical boundary of the FE-grid. This wave then travelled through the grid to the right, thereby being distorted by the obstacle. This report is confined essentially to the analysis of the vertical displacement component, because all important phenomena can be described by considering this component. Details concerning the horizontal displacement component are published elsewhere by the author (Haupt 1978b).

3.1 Rectangular Stiff Obstacles

The figure 5 shows the amplitude of the vertical displacement at the surface depending on the distance over the total length of the FE-grid. The curve $V_{R1}(\xi)$ refers to the case without any obstacle, that is the case of a free Rayleigh wave (see fig. 1b). The attenuation of this amplitude with increasing distance as an exponential function is due to the material damping. If there is an obstacle within the half-space - in the present case with the normalized dimensions $\beta/\tau = 0.2/1.5$ - the curve $V_1(\xi)$ at the surface is obtained.

The effect of the obstacle on the amplitudes of the original wave field can be expressed by the value $\gamma(\xi)$ which is defined as the ratio of the amplitude in the case of distortion by the obstacle to the amplitude of the free, undisturbed Rayleigh wave:

$$\gamma_1(\xi) = \frac{V_1(\xi)}{V_{R1}(\xi)} \qquad (1)$$

As the buildings which are affected by surface-waves are extending only slightly into the half-space, only a small layer of $0.2\ \lambda_R$ thickness at the surface has been considered for the further analysis of the FE-calculations. Therefore, the normalized amplitudes at the surface, $\gamma_1(\xi)$, at a depth of $0.1\ \lambda_R$, $\gamma_2(\xi)$, and at a depth of $0.2\ \lambda_R$, $\gamma_3(\xi)$, have been taken into account.

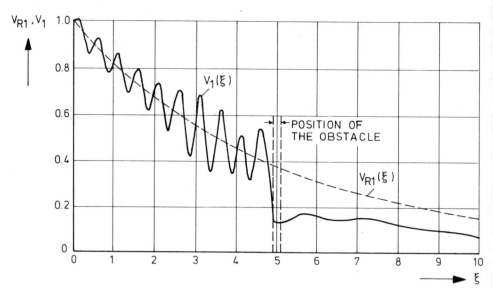

Figure 5. Vertical displ. comp., amplitude at the surface with obst. (V_1) and without obstacle (V_{R1})

In figure 6, typical curves of $\gamma(\xi)$ are presented for a thin, deep obstacle (fig. 6a) and a wide, shallow one (fig. 6b). The location of the obstacle is indicated by the two vertical dotted lines. There are really the three curves $\gamma_1(\xi)$, $\gamma_2(\xi)$ and $\gamma_3(\xi)$ presented in each plot, but they can hardly be distinguished, because they coincide almost perfectly. This fact proves that the function of the vertical displacement near to the surface in the case of distortion is almost the same as that of the free Rayleigh wave.

As may be seen from the two figures, the horizontal distance between two peaks of the distinct interference pattern in the region between $\xi=0$ and $\xi\approx4$ is exactly one half Rayleigh wave length. The attenuation of this regular oscillating curve with distance from the obstacle occurs with twice the attenuation factor of the free Rayleigh wave. Hence, the incoming free Rayleigh wave is partially reflected at the obstacle and travels back as a Rayleigh wave. Beginning at a distance of about one λ_R behind the obstacle, the normalized amplitudes are more or less constant.

At the upper surface of the obstacle, and within it, the amplitudes are very low and almost uniform in case of a thin, deep obstacle. In fig. 6b, however, the amplitudes on the obstacle are not uniform. The position of the minimum, almost at the mid-point, indicates, that this obstacle behaves mainly like a rigid body, vibrating in a rocking mode and in a vertical mode, respectively.

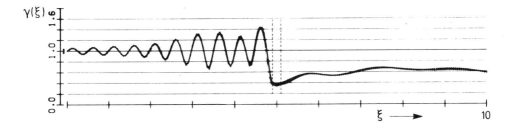

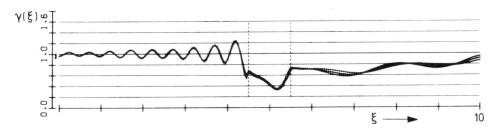

Figure 6. Far-field isolation, γ_1, γ_2, γ_3 depending on ξ, series A, rect. obstacles
a) $\beta/\tau = 0.2/1.5$ b) $\beta/\tau = 1.0/0.2$

Amplitude Reduction Factor

For the further analysis of the FE-calculations a so-called amplitude reduction factor, σ_r, has been computed. This is the average value of the normalized amplitudes of the vertical displacement behind the obstacle within the layer of $0.2 \, \lambda_R$ depth. This value is calculated from the relation:

$$\sigma_r = \frac{1}{4} \int_{\xi=6}^{\xi=10} \frac{1}{3} [\gamma_1(\xi) + \gamma_2(\xi) + \gamma_3(\xi)] \; d\xi . \tag{2}$$

The value σ_r indicates the screening effect of an obstacle in a way, that this value decreases as the screening effect increases.

In fig. 7 σ_r is presented as a function of the obstacle depth τ for different values of its thickness β for an obstacle material corresponding to the series A, which in reality means concrete. Each circle represents the result of one FE-analysis.

343

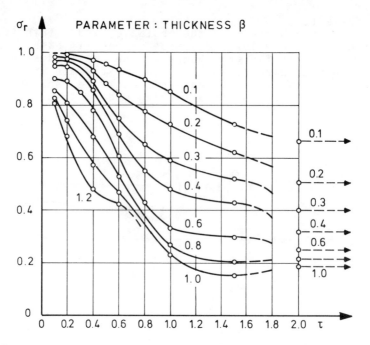

Figure 7. σ_r versus τ, rect. obstacles, series A

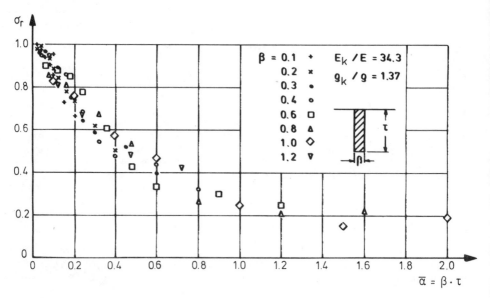

Figure 8.   Far-field isolation, σ_r depending on $\bar{\alpha}$, series A, rect. obstacles

In the cases of $\tau = 2.0$, the lower boundary of the obstacle coincides with the lower boundary of the FE-grid, where the FS-BC has been applied. This condition impedes the displacements of the obstacle at this boundary more than in the case of an obstacle of really infinite depth. This yields an underestimation of the value of σ_r for thin obstacles of a depth $\tau = 2.0$.

For obstacles having a thickness of $\beta = 0.6$ or more the amplitude reduction factor does not decrease furthermore if the depth τ exceeds a value of about 1.4. Obviously for the geometrical system and the material properties described above, the lower limit of σ_r attainable is about 0.15.

It may be seen from this figure, that wide, shallow obstacles - for instance β/τ being 1.0/0.2 - can have the same amplitude reduction factor as thin, deep ones - for instance with the dimensions $\beta/\tau = 0.2/1.0$. Hence, σ_r has been related to the product of β and τ, which is the dimensionless area $\bar{\alpha}$ of cross section of the obstacle:

$$\bar{\alpha} = \beta \cdot \tau = \frac{B \cdot T}{\lambda_R^2} \, . \tag{3}$$

This relation is presented in fig. 8: all values of σ_r are included in a narrow band ranging from $\sigma_r = 1.0$ at $\bar{\alpha} = 0$ to $\sigma_r = 0.2$ at $\bar{\alpha} = 1.4$. In this figure now the individual shape of the obstacle is eliminated. It may be seen that obstacles of totaly different shape have about the same amplitude reduction factor, if only their dimensionless cross section area $\bar{\alpha}$ is the same.

Thus, it can be stated, that the screening effect of rectangular concrete obstacles at the surface against the Rayleigh wave is a very simple function of the dimensionless cross section area of the obstacle and is independent of its shape. This relation now provides the possibility of designing isolation measures in the far-field of a wave source corresponding to the required amplitude reduction, available space and costs.

3.2 Physical Reason of Screening Effect

The results presented here suggest the mass of the obstacle being the essential parameter for the screening effect, because it is proportional to the area of cross section. But this is not the case. The effect observed is due to complicated wave propagation processes caused by the obstacle in its vicinity. Therefore the ratio of the wave velocities is the essential parameter, which will be demonstrated by two extreme examples.

Deep, Thin Obstacle

Figure 9 shows the amplitude function of the vertical displacement component with depth, $V_0(\vartheta)$, in the center plane of a thin deep obstacle like a concrete core wall. The amplitude function of a free, undisturbed Rayleigh wave, $V_{R0}(\vartheta)$, at the same place is plotted as a dashed line. On the right hand side, the respective functions of the phase-angle are to be seen.

The amplitude within the obstacle is almost uniform and the phase-angle does not vary significantly. This means, that the obstacle executes almost a pure rigid-body-motion in vertical direction. In the upper part, the amplitude of this motion is smaller than the free field-amplitude, whereas it exceeds this amplitude considerably in the lower part. Consequently, the upper part of the obstacle impedes the motion of the free half-space, thus generating a partial reflection of the incoming Rayleigh wave. The lower part of the obstacle, however, acts like a new wave source within the half-space,which is only slightly deformed at this depth, and body waves are spreading out in all directions into the half-space. This may be observed also from the phase-function of the vertical displacement component, which indicates a propagation with P-wave velocity in the downward vertical direction.

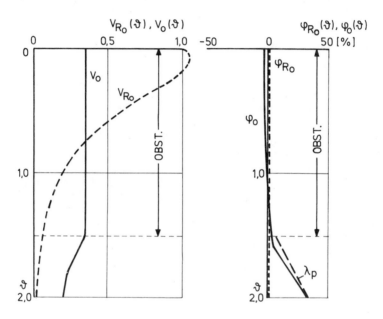

Figure 9. Amplitude and phase-angle of the vert. displ. comp. depending on depth; $\beta/\tau = 0.2/1.5$

These wave propagation processes are considered in more detail in fig. 10, where the central part of the analyzed region is presented with a relatively thin and deep obstacle in the middle. The lines to be observed around the obstacle are so-called isophase-lines, which connect the points having equal phase-angle. They can also be interpreted as wave fronts. The solid lines correspond to the vertical displacement component and the dashed lines to the horizontal component. In this figure, the isophase-lines for phase-angles of 0° and 180° are plotted, which means that the distance between two lines corresponds to one half wave length. If two waves,travelling in the opposite direction,are superimposed, the general direction

346

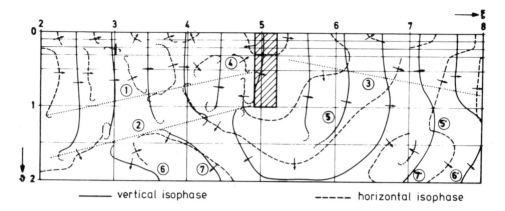

Figure 10. Far-field isolation, isophase-lines, series A,
$\beta/\tau = 0.2/1.0$

of propagation is that of the wave having the greater amplitude.
Hence, in this figure, the isophase-lines indicate only the
region of dominance of one kind of wave.

In front of the obstacle the free Rayleigh wave travelling
to the right is dominant. This can be seen by the vertical solid isophase-lines, which have a distance of exactly one half
λ_R to each other. The horizontal component is out of phase by
90^0 and shows the phase shift of 180^0 at a depth of about
$0.2 \lambda_R$. The isophase-lines of the vertical component reach
down to the dotted straight line number (2). Below this line
the waves are travelling to the left and downwards. Thus in
this lower region, the body waves, radiating from the lower
part of the obstacle, are dominant. A similar line number (1)
can be plotted for the dashed isophase-lines. From the amplitude of the horizontal Rayleigh wave component (fig.1b) and the
position of line number (1) above the line number (2) it can be
shown, that the amplitudes of the vertical displacement component (mainly S-wave) and the horizontal one (mainly P-wave)
of the body waves are almost equal.

In the central part of the figure, the isophase-lines are
passing around the lower end of the obstacle indicating this
point being the wave source. If the FE-grid extended deeper
into the half-space the lines (7) → (7') as well as (6) → (6')
would doubtlessly be connected. Behind the obstacle at the surface, the Rayleigh wave is dominant, which can be seen from the
distance of one half λ_R between two solid lines. In the lower
part of the half-space, however, this distance increases by
about 10 %, see point (5) → (5'). Thus, below the dotted line
number (3) the S-wave is dominant, but above it the Rayleigh
wave.

It has been shown, that the obstacle moves as an almost
rigid body in vertical direction. Consequently it generates a
plane S-wave, propagating to the right parallel to the surface.
As this wave developes to a Rayleigh wave with distance, the
amplitude of the vertical component must increase at the surface, whereas it decreases in the lower part of the half-space.

347

In fig.6a, the increase of the amplitudes near to the surface
within a distance of about one Rayleigh wave length in fact
can be observed.

Summarizing, the isolating effect of this type of obstacle
is due mainly to the transmission of the energy of the Rayleigh
wave from the surface to the inner of the half-space in the
form of body waves. In addition, reflection occurs in the upper
part of the half-space.

Wide, Shallow Obstacle

The other extreme example is a wide,shallow obstacle like a
concrete slab. The amplitude function in the case of distortion,
$V_0(\vartheta)$,indicates a propagation of the Rayleigh wave essentially
underneath the obstacle, see fig.11. Here again body waves
occur,which are generated by the dynamic contact stresses at
the horizontal interface between the obstacle and the half-
space. As the Rayleigh wave is dominant in the upper part of
the half-space, the body waves are indicated by the phase-
function only at a greater depth (see fig.11) and the effect
just described cannot be demonstrated by isophase-lines. In
this case the displacements must be considered directly.

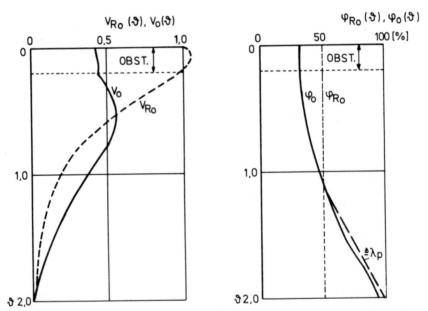

Figure 11.   Amplitude and phase-angle of the vert. displ. comp.
depending on depth; $\beta/\tau = 1.0/0.2$

The main reason for the isolating effect of this obstacle is
the following: A part of the incoming energy is transmitted
through the concrete slab with a higher wave velocity corresponding to the higher modulus of elasticity of its material. Therefore the waves, propagating on different paths, are out of phase to a certain amount behind the obstacle, thus cancelling

each other partially. This is demonstrated in fig.12, which shows the central part of the distorted FE-grid at a particular moment. The obstacle is represented by thick lines. Figure 12a shows the true state of displacement of the wave field, which is disturbed by the obstacle. In fig.12b, the undisturbed free Rayleigh wave field is presented. If these free Rayleigh wave displacements are subtracted from the true disturbed wave field, then those displacements are obtained which are caused only due to the presence of the obstacle, see fig.12c. This means, that fig.12b and fig.12c have to be superimposed to yield fig.12a. All wave fields, of course, have to be considered at the same moment.

In fig.12c, in front of the obstacle, the amplitudes are very small, which indicates a negligible reflection. Behind it, however, the amplitudes are in the order of those of the free Rayleigh wave (fig.12b), but in this case the maxima are shifted by about one half λ_R in positive x-direction. If these two wave fields are superimposed, the waves behind the obstacle will cancel each other partially, due to their phase shift, yielding the small amplitudes to be observed in fig.12a. In fig.12c the body waves within the lower part of the half-space behind the obstacle can also be detected.

Although the wave propagation processes in the vicinity of the obstacles and by this the cause of the amplitude reducing effects are totally different at the two extreme types of obstacles, the interesting, and in fact surprising result of this part of the study is the amplitude reduction factor having the same value and following the same relation with respect to the normalized cross section area.

3.3 Results of Measurements

Most of the measurements have been conducted with an obstacle material yielding a ratio of wave velocities of 5.0 and a ratio of densities of 1.34, thus corresponding to the material of series A in the calculations (table 1). By the procedure of measuring described in chapter 2.2, the same curves $V_{R1}(\xi)$ and $V_1(\xi)$, respectively, were obtained, which have been analysed exactly in the same way as described in the foregoing chapter (Haupt 1978b).

The values of the amplitude reduction factor σ_r, obtained from these experiments, are presented in fig.13. In this figure, the dashed lines indicate the narrow band including all values of σ_r from the calculations. As may be seen almost all reduction factors, received from the experiments, are lying between these two lines.

Some measurements have been performed with an obstacle material where Young's modulus had been chosen such that a ratio of wave velocities of 8 has been obtained (M5/30 and M5/24 in fig.13). At these stiffer obstacles a better screening effect occured, which does agree with corresponding calculations, as will be shown in chapter 4.2.

Summarizing, it can be stated, that the experiments have confirmed the results of the FE-calculations quite well, thus proving the reliability of the theoretical analysis.

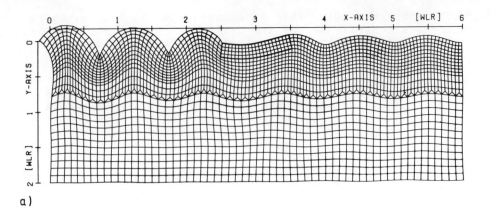

a)

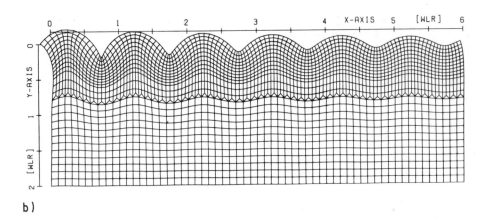

b)

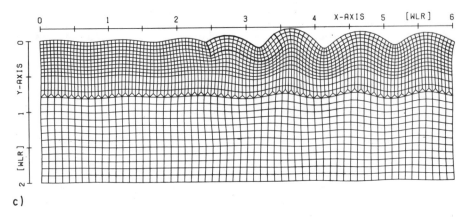

c)

Figure 12. Distorted FE-grids (WLR $\cong \lambda_R$): a) Disturbed wave field b) Free Rayleigh wave field c) Disturbance due to the obstacle

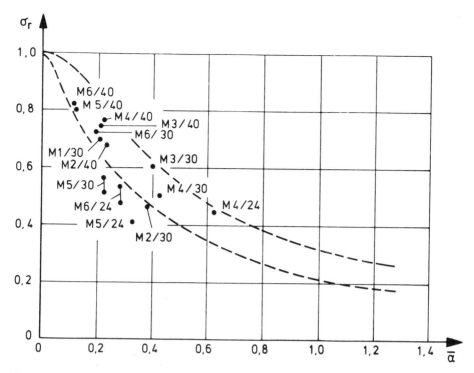

Figure 13. σ_r obtained from experiments depending on $\bar{\alpha}$

4. PARAMETER VARIATION

4.1 Variation of the Obstacle Shape

The physical interpretation of the isolation effect of different
rectangular obstacles suggests the consideration, that a special shape of the obstacle could possibly combine the effects
of the thin, deep obstacle and the wide, shallow one, thus en-
hancing the screening effect. Therefore, in total 19 different
shapes of the obstacle have been investigated, see fig.14. In
this figure, the dimension b is equal to λ_R and the angle of
inclination of the tilted obstacles is 45° in the direction of
the Rayleigh wave propagation, which is from the left to the
right, as well as against it, respectively.
   As may be seen in fig.15 almost all values of σ_r are included within the narrow band indicating the range of the results obtained at rectangular obstacles. In this figure, $\bar{\alpha}$
represents the dimensionless area of the concrete cross sections. There are only two obstacles showing a considerably
better screening effect, which are number 15 and number 17. Due
to their shape, these obstacles really combine the effects of
the thin, deep and the wide, shallow obstacle. In practice
their construction would be quite expensive and thus they are
not of great interest as an isolating measure.

351

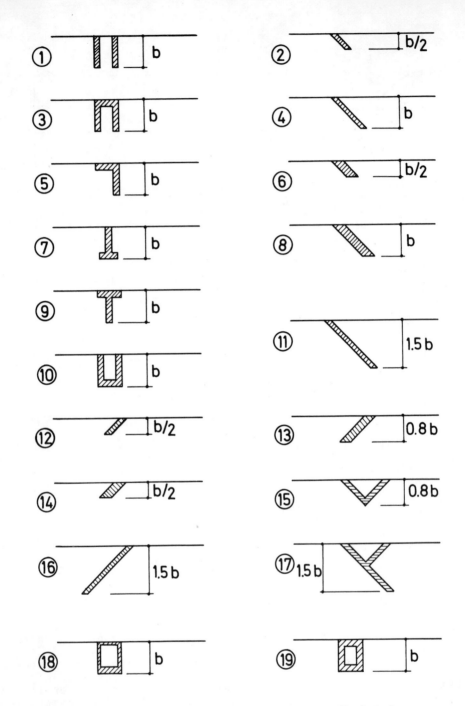

Figure 14. Variation of the obstacle shape (b $\hat{=} \lambda_R$)

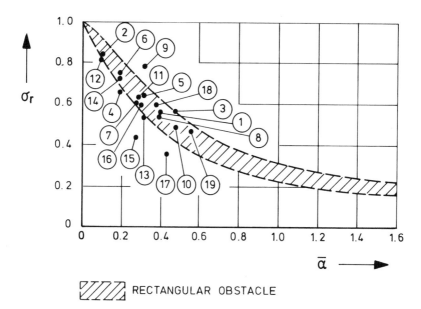

RECTANGULAR OBSTACLE

Figure 15. σ_r versus α for different obstacle shapes compared to rectangular obstacles

From this study it can be concluded, that a special shape of the obstacle in general will not enhance the screening effect as against that of rectangular ones, if the required amount of concrete and the costs of construction are taken into account.

4.2 Variation of the Obstacle Material

As may be seen from table 1, there is - besides series A - a group of 4 series (B through E) with a variation of the ratio of Young's moduli and another group, where the ratio of the densities has been varied (F through J). For a better presentation, the narrow bands of the σ_r-values obtained in each series are replaced in the following by their mean curves, see fig.16.

Stiff Obstacle Material

In series B the stiffness of the material has been chosen such that $v_k/v = 5.0$, as it is in series A. The values of σ_r of this series are only slightly greater than those of series A (fig.16), thereby indicating only a small reduction of the screening effect. On the other hand, in series F, the density of the obstacle material has been chosen to be that of the series A, but the Young's modulus is the same as that of the surrounding material. In this case, the amplitude reduction factors are all in the range of 0.95 to 1.0 (fig.16). This result confirms the

conclusion of the previous discussion: the screening effect of
stiff obstacles is governed mainly by its material stiffness
and the - only small - increase of the obstacle material density is of no significance.
    If the ratio E_k/E is chosen to be 400, as in series C, the
values of σ_r are only slightly smaller than those of the series
A (fig.16). Hence, it is not worth-while to use a concrete of
high quality for the construction of the isolating measure. A
very thin wall of this material is comparable in its behaviour
to a sheet pile wall. As such a wall will have a normalized
cross section area $\bar{\alpha}$ of 0.1 or less, the amplitude reduction
factor is about 0.9 or more. This very small screening effect
of sheet pile walls agrees well with the results of experiments
conducted by the author (Haupt 1978b) and by other investigators (Woods 1968).

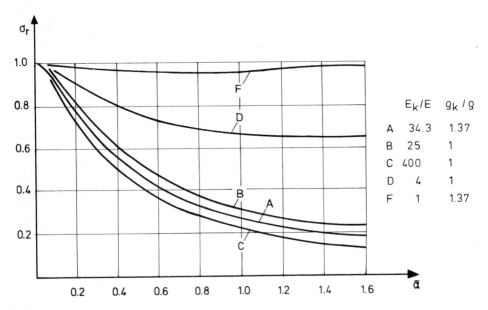

Figure 16. σ_r depending on $\bar{\alpha}$ for different obstacle material

    If Young's modulus is only 4 times higher than that of the
half-space material (series D) the relationship between the
amplitude reduction factor and $\bar{\alpha}$ is similar to that of series
A, B or C, respectively. But in this case, σ_r is about 0.6 at
minimum, therefore this material does not provide sufficient
isolating effect (fig.16).

Soft Obstacle Material

If the reflection of an incoming wave at an interface is considered, an obstacle material of a smaller Young's modulus than
that of the half-space could be of interest. In fig.17 the

results obtained for series E are presented. This material is referred to as soft. Here the effectiveness of the shallow obstacles is about the same as in the case of a stiff obstacle material. This is due to the mechanism described before: as there is again a difference in wave velocity of the obstacle material and the half-space material, a partial extinction of the waves propagating on different paths occurs behind the obstacle. It seems that at an obstacle-depth of $\tau \approx 0.6$, the partition of the incoming energy on the two different paths of propagation is about equal, thus yielding a maximum cancelling effect. As the soft material obstacle does not move as a rigid body, no energy can be transmitted into the inner of the half-space. Therefore, at the deep obstacles the amplitude reducing effect is due only to the reflection of the waves at the two vertical interfaces. Consequently σ_r becomes independent of τ, having a value of about 0.7 to 0.8. For this obstacle material no systematic relationship between the amplitude reduction factor and $\bar{\alpha}$ could be established.

The series G and H, where the ratios of the densities are 4 and 25 respectively, are only of academic interest. It should be noted, however, that in the last case for some special obstacle dimensions a value of σ_r being about 0.1 was obtained (Haupt 1978b).

In practice, the ratio of the densities cannot much exceed the value of 1.6. Therefore the last case of material parameter variation is of interest, dealing with a ratio of the densities of 0.2, see fig.18. The screening effectiveness of wide, shallow obstacles made of this material is in the same order of that of concrete slabs or even better. Thus, a very light-weight and soft concrete or any artificial material could be used instead of a normal concrete, the necessary space provided. In cases, where only transient isolation measures are required, this material could be applied with advantage, because it would be easy to remove. Here again the maximum screening effect due to the partial substraction of the waves propagating on different paths occurs at an obstacle depth of about 0.6. As the material stiffness is not increased as against the surrounding material, the deep obstacle screening mechanism does not take place.

This study of the influence of the obstacle material properties - which is not complete at all - gives an idea in which direction and to which extent the variation of a material parameter influences the effect of the obstacle on the wave field. It should be noted, that the variation of two parameters at the same time does not necessarily result in an addition of the two influences, caused by each of them separately.

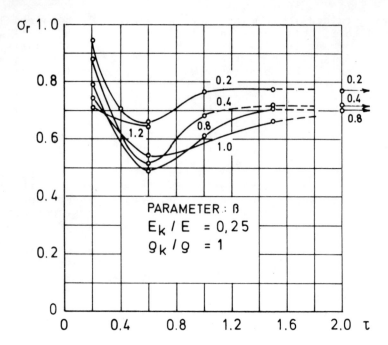

Figure 17. σ_r depending on τ, series E

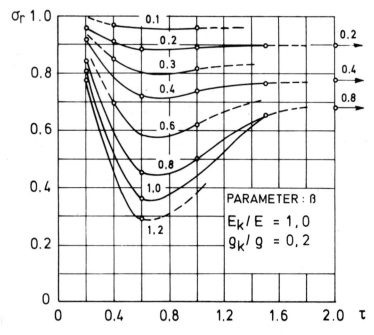

Figure 18. σ_r depending on τ, series J

5. NEAR-FIELD ISOLATION

In a Rayleigh wave field, the dynamic stresses causing the vertical motion of a deep obstacle are in phase. In contrast to this, in the vicinity of a wave source, the phase-angle of the stresses along a vertical plane extending from the surface into the half-space varies with depth. The effects of the stresses on a stiff obstacle, therefore, will partly cancel each other, thereby reducing the motion of the obstacle. From this a decrease of the amplitude reduction factor can be expected.

Therefore this part of the study deals with the near-field isolation, i.e. the influence of the normalized distance ζ between a vibrating strip foundation at the surface and a thin, deep rectangular obstacle. The properties of the foundation material were those of concrete (series A). The thickness β of the obstacles has been chosen to be 0.1 and 0.2, respectively, and as obstacle materials only that of series A and that of series E have been considered. The functions $\gamma(\xi)$ now relate the amplitudes of the disturbed wave field to the undisturbed near-field.

Isophase-lines

For a better understanding of the wave propagation processes in the near-field of a wave source, the isophase-lines are considered in fig.19. The oscillating strip foundation is located in the upper left corner of this figure and the vertical axis of symmetry is passing through the middle of the foundation.

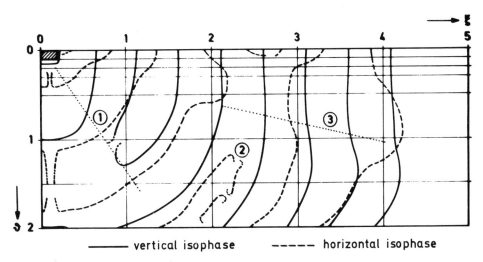

———— vertical isophase   – – – – – horizontal isophase

Figure 19.   Isophase-lines, oscillating strip foundation at the surface

At the surface, the Rayleigh wave is dominant already near to the wave source, which is indicated by the vertical solid

lines and the spacing of one half λ_R between two successive
lines. There are two isophase-lines of the vertical component
terminating at the line number (1). From the spacing between
these lines and between the dashed ones, respectively, it can
be concluded that above the line number (1) the propagation of
the vertical displacement component is governed by the S-wave,
whereas below it by the P-wave. This line number (1) indicates
the location of zero amplitude of the shear wave, radiating
from a point source at the surface, as it has been calculated
by Miller and Pursey (1954). It seems that their analytical results agree quite well with those obtained by numerical procedure.

In the near-field again, the region of dominance of the
Rayleigh wave and that of the body waves can be distinguished
clearly: They are separated by the line number (3). It must be
admitted, however, that some phenomena related to the horizontal
displacement component - see location number (2) - are not yet
fully understood.

Amplitude Reduction Factors

The results of this part of the study are presented in fig.20
and fig.21 respectively, where the amplitude reduction factors
are plotted against ζ for the obstacle depth τ = 0.5, 1.0, 1.5
and 1.83,respectively. In fig.20, which shows this relation for
the obstacle material corresponding to series A, the values of
σ_r decrease considerably, if the obstacle comes close to the
foundation. This effect is more pronounced for τ being great.
In the case of β = 0.2 (fig.20b), the amplitude reduction factors are throughout smaller than the corresponding values for
β = 0.1 (fig.20a). The minima, which can be observed in the
curves corresponding to τ equal 1.0 and more, are due to the
interference phenomena within the half-space mentioned earlier
(Haupt 1978b).

The curves corresponding to an obstacle depth of 0.5 are
almost horizontal from $\zeta \approx 0.6$ onwards. The reason is the development of the Rayleigh wave from the body waves near the surface (see fig.19). Hence, from some distance on, an obstacle
of small depth is acting like a far-field isolation measure,
for which the distance to the wave source is not relevant.

In the following figure 21, the amplitude reduction factors
of soft material obstacles are presented for β = 0.1 (fig.21a)
and for β = 0.2 (fig.21b). In this case, a systematic relation
between σ_r and ζ for the different values of τ, analogue to
fig.20, cannot be observed. Contrary to the results obtained
at the stiff material obstacle here σ_r increases with the obstacle approaching the foundation and can even exceed the value
of 1.0.

This effect can be explained in the following way: beyond a
certain angle of inclination of the direction of propagation,
the body waves do not further contribute to the development of
the Rayleigh wave. If these waves encounter an obstacle, they
will be refracted partly towards the surface and do now contribute to the Rayleigh wave behind the obstacle, see fig.22.
In the case of a soft material obstacle, this effect can do-
minate the screening effect, which is rather small for thin,

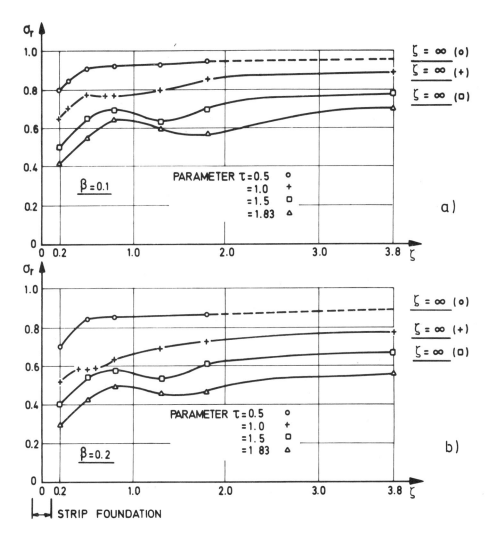

Figure 20.   Near-field isolation, σ_r depending on ζ, series A
a) ß = 0.1   b) ß = 0.2

deep obstacles, see fig.17.
   This analysis has been performed as a plane-strain problem
with the wave source being a strip foundation of infinite
length. In practice, the dimensions of the wave source are
often small compared to the length of, for instance, a concrete
core wall. For this plane structure within a rather axisymmetric
wave field, the same considerations concerning the variation of
the phase-angle with depth are valid in the horizontal direction. Thus, the screening effect of stiff obstacles, obtained
from the plane-strain analysis, will be underestimated, which
fact has also been proved by experiments (Haupt 1978b).

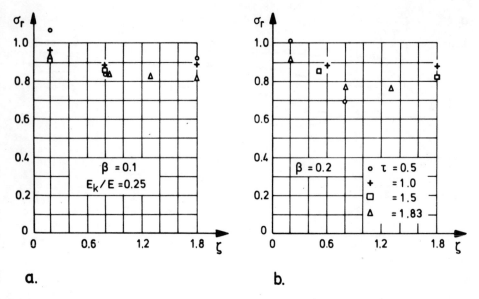

a.

b.

Figure 21. Near-field isolation, σ_r depending in ζ, series E
a) $\beta = 0.1$ b) $\beta = 0.2$

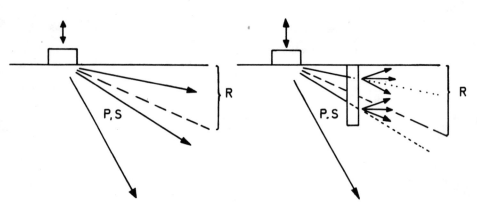

Figure 22. Amplitude enhancement effect of soft material
obstacle near to the wave source

Foundation - Obstacle - Interaction

It should be noted, that in the case of near-field isolation,
the presence of the obstacle can affect the vibration of the
oscillating foundation as well as the wave field at the opposite side. In fig.23, the two cases of near-field isolation by
a concrete core wall, having the dimensions $\beta/\tau = 0.2/1.5$, at
a normalized distance of $\zeta = 0.5$ (fig.23a) and of $\zeta = 1.8$
(fig.23b) respectively are presented. The position of the

foundation is indicated by the two vertical solid lines, that of
the obstacle again by the dotted lines.

Between the foundation and the obstacle a distinct reflection pattern occurs. If the oscillating foundation is situated
at a minimum point of this pattern, its amplitude is reduced,
see fig.23a. As the wave generated by the foundation and propagating to the left of it and the wave reflected at the obstacle in this case are out of phase, a constant reduction of
the amplitude takes place.

In contrast to this, fig.23b shows the case of the foundation being positioned at a maximum point of the reflection
pattern. Here a considerable increase of the amplitude of the
foundation vibration and of the wave field on the left hand
side occurs. Thus,if one deals with near-field isolation measures the possible increase or decrease of the amplitudes on
the opposite side of the wave source has to be taken into
account.

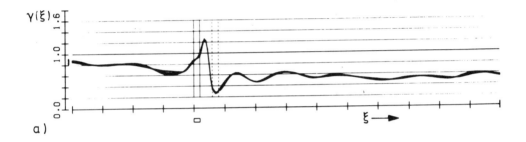

a)

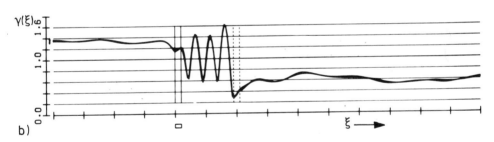

b)

Figure 23.   Near-field isolation, series A, β/τ = 0.2/1.5,
γ_1, γ_2, γ_3 depending on ξ
a) ζ = 0.5   b) ζ = 1.8

Film Presentation

The discussion of the physical reason of the screening effect
has been done by use of graphs, which cannot include at the
same time the space variables as well as the time variable.
Therefore, a little computer film has been prepared, showing
the distortion of the FE-grid during the propagation of the

waves. Some typical pictures of this film and the corresponding comments are presented in the appendix.

The moving picture film is an appropriate way of regarding dynamic problems, because the visualization of the displacements and strains below the surface and around the obstacle provides a better understanding of the wave propagation processes within the half-space and enables the engineer to arrive at an effective design of screening measures.

6. SUMMARY AND CONCLUSION

The problem of the screening effect of solid obstacles within an homogeneous half-space has been solved theoretically by the use of the FE-method and experimentally in a model scale. The application of the influence-matrix procedure has permitted to calculate a large number of cases, thus enabling the use of the FE-method as an investigation method which yields general relations instead of special case results.

For rectangular stiff material obstacles at the surface and in the far-field of a wave source, a very simple law has been found, relating the amplitude reduction factor to the area of cross section of the obstacle in a way that this factor decreases with increasing area. Thus, the actual shape of the obstacle is of no significance for the screening effect. This is in general also true for non-rectangular stiff material obstacles. If, however, the obstacle material is soft compared to that of the surrounding half-space the amplitude reduction factor depends on the actual shape of the obstacle: a satisfactory screening effect is obtained only at rather compact obstacles.

The near-field isolation by concrete core walls is more effective than the far-field isolation, whereat this enhancement of the effect is pronounced at greater obstacle depth. At soft material walls, however, the amplitude reduction factor increases if the obstacle approaches the wave source and can even exceed the value of 1.0.

By detailed analysis of the subsurface wave propagation processes within the vicinity of inhomogeneities of different shape the amplitude reducing effect of such obstacles could be related to physical phenomena, which can be visualized by a computer film.

REFERENCES

Dolling, H.J. 1970, Abschirmungen von Erschütterungen durch Bodenschlitze, Die Bautechnik 5/6
Haupt, W.A. 1976, Influence-Matrix Boundary Condition for the Analysis of Dynamic Problems by FE-Method, Proc.Int.Symp. Num.Meth. in Soil and Rock Mech. (NMSR 75), Karlsruhe
Haupt, W.A. 1978a, Numerical Methods for the Computation of Steady-State Harmonic Wave Fields, Proc.Dyn.Meth.in Soil and Rock Mech. (DMSR 77), Vol.I, Rotterdam, A.A. Balkema Publ.

Haupt, W.A. 1978b, Verhalten von Oberflächenwellen im inhomo-
genen Halbraum mit besonderer Berücksichtigung der Wellenab-
schirmung, Veröffentlichung des Instituts für Bodenmechanik
und Felsmechanik (IBF), Universität Karlsruhe, Vol.74, 1978
Lysmer, J., R.L. Kuhlemeyer 1969, Finite Dynamic Model for In-
finite Media, Proc. ASCE, No. EM4, Aug. 1969
Rayleigh, Lord J.W.S. 1885, On Waves Propagated along the
Plane Surface of an Elastic Solid, London Math. Soc.,
Proc. 17
Woods, R.D. 1968, Screening of Surface Waves in Soils, Proc.
ASCE, No. SM4, July 1968
Woods, R.D., N.E.Barnett and R. Sagesser 1974, Holography -
A New Tool for Soil Dynamics, Proc. ASCE, No. GT11, Nov.1974

APPENDIX

The following eight fig.A2 through A9, which are presenting
the central part of the distorted FE-grid, are taken from the
film. The fig.A2 through A4 deal with the near-field isolation,
the obstacle material being that of series A. For the system
see fig.A1a. In fig.A5 through A9 different cases of the far-
field isolation are presented,whereby in the first three of
these figures the obstacle material is again that of series A
whereas in fig.8 and fig.9 a material corresponding to series E
has been used. In these figures WLR means λ_R.

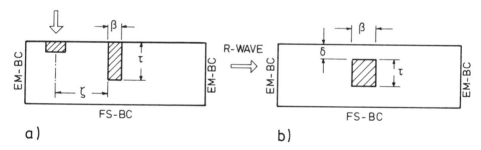

a)   b)

Figure A1. FE-system   a) Near-field isolation
b) Far-field isolation

Figure A2 presents the case where there is no obstacle within
the half-space besides the oscillating strip foundation, i.e.
the free near-field. The typical deformation field can be observed: Below and to the side of the foundation body waves are
spreading out into the half-space. At the surface the Rayleigh
wave is dominant even at short distance from the foundation.
   The next case, which is presented in fig.A3, deals with the
screening of the waves by an obstacle, which can be considered
as a concrete core wall. The normalized dimensions are
$\beta/\tau/\zeta$ = 0.2/1.5/1.2. At the surface almost a standing wave
is generated, due to the reflection of the waves at the wall.
Within the half-space the reflection of the incoming waves can
be observed directly. The isolating effect is quite good
($\sigma_r \approx 0.5$).

In fig.A4 the case of a thin wall near to the foundation is presented. This can be considered approximately as a sheet pile wall, although that would even be thinner. This wall follows quite well the motion of the surrounding half-space. Therefore the screening effect of this sheet pile wall is small. Note the vibration of the foundation being influenced by the obstacle.

In the following some cases of obstacles in the far-field of a wave source are presented. For the system see fig.A1b. The Rayleigh wave in its theoretical mode penetrates from the left into the FE-grid.

Fig.A5 shows the distortion of this Rayleigh wave field by a wall-shaped obstacle with the dimensions of $\beta/\tau = 0.2/1.0$. At the surface a partial reflection of the incoming waves can be observed. In the lower part of the half-space body waves are spreading out from the end of the obstacle similar to the case of the oscillating foundation. Thus, the obstacle acts like a new wave source within the half-space. Behind the obstacle the Rayleigh wave is dominant within some distance.

In the case of a wide, shallow obstacle, which can be seen in fig.A6, body waves also occur. They are generated at the horizontal interface between the obstacle and the half-space. Due to interference phenomena, which cannot be observed in this figure, the screening effect of this obstacle is as good as that of the foregoing one.

Next a square bloc at the depth of 0.3 Rayleigh wave lengths below the surface is considered, see fig.A7. Although the obstacle does not reach up to the surface the reflections are strong and - due to the rigid-body-motion of the bloc - also strong body waves are generated in the lower part of the half-space. This obstacle has a very good screening effect.

In the following, two examples of soft material obstacles (series E) within the far-field are presented. Fig.A8 shows the case of a thin, deep obstacle with the dimensions $\beta/\tau = 0.2/1.5$. Due to the low stiffness of its material the displacements within the obstacle are very strong, especially at the surface, where the amplitudes of the half-space displacements are great. In this case no body waves are generated by the lower end of the obstacle, therefore the screening effect is poor.

The last case deals with a more shallow and wide obstacle: the dimensions are $\beta/\tau = 1.0/0.4$, see fig.A9. Here the scale factor of the displacements is half of that of the foregoing figures to avoid overlapping of the elements. Within the obstacle the enhancement of the amplitudes as well as the short wave length can be observed clearly. From the latter results the partly extinction of the two waves, which travel on different paths, thus yielding the small amplitudes behind the obstacle.

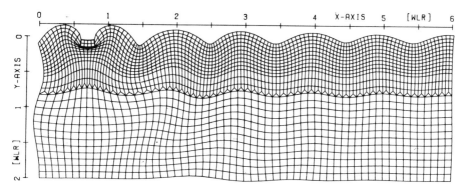

VIBRATING STRIP-FOUNDATION AT THE SURFACE

Figure A2.   Vibrating strip foundation without obstacle

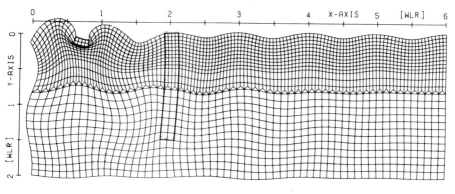

VIBRATING STRIP-FOUNDATION AND WAVE BARRIER

Figure A3.   Near-field isolation, $\beta/\tau/\zeta$ = 0.2/1.5/1.2

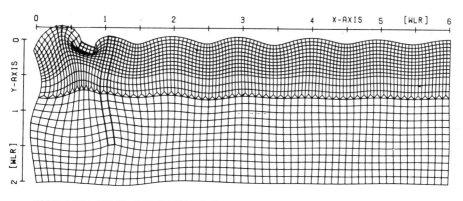

VIBRATING STRIP-FOUNDATION AND WAVE BARRIER

Figure A4.   Near-field isolation, $\beta/\tau/\zeta$ = 0.1/1.5/0.3

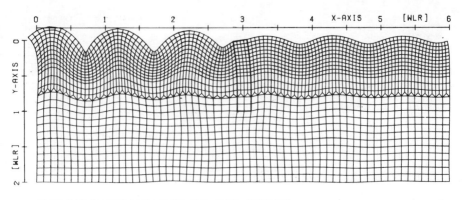

FREE FIELD RAYLEIGH-WAVE AND WAVE BARRIER

Figure A5. Far-field isolation, $\beta/\tau/\delta$ = 0.1/1.0/0.0, series A

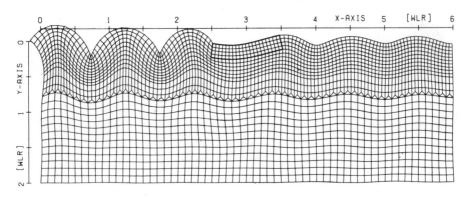

FREE FIELD RAYLEIGH-WAVE AND WAVE BARRIER

Figure A6. Far-field isolation, $\beta/\tau/\delta$ = 1.0/0.2/0.0, series A

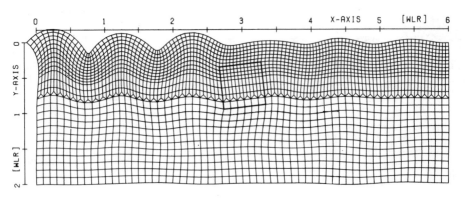

FREE FIELD RAYLEIGH-WAVE AND WAVE BARRIER

Figure A7. Far-field isolation, $\beta/\tau/\delta$ = 0.6/0.6/0.3, series A

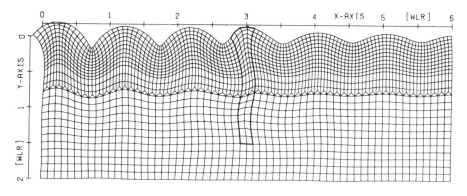

FREE FIELD RAYLEIGH WAVE AND WAVE BARRIER

Figure A8.   Far-field isolation, ß/τ/δ = 0.2/1.5/0.0, series E

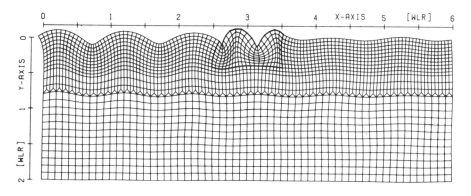

FREE FIELD RAYLEIGH WAVE AND WAVE BARRIER

Figure A9.   Far-field isolation, ß/τ/δ = 1.0/0.4/0.0, series E

Measurement of vibrations

B. PRANGE
University of Karlsruhe, Karlsruhe, Germany

SYNOPSIS

Problems of vibration transducer response and transducer/object interaction are investigated. The different instrumentation methods are outlined in the context of dynamic data acquirement and evaluation. It is shown how inertia effects of transducers can be compensated electronically to obtain the undistorted freefield data.

INTRODUCTION

The measurement of vibrations is an important part of Soil and Rock Dynamics. It is the basis of dynamic experiments to investigate the respective dynamic material parameters and is essentially connected to model testing in the laboratory and investigations in the field regarding the dynamic behavior of structures and foundations as well as the propagation of waves.

Human beeings are very sensitive to vibrations:

We feel body vibrations having an acceleration of less than 10^{-3} g.

We are able to distinguish easily between two notes played on two adjacent keys on the piano, a frequency step of $\sqrt[12]{2}$ = 1.059.

The human brain is capable to register a time difference between the arrival of acoustic signals at the right and left ear of the order of 10^{-5} sec, the reason for stereo listening and acoustic direction finding.

The minimum power needed by the ear to perceive an acoustic signal is of the order of 10^{-17} Watt.

We are, however, not able to measure with our senses quantitatively the three basic parameters of vibrations:

Amplitude, Phase, Frequency.

To gain these parameters, we must take advantage of instruments capable of performing the two basic steps in dynamic instrumentation:

Transformation of mechanical vibration data into electrical data

Indication and evaluation of these data for analysis, recording and control.

369

Consequently, this paper is subdivided into three parts
1. DATA ACQUIREMENT
2. DATA EVALUATION
3. DATA CONTROL

The interconnection may best be understood by looking at the diagram showing the data flow, fig.1:

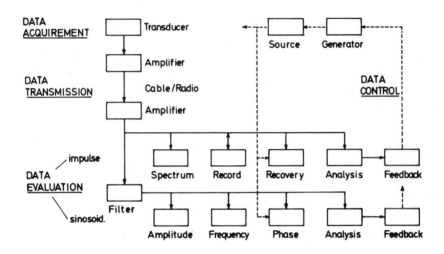

Fig.1   Data flow diagram for measurement of vibrations

At the top level, the mechanical vibration data (displacement, velocity or acceleration) are transformed into electrical data.
All problems of transducer-object-interaction and transducer response have to be dealt with in this part.
At the next level the electrical signal is treated for transmission from the location of the transducer to the location of the evaluation instruments.
At the lower levels, the signals are analyzed, indicated and recorded.
In some cases, the signals are reconnected to the vibration source by means of feedback processes to control the dynamic behavior of the system under investigation.
The various instruments will be dealt with in the resp. sections below.

1. DATA ACQUIREMENT

The first question we have to ask ourselves in the context of vibration measurements is wether we change the dynamic behavior of our object-system by the very procedure of measuring. If in most the answer cases is yes, we have to investigate the

370

possibilities of reducing this effect which immediately leads us to the problems of contact transducers and contactless transducers. Closely related to this problem is the question wether we want to measure the relative vibrations between two objects or the absolute vibrations of one object. If, in the former case, one of the two objects is a fixed reference point, relative to which we want to measure the vibrations of the other object, we of course measure absolute vibrations, depending, however, on the fact that all translations and rotations in the six degrees of freedom of this fixed point must be null.

Contactless Transducers

These are transducers which do not affect the vibrating object by inertia effects due to transducer mass and where other contact forces (or moments) are either zero or neglegible. The following transducers fall in this group:

Capacitive Transducers: The vibrations of the object change an electric field sensed by the transducer.

Magnetic Transducers: The vibrations of the object change a magnetic field sensed by the transducer.

Both types of transducers are basically non-linear, unless the amplitude of the vibration is small in comparison to the field-gap between object and transducer.

Electro-Magnetic Transducers: A carrier-frequency is modulated by the vibrations of the object. The frequency may be in the VHF range in the case of Radar-transducers or in the frequency range of light in the case of optical transducers. We then have the following types of transducers:
Displacement Follower: An electron beam follows the image of a vibrating target on a vidicon.
Laser Holography: The object is illuminated by a pulsed laser beam, which, together with a reference beam "freezes" the vibrations of the object in a stroboscopic hologram. This very interesting technology is described in detail by Woods (1978).

All the above mentioned transducers have in common that they supply vibration data relative to a fixed point, the position of which has to be determined separately. It may be subject to vibrations itself (building vibrations, wind, microseismics etc.), a condition which might be very difficult to control.

Contact Transducers

These are transducers which are directly connected to the object and therefore give rise to inertia effects. They may measure the relative vibrations between two objects or, most commonly, the absolute vibrations of one object. They have the advantage of not depending upon a reference point but the above mentioned disadvantage of inertia effects.

Absolute transducers measure displacement, velocity or acceleration of translations or rotations, respectively. In principle, the absolute vibrations of an object are gained by measuring the relative motion between an inert mass and the casing of the transducer excited by the object.

Transducer Response

The transducer response shall be investigated on the example of a translation transducer, fig.2:

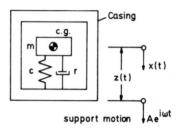

Fig.2   System of absolute dynamic transducer (translation)

where $Ae^{i\omega t}$ = motion of the object, transferred to the transducer-case
$x(t)$ = absolute motion of inert mass m
$z(t)$ = relative motion between inert mass m and transducer-case

The condition of equilibrium at c.g. leads to

$$-m\ \ddot{x}(t)\ -\ r\ \dot{z}(t)\ -\ c\ z(t)\ =\ 0 \tag{1}$$

with $x(t)$ $=\ z(t)\ +\ A\ e^{i\omega t}$
$\dot{x}(t)$ $=\ \dot{z}(t)\ +\ i\omega A\ e^{i\omega t}$
$\ddot{x}(t)$ $=\ \ddot{z}(t)\ -\ \omega^2\ A\ e^{i\omega t}$ $\tag{2}$

Putting equ.2 into equ.1 we find the differential equation

$$z(t)\ +\ r/m\ z(t)\ +\ c/m\ z(t)\ =\ \omega^2\ A\ e^{i\omega t} \tag{3}$$

where the exciter term on the right-hand side is due to the transducer-case motion = object motion. The solution of equ.3 is well known, e.g. Harris and Crede (1961):

$$z(t)\ =\ Z_0\ e^{i(\omega t+\varepsilon)} \tag{4}$$

where

$$Z_0\ =\ A\ \frac{n^2}{\sqrt{(1-n^2)^2+n^2(2\ D)^2}}\ =\ A\ V_4 \tag{5}$$

$$\tan\varepsilon\ =\ -2\ Dn/(1-n^2) \tag{6}$$

with normalized frequency $\eta = \omega/\omega_E$; $\quad \omega_E = \sqrt{c/m}$

$$(7)$$

damping ratio $\qquad D = r/r_c = r/(2\sqrt{cm})$

The graphical representations of equ.5 and equ.6 are given in fig.3a and 3b.

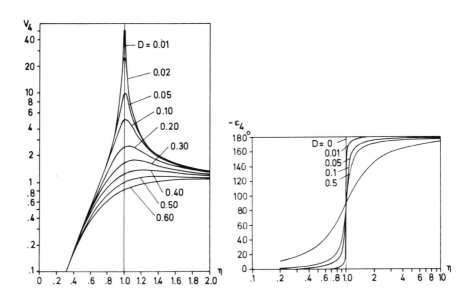

Fig.3a Amplification function V_4 versus normalized frequency η

Fig.3b Phase angle ε_4 versus normalized frequency η

In equ.5, we can distinguish between two cases:

$\eta \ll 1$; the frequency range investigated by the transducer lies very much below the resonance frequency of the inert mass m (high-tuned transducer). The response function V_4 reduces to η^2 and therefore

$$Z_o \approx A\eta^2 \qquad\qquad (8)$$

describing an acceleration transducer.

$\eta \gg 1$; the frequency range investigated by the transducer lies very much above the resonance frequency of the inert mass m (low-tuned transducer). The resonance function V_4 reduces to 1 and therefore

$$Z_o \approx A \qquad\qquad (9)$$

describing a displacement transducer.

373

Signal Distortion

We now have to investigate the effect of the phase-shift ε given by equ.6 on the distortion of the vibration signal.

At the output of the transducer, the image of the original signal to be measured is only correct if all components of the original signal are recombined in their correct sequence, adding only a constant delay time. Only then the relative location of the different spectral parts of the original signal along the time axis remains unchanged, beeing only shifted by the delay time (transfer time in the transducer).

If the delay time is denoted t_0, we find for the original signal and the image signal under those conditions:

$$\text{original:} \quad A(t) = A\, e^{i\omega t}$$
$$\text{image:} \quad z(t) = A\, V_4\, e^{i\omega(t+t_0)}$$
$$z(t) = A\, V_4\, e^{i(\omega t + \omega t_0)}$$

and from equ.4:

$$z(t) = A\, V_4\, e^{i(\omega t + \varepsilon)}$$

resulting in

$$\varepsilon = t_0\, \omega \tag{10}$$

Equ.10 indicates, that signal distortion does not take place if the phase-shift ε is proportional to frequency.

If we relate the criteria "high-tuned" or "low-tuned" trans-ducer of equ.8 and equ.9 to the vibration-signal component with the lowest frequency (the other components beeing e.g. harmonics), we can illustrate the different types of transducers by fig.4.

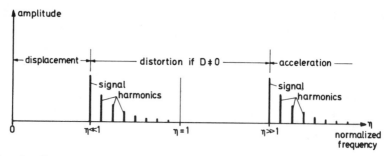

Fig.4 Frequency range of signal (plus harmonics) and dynamic transducer (displacement or acceleration)

In the case of equ.8, equ.6 reduces to

$$\tan\varepsilon \approx -\,2\,D\eta \tag{11}$$

and in the case of equ.9 to

$$\tan\varepsilon \approx 2\,D/\eta \tag{12}$$

If the above criterion $\eta \ll 1$ is assumed to be valid for $\eta = 0.1$

and similarly $\eta \gg 1$ valid for $\eta = 10$, equ.11 and equ.12 reduce to

$$\tan\varepsilon = 0.2 \, D \qquad\qquad (13a)$$

If we further assume that the transducer is critically damped, $D = 1/\sqrt{2}$, we have

$$\tan\varepsilon = 0.1 \sqrt{2} \qquad\qquad (13b)$$

For this value, tan and argument ε are almost equal:

$$\frac{0.1/\sqrt{2}}{\varepsilon} = 1.00663 \qquad\qquad (14)$$

the error beeing approx. 7 ^{o}/oo.
The respective error in equ.8 and equ.9 is then 5.10^{-5}.
This means, that for $D \leqslant 1/\sqrt{2}$ and $\eta \leqslant 0.1$ and $\eta \geqslant 10$ resp. the phase shift is proportional to frequency for acceleration transducers and for displacement transducers.

However, as seen from fig.4, in the case of displacement transducers, the vibration-signal components of higher frequency (harmonics) violate the condition $\eta \ll 1$ and therefore displacement transducers with $D \neq 0$ will always distort a vibration signal consisting of a number of components unless the highest frequency-contribution is still lower than $\eta \ll 1$.
Table 1 summarizes the above results:

Table 1   Signal Distortion

Transducer type	D	ε	signal	phase error
acceleration $\eta \ll 1$	0	0	no distortion	0
acceleration $\eta \leqslant 0.1$	$\leqslant 1/\sqrt{2}$	$t_o\omega$	no distortion	<7 ^{o}/oo
displacement $\eta \gg 1$	0	-180°	no distortion	0
displacement $\eta \gg 1$	$\neq 0$	$\neq t_o\omega$	distortion	$\neq 0$

Transducer-Soil-Interaction TSI

So far we have covered the response of the internal transducer system, assuming that the transducer case motion and the free undisturbed object motion are identical. This may be correct for measuring the vibrations of e.g. machinery, structures or foundations (due to the small mass ratio transducer mass/object-mass). It is certainly not correct if the measurement of free wavefields of the subsoil is concerned.
In this case, the transducer due to inertia effects draws energy from the wavefield and returns it with a phase difference. Thus, the relevant parameters of the wavefield are changed and the measured data are no more those of the freefield.
The interaction relations between a transducer and a wave-field can readily be extended to the interaction between a mass and a multi-degree-of-freedom system (transducer-structure-interaction). The relevant equations will, however, be analyzed for the case of wavefields only.

375

The rigid case of the transducer has 6 degrees of freedom on the soil surface (3 translations, 3 rotations). In each degree of freedom, the inertia mass of the transducer, the spring reaction and the damping reaction of the subsoil form a dynamic system with the resp. resonance curves. The following analysis concerns only one degree of freedom (vertical translation) to clarify the TSI relations.

If the strain-level in the soil due to wave propagation is small, the stress-strain relations become linear and the principle of superposition is valid. Hence, the interaction between the transducer mass m and the free wavefield may be subdivided into two states, following fig.5:

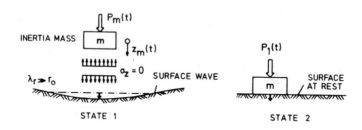

Fig.5 Superposition of two states of transducer on wavefield

In state 1, the transducer mass vibrates synchronous to the motion of the freefield at the soil surface. No dynamic stresses in the contact area are generated. To realize synchronism, a fictitious force $P_m(t)$ is necessary.

In state 2, this fictitious force $P_m(t)$, which does not exist in reality, is compensated by an identical negative force $P_1(t)$. This force $P_1(t)$ acts on the soil surface at rest (the dynamic deformation due to the free wavefield beeing the "zero"-reference).

The real motion of the transducer then results from the vectorial addition of the motions in state 1 and state 2.

In state 1, the motion of the transducer $z_m(t)$ is identical to the motion of the free wavefield $z_0(t)$.

Assuming harmonic motion with angular frequency ω we put

$$z_m(t) = z_0(t) = Z_0 \, e^{i\omega t} \qquad (15)$$

The fictitious force $P_m(t)$ for synchronism is gained from the equilibrium at node m:

$$-m \, \ddot{z}_m(t) + P_m(t) = 0; \quad P_m(t) = -Z_0 \, m \, \omega^2 \, e^{i\omega t} \qquad (16)$$

We find the compensating force $P_1(t)$ from

$$P_m(t) + P_1(t) = 0; \quad P_1(t) = Z_0 \, m \, \omega^2 \, e^{i\omega t} \qquad (17)$$

$P_1(t)$ causes the soil surface in state 2 to vibrate with the motion

$$z_1(t) = Z_1 \; e^{i(\omega t + \varepsilon_1)} \tag{18}$$

where in similarity to equ.4-7

$$Z_1 = Z_0 \; V_4 \tag{19}$$

$$\tan\varepsilon_1 = -2 \; D\eta/(1-\eta^2) \tag{20}$$

We must realize, however, that m, c and r are now the total transducer mass, the soil spring reaction and the soil damping reaction, respectively.

Vectorial addition of the motions in state 1 and state 2 leads to

$$z_2(t) = z_0(t) + z_1(t) = Z_0 \; e^{i\omega t}(1 + V_4 \; e^{i\varepsilon_1})$$
$$= Z_2 \; e^{i(\omega t + \varepsilon_2)} \tag{21}$$

The ratio between the distorted and the freefield amplitude is then

$$Z_2/Z_0 = V_2 = \sqrt{1 + 2\,V_4 \; \cos\varepsilon_1 + V_4{}^2} \tag{22}$$

The phase angle ε_2 is found from

$$\tan\varepsilon_2 = V_4 \; \sin\varepsilon_1 \; / \; (1 + V_4 \; \cos\varepsilon_1) \tag{23}$$

If we put

$$\cos\varepsilon_1 = 1 \; / \; \sqrt{\tan^2 \varepsilon_1 + 1}$$
$$\sin\varepsilon_1 = 1 \; / \; \sqrt{\cotan^2 \varepsilon_1 + 1} \tag{24}$$

into equ.22, we find

$$V_2 = \sqrt{1 + 2\,V_4 \; / \; \sqrt{[2\,D\,/(1-\eta^2)]^2 + 1} \; + V_4{}^2}$$

and inserting V_4 from equ.5 we arrive at

$$V_2 = Z_2/Z_0 = \sqrt{\frac{1 + (2\,D\eta)^2}{(1-\eta^2)^2 + (2\,D\eta)^2}} \tag{25}$$

In the same way the phase angle ε_2 can be determined by putting equ.24 into equ.23:

$$\tan\varepsilon_2 = \frac{V_4 \; / \; \sqrt{[\,(1-\eta^2)/2\,D\eta]^2 + 1}}{1 + V_4 \; / \; \sqrt{[2\,D\eta/(1-\eta^2)]^2 + 1}}$$

Inserting equ.5 we find

$$\tan\varepsilon_2 = \frac{-2\,D\eta^3}{1-\eta^2 + (2\,D\eta)^2} \tag{26}$$

Equations 25 and 26 describe the amplitude- and phase-distortion of the freefield motion due to the presence of the inertia mass of the vibration transducer. Fig.6 shows the response curves equ.25 and equ.26.

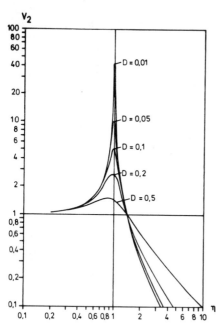

Fig.6a Amplification function V_2 versus normalized frequency η

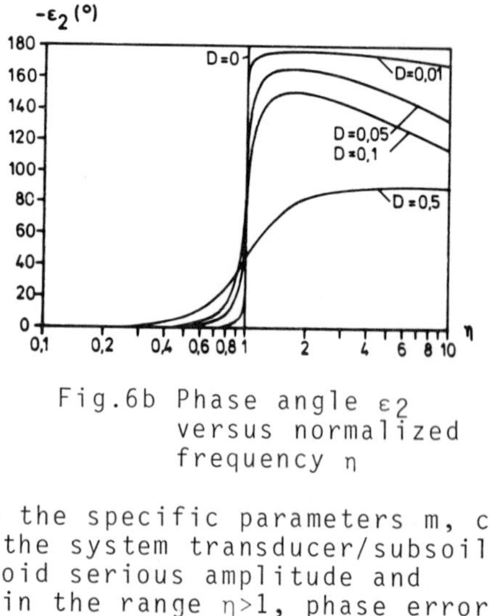

Fig.6b Phase angle ε_2 versus normalized frequency η

It is seen from fig.6 that unless the specific parameters m, c and r controlling STI are known, the system transducer/subsoil should be high-tuned ($\eta \ll 1$) to avoid serious amplitude and phase distortion. Depending on D in the range $\eta > 1$, phase errors may be introduced by STI to such an extend, that the freefield motion e.g. is downward, when the transducer indication is al-most upward, the worst possible error.

We could have arrived at the same results if we had looked at the system like fig.7.

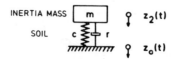

Fig.7 Support excitation of lumped parameter system

378

m, c and r are the lumped parameters of the system transducer/ subsoil, the excitation beeing the support motion $z_0(t)$ resulting in the motion $z_2(t)$ of the inertia mass:

$$z_1(t) = z_2(t) - z_0(t) \tag{27}$$

Equilibrium around node 2 yields

$$-m\,\ddot{z}_2(t) - r\,\dot{z}_1(t) - c\,z_1(t) = 0 \tag{28}$$

From equ.27 we find

$$z_2(t) = z_1(t) - Z_0\,\omega^2\,e^{i\omega t} \tag{29}$$

Putting equ.29 into equ.28 we find

$$m\,\ddot{z}_1(t) + r\,\dot{z}_1(t) + c\,z_1(t) = Z_0\,m\,\omega^2\,e^{i\omega t} \tag{30}$$

the solution beeing identical to equ.19 and equ.20. Vectorial addition of $z_0(t)$ and $z_1(t)$ leads to the already known solutions of equ.25 and equ.26.

Transducer technology

Two transducers representing the groups of displacement and acceleration transducers shall be presented here which are most commonly in use:

The electro-dynamic transducer, fig.8, consists of a coil moving in the gap of a permanent magnet N-S. The coil body m is supported by guide springs c, the resistor R across the voltage output functions as the damping coefficient r. Thus, we find the values m, c, r of fig.2 again in fig.8:

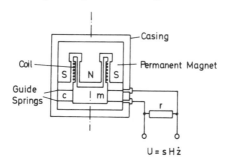

Fig.8 System of electro-dynamic vibration transducer

Due to induction, the output voltage U(t) is

$$U(t) = s\,H\,\dot{z}(t) \tag{27}$$

where H = Intensity of magnetic field
      s = geometric coil constant

For $\eta \gg 1$, the displacement transducer by equ.27 is transformed
into a velocity transducer. The damping ratio D governing equ.5
and equ.6 can easily be selected by chosing the proper value R
of the resistor across the output (current I_R proportional to
velocity, see equ.27 and current analogy below). For real-time
measurements, the velocity voltage U(t) may be integrated or
differentiated by electronic circuits to gain displacement or
acceleration, respectively.

Most geophones are based on this system, the output sensitivity (and the impedance) beeing governed by the number of
windings on the coil.

The piezo-electric transducer consists of a crystal of active
piezo-electric material (e.g. quartz) subjected to compression
or shear due to the vibrations, fig.9.

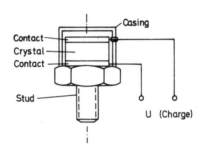

Fig.9 System of piezo-electric vibration transducer

In the undeformed state, the charges of the spatial molecule
structure balance each other (e.g. quartz SiO_2; $S_i/4+$, $O/2-$).
The deformation of the molecular structure gives rise to sur-
charges on the contacts resulting in a voltage U across the
output. The parameters m, c and r of fig.2 are represented by
the continuous mass and elasticity of the crystal together with
the mechanical damping properties of the crystal fixture. The
resonance frequency of the crystal (e.g. height $\leqslant$ quarter wavelength in fig.9) is usually very high, hence the piezo-electric
transducer in most cases is an accelerometer ($\eta \ll 1$). The charge
sensitivity is governed by the arrangement of the crystals and
their piezo-electric constants.

Both the electro-dynamic and the piezo-electric transducers
are active transducers. They generate an electric power which
is drawn from the object in the form of mechanical power.
Consequently, they can not operate at f = 0, because a static
displacement cannot produce power continously. This is reflected by the fact that the output of velocity or acceleration
transducers is zero for f = 0.

2. DATA EVALUATION

We now leave the top level fig.1 and focus our attention to the
further treatment of the signal in the course of evaluation.

380

Amplifier

The voltages (or charges) at the output are usually very small
and have to be amplified. Furthermore, the impedances of the
transducer and the transmission line have to be matched. Consequently, the first amplifier is usually placed close to the
transducer, sometimes even integrated into the transducer itself. There are a number of specifications we have to observe:

Signal distortion: The input signal will not be distorted at
    the output if the amplifier adds only a constant delay time
    (see equ.10).
Frequency response: The amplifier response must be flat within
    the frequency range investigated.
Overload: The input signal must not exceed the maximum input
    level, otherwise the output signal will be heavily distorted
    due to nonlinear amplification.
Interference: Common mode rejection and signal/noise ratio shall
    be large to avoid other than the amplified input signals to
    appear at the output.
The same considerations hold for the amplifier at the end of the
transmission line (amplification/impedance matching).

    In the lower levels of fig.1 we distinguish between the
evaluation and analysis of
nonperiodic impulse signals (stationary, nonstationary;single
    or multiple pulses; stochastic pulse sequence, noise etc.)
    and
periodic sinosoidal signals (monochromatic or multichromatic),
though the boundary between the two cannot be defined exactly
as far as instrumentation is concerned.

NON-PERIODIC IMPULSE-SIGNALS

Recovery

Signal recovery is achieved by sampling the signal elements and
averaging the samples over a prolonged period of time. Thus,
noise contributions will decrease with increasing sampling time
improving the signal to noise ratio. Fig.10 illustrates the
recovery procedure:

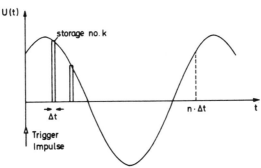

Fig.10 Sampling and storage/average-procedure in signal
        recovery operation

381

The periodic signal is subdivided into n elements of duration
Δt and stored in the resp. storages K = 1...n. The content of
the storage is averaged with the new sample every sampling cycle. The improvement of the signal/noise ratio is $\sqrt{N}$, N beeing
the number of sampling cycles.

The sampling cycle is initiated by a trigger impulse which
must be derived from the signal source. Consequently, recovery
mode operation is only possible in the case of vibrations
excited artifically and only if such vibrations are caused by
recurrent events (yielding a trigger impulse).

Spectrum

To study the particular characteristics of the vibration signals,
the different frequency-contributions and their respective phase relation must be analyzed. In case of the 1/3 octave-spectrum,
the interesting frequency range is divided into sections representing 1/3 of the octave:

octave = frequency ratio 2

1/3 octave (terz) = frequency ratio $(\sqrt[12]{2})^4 = \sqrt[3]{2} = 1.2599$

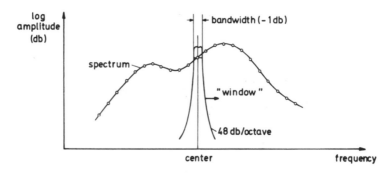

Fig.11 Measurement of signal spectrum by bandpass filtering

Fig.11 illustrates the operation of a 1/3 octave filter having
a bandwidth of .2599 of the center (terz) frequency. If this
"window" is drawn across the interesting frequency range, the
signal contributions falling within the bandwidth are gained
from the output.

By means of more sophisticated analysis instruments (see
below) the real and imaginary parts of the spectrum (vector
components) or the resp. vector amplitude- and phase-spectrum
can be measured. This allows the calculation of the response of
a complex system to an non-sinosoidal excitation.

The spectrum analysis may concern the response spectrum of
systems or the spectrum of the vibration source. In the latter
case it is in fact the finger print of the vibration source
(e.g. train vibrations, traffic, vibrating machines etc.).

Recording

The purpose of recording is threefold:
   analysis of vibration data
   repetition of vibration tests
   documentation

   In the case of analysis (qualitative or quantitative), we
may employ the following methods:
Storage Oscilloscope
   Signal storage for some hours; split-screen for comparison
   of different signals
Oscilloscope Photography
   for evaluation and documentation
Light Recording
   Time axis represented by variable speed of moving paper;
   single- or multi-trace generated by light spots (UV-light)
   deflected by galvanometers. The total paper width is
   available for each trace. Small galvanometer mass allows
   high upper limiting frequencies.
Pen Recording
   Time axis as above, single- or multi-trace by electro-
   magnetic or electro-dynamic drive of pens. Usually no overlap of traces is possible; due to relatively large inertia
   forces the upper limiting frequency is generally low.
Level Recording
   Here instead of the actual signal its level is recorded,
   usually by pen-recorders. The level may consider peak,
   average of effective values and may be recorded as linear
   or logarithmic level (decibels). The levels depend on the
   time-constant of the recording instrument, hence the lower
   and upper limiting frequency of the signal, the writing
   speed of the pen and the damping characteristics of the recording system have to be considered. If the level is known,
   it can be compared with the maximum amplitudes of single
   events in the signal recorded e.g. by a light recorder and
   the resultant crest-factor can be calculated.

   All above records serve the purpose of documentation as well,
provided all relevant test data are preserved on the record.
   In the case of repetition of vibration tests, magnetic tape
recorders are used exclusively. The great advantage is, that the
data can be reproduced as often as desired. The evaluation can
be performed versus time by the recording instruments mentioned
above: a = f(t). In addition, the data may be evaluated in
dependency upon each other: a = f(b,c,d...). Multi-channel recording preserves the correct phase relation of the different
data and allows to register reference time signals as well as
comments of the operator concerning the test situation.
To record low frequency or static data, FM or PCM modulation
is to be preferred.

Analysis

Mathematical operations of vibration data analysis can be performed by advanced electronic analog/digital analyzers. To

mention some of the operations necessary in system analysis, the analyzers may be capable to compute and graphically present the following functions:

Auto and Cross Correlation Functions
Impulse Response, Weight Function
Amplitude Histogram, Amplitude Probability Density
Auto and Cross Power Spectrum
Power and Energy Spectral Density
Transfer Function
Coherence Function
Signal Recovery (in the case of recurrent events)

Correlation methods in particular are a very powerfull tool in soil dynamics, Roesler (1978). They enable the measurement of vibrations at extremely low signal/noise ratios (in some cases 1:100) and allow to distinguish between the different wave types arriving at the transducer. Dispersion effects can be determined by exciting wavefields with narrow-band noise.

In addition to the presentation on a CRT-screen, the results may be plotted on a x-y-plotter or registered on tape for further evaluation on a computer.

An other feature of analysis by electronic instruments is the possibility of inserting a feedback loop into the vibration system under consideration. This allows the automatic control of certain dynamic conditions of the system in real time. An example is the inertia compensation of vibration transducers mentioned below.

PERIODIC SINOSOIDAL SIGNALS

Filter

To suppress undesired signals like transients, noise or inter-ference (mains, machinery etc.), a filter is inserted between the incoming signal and the evaluation instruments. It may be a low pass, high pass or band pass (or band rejection) filter depending on the frequency range of the interference and wether the vibration signal is monochromatic or multichromatic. A possible amplitude- and phase-distortion of the vibration-signal due to the filter must be taken into account.

The following instruments measure the basic properties of a sinosoidal signal: amplitude, phase, frequency:

Amplitude

Either by analog or digital instruments the amplitude is measur-ed, usually in the form of the rms-value beeing $1/\sqrt{2}$ of the peak value for a sinosoidal signal. Attention should be paid to the frequency response of the rms- (or peak-)rectifier with respect to the signal frequency.

Phase

To measure the phase angle, we need a reference signal derived either from the vibration source or from a reference time signal.

Besides analog or digital instruments we have two possibilities to establish the phase relation:

Vector Measurement

In this case, the real and imaginary component of the vibration vector is measured and the phase angle determined from tan ε = IM/RE. The vector components and the phase angle is indicated including the distortion due to an interference signal still present at the input.

Phase Measurement

In this case, the vibration signal is directly compared with a reference signal with respect to phase difference. This can be achieved on an oscilloscope by triggering the vibration signal by the reference signal or by displaying both signals. The phase angle may then be read as time difference on the screen. The accuracy may be of the order of 1 %, depending on the interference (instable zero intersection). We may also employ the z-coordinate of the oscilloscope using a light spot on the screen positioned upon the vibration signal by an adjustable time delay.

The highest accuracy can be achieved by presenting the vibration signals as a function of the reference signal, thus obtaining a Lissajous-pattern. If the phase of the reference signal is adjusted such that the Lissajous-pattern becomes a straight line (e.g. inclination 45°), the accuracy of the phase measurement may be of the order of 0.5° (1°/oo) even in the presence of some interference signals. This is due to the fact that the human eye is very sensitive to the symmetry of the Lissajous-pattern (beeing maximum for ε = 0). The reference signal is taken from a variable phase generator which may control the vibration source as well. If access to the vibration source is not available, the variable phase generator must be phase-locked to a separate transducer signal, to which all other data are than compared with respect to phase.

Phase measurements are essential e.g. in problems of
  wavefields
  resonance corves
  energy dissipation
where the accuracy of the analysis depends much more on phase than on amplitude.

Frequency

The frequency is usually read from digital frequency meters, presenting the data directly in Hz. This implies, that we need a duration of the measurement of 1 sec (Hz = no. of cycles per second). The measurement of low frequencies would therefore result in a very low accuracy because only a few digits would be indicated with an error of $\pm$ 1 digit. In this case, period times should be measured with $\overline{a}$ much greater accuracy (reading in ms or μs).

385

Recording

The recording procedures are similar to the ones mentioned for non-periodic impulse signals with all the considerations above.

Analysis

If we switch off the filter, periodic sinosoidal signals may be studied by means of analysis instruments resulting in e.g.:
    Signal Recovery
    Fourier Analysis
    Resonance Curves, Nyquist Diagram
    Transfer Function
    Auto and Cross Power spectrum, Power and Energy Spectral-
    Density in energy and hysteresis problems
    Auto- and Cross Correlation Functions
Again, it is possible to insert a feedback loop into the vibration system, connecting the output of an analyzer with the vibration source, thus obtaining a certain dynamic condition of the system.

3. DATA CONTROL

The possibilities of data control in the form of a feedback loop within the dynamic system under consideration shall be illustrated on the example of the electronic inertia compensation of vibration transducers.

TRANSDUCER INERTIA COMPENSATION

Three possibilities are available to compensate inertia effects besides the actual calculation of the TSI-functions described by equ.25 and equ.26 and fig.6:

Analog Computer

In this case, we take advantage of the well-known electrical current-analogy for vibrational systems, Harris and Crede (1961), where we find the electrical circuit analogous to fig.7 in fig.12b:

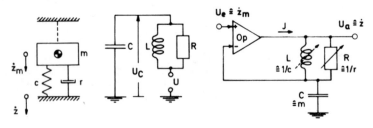

Fig.12 Electro-mechanical equivalent: a) mechanical system
        b) electrical analog system   c) analog computer

386

The capacitor C represents the transducer mass m, the inductance L represents the inverse spring constant 1/c and the resistor R the inverse damping constant 1/r. At all points of the system, currents are analogous to forces and voltages to velocities. If the model factors are chosen for real-time, the analog computer in fig.12c provides the inverse functions of equ.25 and equ.26 as follows:

The velocity signal $\dot{z}_m \cong U_e$ of the transducer motion on the object (wavefield) is fed into the non-inverting input of an operational amplifier OP. The output currents into L and R represent the coupling forces through c and r to the motion of the free object (freefield) in the mechanical analogon (trans-ducer). The voltage U_c across the capacitor C must therefore be analogous to the velocity of the transducer mass. U_c is fed into the inverting input of the operational amplifier OP. If $U_c \neq U_e$, the output current I will immediately change until the differen-ce becomes zero. The output voltage U_a is than analogous to the velocity $\dot{z}$ of the free object (freefield). The same holds true for the respective accelerations. L and R must be set to represent the respective object (or soil) data c and r. For calibration, see Prange (1978).

The feedback loop in the case of the analog computer is realized electronically by the inverting input of the operatio-nal amplifier; in the following examples it will be a mechanical feedback.

Manual Feedback Control

In the case of monochromatic sinosoidal vibrations, the ficti-tious force $P_m(t)$ in eq.16 can be realized by placing an auxiliary vibrator on top of the vibration transducer. Thus, synchronism between the free object (soil) motion and trans-ducer motion can be achieved avoiding the contact stresses due to inertia effects. The combined system transducer/vibrator is shown in fig.13.

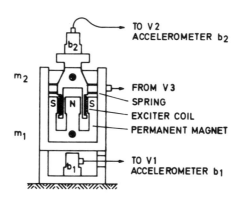

Fig.13 Inertia compensation: Combined Transducer/Vibrator

The auxiliary vibrator consists of an electro-dynamic system generating the force $P_m(t)$ by the acceleration of the top mass m_2, measured by accelerometer b_2. If no contact forces are present in the contact between transducer and object, the system transducer/vibrator is a free system with the center of gravity at rest.
If the mass ratio is denoted

$$m_1/m_2 = a \qquad\qquad (28)$$

we find the accelerations of the free system to be

$$b_1/b_2 = -1/a \qquad\qquad (29)$$

The manual feedback control therefore must be capable to regulate the current through the electro-dynamic auxiliary vibrator in such a way that the acceleration b_2 is a-times greater than b_1 and the phase difference between the two is always 180°. The electronic circuit to achieve this condition is found in fig.14.

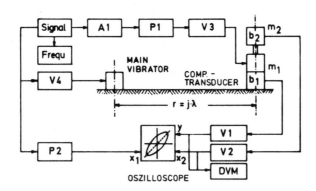

Fig.14 Electronic circuit for manual control of transducer
       inertia compensation

The sinosoidal signal is generated in a signal generator and amplified by V4. It drives the main vibrator exciting the object vibrations. The same generator signal in controlled manually in amplitude by A1 and in phase by P1 and then amplified by V3. The current output of V3 drives the exciter coil of the auxiliary vibrator, generating the force $P_m(t)$. Both acceleration signals b_1 and b_2 are amplified by amplifiers V1 and V2, the amplification ratio being a. The output voltages are connected to the x- and y-input of an oscilloscope to obtain a Lissajous-pattern, which in the case of full inertia compensation is a straight line of 45°, as indicated by equ.29.
   Amplitude, phase and frequency of the free object (free-field) vibrations are then measured as usual, e.g. by oscilloscope, variable phase generator P2 and digital frequency meter.

Automatic Feedback Control

The manual procedure of above can be performed automatically, if we write equ.29 in the form

$$a\, b_1 + b_2 = 0 \qquad (30)$$

The electronic circuit to achieve the condition of equ.30 is given by fig.15.

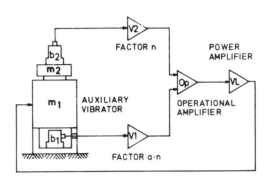

Fig.15   Automatic feedback control for transducer inertia compensation

The output voltages of the acceleration transducers b1 and b_2 are fed into amplifiers V1 and V2, the amplification ratio again beeing a. Any deviation from equ.30 will instantaneously produce an output voltage of the operational amplifier OP, resulting in an output current of the power amplifier VL, driving the exciter coil of the auxiliary vibrator. The resultant effect is, that the center of gravity of the system transducer/vibrator remains at rest and the transducer vibrates in synchronism with the free object.
   The advantage of the automatic feedback control is the application to non-sinosoidal impulse vibrations within the upper limiting frequency range of the electro-mechanical feedback system.

CONCLUSIONS

The measurement of vibrations involves in most cases inertia effects in transducers as well as in the combined system transducer/object. Hence, the dynamic response of the transducer and the system transducer/object must be considered to obtain the correct vibration data. This can be achieved by applying the inverse complex response functions to the measured data or by compensating the inertia effects by electronic feedback circuits. In the former case, all distortion contributions (amplitude and phase) of the instruments involved in the data flow must be considered.

389

In the latter case, care must be taken to avoid instability in the feedback loop of the control system.

Advanced electronic analysis equipment enables the computation of dynamic system parameters in real time by comparing the dynamic input and output of the system.

REFERENCES

Harris, C.M. and Crede, C.E. 1961, Shock and Vibration Handbook, McGraw-Hill, New York
Prange, B. 1977, Inertia Compensation of Vibration Transducers. Proceedings I.C.S.M.F.E. Tokyo
Prange, B. 1978, On the Interaction between a Body and the Wavefield. Publ. Institute Soil and Rock Mech., University of Karlsruhe
Prange, B. 1978a, Primary and Secondary Interferences in Wavefields. Proc. Dynamical Methods in Soil and Rock Mechanics, DMSR 77; Balkema, Rotterdam
Richart, F.E., Hall, I.R. and Woods, R.D. 1970, Vibrations of Soils and Foundations, Prentice Hall Inc., Englewood Cliffs, N.J.
Roesler, S.K. 1978, Correlation Methods in Soil Dynamics, Proc. Dynamical Methods in Soil and Rock Mechanics, DMSR 77, Balkema, Rotterdam
Woods, R.D. 1978: Holographic Interferometry in Soil Dynamics, Proc. Dynamical Methods in Soil and Rock Mechanics, DMSR 77, Balkema, Rotterdam

Proceedings of DMSR 77 / Karlsruhe / 5-16 September 1977 / Volume 1

Holographic interferometry in soil dynamics

RICHARD D. WOODS
University of Michigan, Ann Arbor, Mich., USA

SYNOPSIS

The relatively new technique of stroboscopic, holographic interferometry has been used to study the isolation effectiveness of cylindrical hole barriers and slurry filled trench
barriers. Also, the effectiveness of trench barriers in
screening waves generated by embedded footings was studied.
The holographic interferometry technique was extended to obtain two vectors of particle motion at the surface of the half
space. Rows of cylindrical holes and slurry filled trenches
were both found to be effective wave barriers if properly pro-
portioned, but trench barriers are not effective for waves
generated by footings embedded more than about 1/3 wavelength
of the Rayleigh Wave.

1 INTRODUCTION

Holography is a kind of photography which records wave fronts
rather than focused images. By coupling holography with the
well known principles of interferometry, it is possible to
easily obtain "contour maps" of the surface displacements of
objects of any shape for either static or dynamic deformations.
The combined technique is called "holographic interferometry"
or "hologram interferometry".

Holographic interferometry has been used often to study deformation of solid objects (Ennos, et al, 1968, Powell and
Stetson, 1965, Sampson, 1970) and as described herein and in
Woods and Sagesser (1973) and Woods, et al, (1974) has been
extended to study static and dynamic deformations of a sand
half-space model. In static tests the settlements around
footings of various shapes were studied. In dynamic tests the
seismic wave screening effectiveness of several types of
barriers was studied.

2 HOLOGRAPHIC INTERFEROMETRY

2.1 Holography

Holography is often called "lensless" photography or photography by wavefront reconstruction because an entire wave-

front rather than a focused image is captured in the photographic emulsion, Leith and Upatnieks (1965). To make a hologram, a coherent light beam, produced by a laser, is divided
into an object beam and a reference beam by a beam splitter,
Fig. 1a. The object beam OB is reflected from the object onto
a photographic plate where it intersects the reference beam
RB, and a standing wave interference pattern is formed. The
developed photographic plate is the hologram. The captured
wavefront interference pattern is a complex pattern of small,
light and dark zones which show no obvious relationship to the
object when viewed in normal light. However, if this hologram
is illuminated only by the reference beam under the same conditions as the original exposure, Fig. 1b, the hologram will
reconstruct the original object beam and a virtual image of
the object can be observed by looking through the hologram.

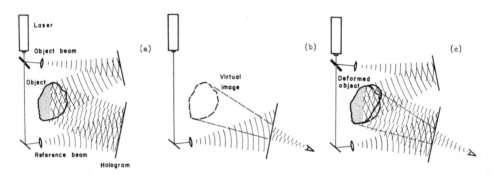

Fig. 1 - Making a hologram and interferogram. (from Woods,
et al, 1974).

2.2 Holographic Interferometry

Using holography, it is possible to store a wavefront and at
any time reconstruct that wavefront. This stored wavefront
can be reconstructed in the original optical setup and be
caused to interfere with a new wavefront from the same object
which has subsequently been deformed, Fig. 1c. Now the wavefront reconstructed from the hologram will not exactly match
the current wavefront from the deformed object and the interference of these slightly different wavefronts will cause dark
and bright fringes to appear on or near the object. The
fringes are essentially contours of equal displacement and may
be viewed by eye or photographed by conventional photography.
A photograph of a hologram on which a fringe pattern exists is
called an interferogram. The interferogram produced by the
above process is called a real-time interferogram. If the object was deformed only in the elastic range, the fringes will
disappear if the force causing deformation is removed.
   With only a slight modification of the sequence of operations, the above procedure can yield another type of interferogram. An exposure of the photographic plate can be made with
conditions as in Fig. 1a, however, the plate is not developed.

392

Then the object can be deformed as in Fig. 1c and the same plate re-exposed. Now, when the plate is developed, both wavefront interference patterns, undeformed and deformed, are stored in the double exposure hologram. The double exposure hologram can be illuminated by the reference beam as in Fig. 1b and an observer can see a virtual image of the object with fringes representing contours of equal displacement superimposed. A photograph of the double exposure hologram is also an interferogram.

The preceding described the process of obtaining interferograms of statically deformed objects. Several possibilities exist for making interferograms of vibrating objects as described by Ennos and Archbold (1968), Power and Stetson (1965), Sampson (1970, 1971), and Brooks, Heflinger and Wuerker (1966). For the dynamic tests reported here, the double exposure, stroboscopic technique as described by Brooks et al (1966) was used. In this technique the object of interest is stroboscopically illuminated with a pulsed laser at the same frequency at which it is excited. The stroboscopic technique was coupled with the double exposure technique to capture "stopped motion" contours of traveling waves.

2.3 Interpretation of Interference Fringes

A fringe is generated whenever a particular point on an object is displaced by about one-half the wavelength of the laser light in the direction of a "sensitivity vector". The "sensitivity vector" as used here is a slight extension of the concepts presented in Orr, Tehon and Barnett (1968). Every point in the field of view in an interferogram has associated with it a sensitivity vector which is a function of the geometry of the optical setup and the wavelength of the laser light. The sensitivity vector has a direction normal to the undeformed surface and the dot product of this sensitivity vector with the point displacement vector prescribes the number of fringes which will be observed. If the optics are arranged appropriately, the fringes represent lines of equal vertical displacement; therefore, a hologram or interferogram may be read like a contour map with a contour interval of about 1/2 the wavelength of the laser light. With other arrangements of the optics, displacements in other directions can be obtained as will be described subsequently.

The laser used in these studies was a pulsed Argon laser operating at 5.145×10^{-4} mm wavelength. This gave a contour interval of about 2.5×10^{-4} mm.

3 MODEL HALF-SPACE AND STATIC TESTS

A model half-space was prepared by showering dry Agsco No. 16 fine sand from a height of about one meter into a box 100 cm x 100 cm x 30 cm. This technique produced a sand with a void ratio of 0.71 and a friction angle of 38 degrees. The sand was moistened to a water content of about 10% to develop apparent cohesion. The apparent cohesion permitted excavation of trenches or open holes without support, and also permitted upward particle accelerations near the vibrating footing to

Fig. 2 - Model half-space
and holography apparatus
(from Woods, et al, 1974).

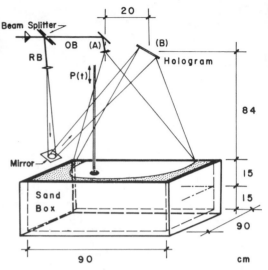

Fig. 3 - Schematic of half-space
model and holography apparatus (from
Woods and Sägesser, 1973).

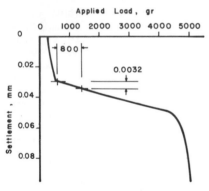

a) load-displacement curve

b) Interferogram for load increment of 1/5 ultimate

Fig. 4 - Static footing load test (from Woods, et al, 1974).

exceed 1g. The model half-space box can be seen in Fig. 2.
Figure 3 shows schematically the optics and hologram plate
arrangements for the half-space model of Fig. 2.

The central cylindrical zone (90 cm dia. x 30 cm deep) of
the sand box was left in the dense condition, but the peri-
meter zone between the central cylinder and the square box was

loosened. The loosened zone proved to be an effective energy
absorber for the steady state tests.

The load-deflection curves for statically loaded model
footings resting on sand were obtained. Vertical displacements
of the footings were measured with a dial gage and the applied
load was measured with a load cell. Simultaneously, the vertical displacement of the footing and the surrounding sand was
obtained using static, real-time holographic interferometry.

A load-displacement curve for an actual load test to failure is shown in Fig. 4a. Figure 4b is a double exposure hologram showing the displacement under a load increment of 800 gm
(7.86 N), about 1/5 of the ultimate load. The nearly parallel
fringes on the footing in Fig. 4b represent rigid body rotation of the footing while the concentric fringes around the
footing represent contours of ground surface displacement.
The fringe numbers of the upper and lower edges of the footing
are 12 and 14 respectively. The average settlement, Δ, is:

$$\Delta = 13(2.5 \times 10^{-4})\text{mm} = 0.0032 \text{ mm.}$$

4   ISOLATION BY BARRIERS

The effectiveness of open trench barriers in reducing the amplitude of ground transmitted vibrations has been reported by
Woods (1968), Dolling (1970) and Barkan (1962). These previous studies showed that the effectiveness of open trenches
was directly related to the dimensionless depth of the trench,
i.e., the depth (H) divided by the wavelength of the Rayleigh
wave (λ_R). The minimum effective H/λ_R is about 0.67.

For Rayleigh waves of long wavelengths, a trench of sufficient depth to be effective may be difficult to provide both
technically and economically. Two potential alternatives to
the open trench have been studied using holographic interferometry. One alternative is a row or rows of open or thinly
lined cylindrical holes, and the other is a heavy fluid filled
barrier.

4.1   Rows of Cylindrical Holes

Figure 5 shows schematically the configuration of a barrier
composed of a row of open cylindrical holes. A series of 40
open cylinder tests were performed in the model half-space
facility to study the geometrical parameters influencing their
effectiveness as wave barriers. The critical parameters for
the cylindrical hole barriers are shown in Fig. 5; R is the
distance of the barrier from the source, L is the length of
the barrier, H is the depth of the barrier, D is the diameter
of each cylindrical hole, and S is the center to center spacing of the holes.

After each test with rows of holes, the holes were connected by removing the separation walls to make a trench. The
results of the trench tests could then be used to correlate

with previous studies by Woods (1968) and Dolling (1970). These investigators found that the effectiveness of a trench was significantly influenced by its depth, H. For depths less than about 2/3 wavelength, the effectiveness was minimal and for depths greater than about 1 1/4 wavelengths, little additional effectiveness was achieved.

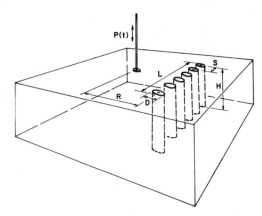

Fig. 5 - Schematic of cylindrical hole barrier (from Woods, et al, 1974).

The amplitude reduction factor for <u>depth</u> <u>only</u> as a variable, as reported by Dolling (1970), is represented by F_A'' and is defined as:

$$F_A'' = \frac{\text{amp w/trench barrier}}{\text{amp without barrier}}$$

Figure 6 shows F_A'' as a function of H/λ_R.

To eliminate depth as a variable in the tests of rows of cylindrical holes, an H/λ_R of about 1.4 was used. To eliminate barrier length as a variable, an L/λ_R of about 2.5 as suggested by Woods (1968) was used. The preceding constraints eliminated

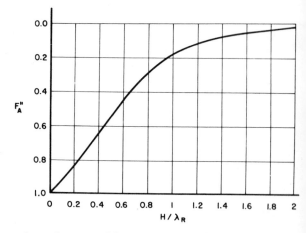

Fig. 6 - Amplitude reduction factor for depth, F_A'', (After Dolling, 1970, for $\nu = 1/3$).

all variables except D, S and R. The distance to the barrier, R, was limited by the size of the model half-space and was also held constant, however, R/λ_R varied slightly because various excitation frequencies and consequently various λ_R's were used.

To determine the effectiveness of a particular barrier using the double exposure holographic technique, two double exposure holograms were required. The first double exposure was made with no barrier, i.e., unobstructed half-space. An exposure was made with the half-space at rest, then a second exposure was made with the half-space being excited by a vertical harmonic force applied to a circular footing on the surface. The laser was pulsed stroboscopically with the exciting force. The developed double exposure hologram resulting from

396

this technique contained concentric fringes representing "stopped", traveling waves decaying with distance from the source. After excavating a barrier (row of holes), another double exposure hologram was made, one exposure static, one exposure with the circular footing being vibrated. This double exposure hologram showed the alteration of the traveling wave pattern caused by the barrier.

Each pair of interferograms made as described above and shown in Figs. 7 and 8 illustrate the screening effect of a barrier composed of a row of cylindrical holes. The left interferogram of each figure shows the free-field, unobstructed surface displacement pattern generated by a circular footing excited at about 1400 Hz. The right interferogram shows the area and extent of surface wave screening provided by a barrier consisting of cylindrical holes. The right interferogram clearly shows the diffraction around the corners and the wave interference caused by the reflected portion of the energy. At some points the wave interference results in an increase in amplitude and in some shortening of the wavelength.

Because the half-space model was not perfect, reflections off the bottom and sides occurred and the fringe pattern was not perfect. This resulted in a "wood-grain" appearance as in Figs. 7 and 8. Therefore, the numerical evaluation of barrier effectiveness was done by obtaining an average effect from several lines in a sector $\pm$ 15 degrees both sides of an axis through the source and perpendicular to the barrier. The numerical analysis consisted of plotting diagrams of amplitude vs. distance from the source for <u>before</u> barrier and <u>after</u> barrier conditions. The ratio of <u>after</u> barrier amplitude to <u>before</u> barrier amplitude in the region beyond the barrier was called the amplitude reduction factor, F_A', for parameters S and D.

The net space available for energy to penetrate the barrier (S-D) (see upper right corner of Fig. 9) was found to be the key spacing parameter. Figure 9 shows the relationships between net scaled spacing, $(S-D)/\lambda_R$, scaled diameter, D/λ_R and a factor called isolation "Effectiveness," $(1 - F_A')$. An Effectiveness of 1 indicates that the barrier stops all Rayleigh wave energy, and an Effectiveness of 0 indicates all energy passes the barrier. For a trench of the same depth as the holes, $H/\lambda_R = 1.4$, the Effectiveness according to Dolling, Fig. 6, would be about 0.96. This is shown as the point on the Effectiveness axis, Fig. 9.

It was found that for barriers with hole diameters between 1/3 and 1/6 wavelength, no difference in isolation effectiveness was observed for constant net scaled spacing. But, for hole diameters less than 1/6 wavelength, the effectiveness drops off quickly at all hole spacings.

The shaded zone in Fig. 9 was well established, however, the dashed lines for D/λ_R < 0.17 are speculative and are based on only a few tests. As a practical matter, it was very difficult to excavate closely spaced, small diameter holes without soil failures or interconnecting the holes. Therefore, only a few successful tests of that type were accomplished. The general trend from all tests however, showed that as D/λ_R decreases, $(S-D)/\lambda_R$ must also decrease to accomplish satisfactory isolation.

4.2 Multi-row Barriers

If, because of certain limitations in the field, cylindrical
holes cannot be spaced closely enough and/or the depth is not
optimal, the effectiveness of a barrier may be insufficient.
A logical question then, is whether more than one row of holes
would provide any improvement, and tests were performed to
evaluate this proposition.

All barriers with a second row were more effective. In
some tests the amplitudes of the wavefronts were simply reduced further, while in others wavefronts beyond the barrier
no longer existed but an erratic pattern caused by construc-
tive and destructive interference resulted. Figure 10 is an
example where the spacing between rows of cylindrical holes is
2D. Some improvement in effectiveness is demonstrated by the
double row but the same improvement might also be achieved by
closer spacing of the holes in the single row.

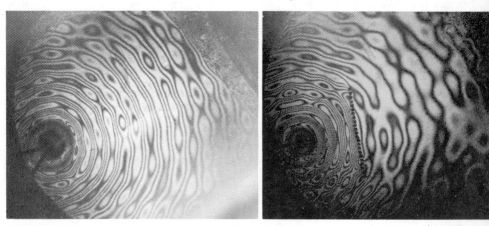

Fig. 7 - Comparative Pair: left without barrier, right with
barrier (after Woods, et al, 1974).

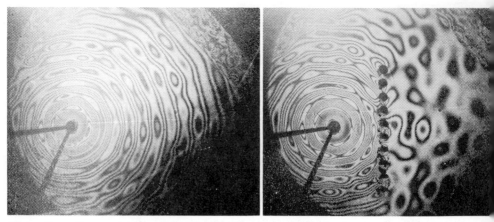

Fig. 8 - Comparative Pair: left without barrier, right with
barrier (after Woods, et al, 1974).

398

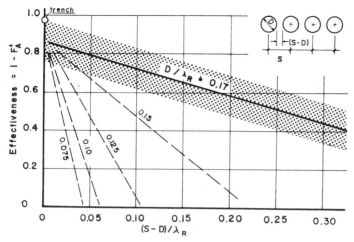

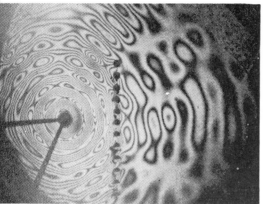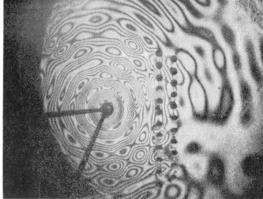

Fig. 9 - Isolation effectiveness as a function of hole diameter and spacing (from Woods, et al, 1974).

Fig. 10 - Comparative pair: left single row, right double row (from Woods, et al, 1974).

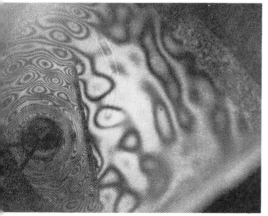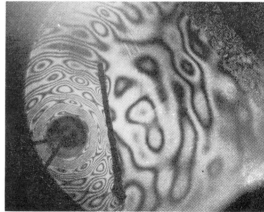

Fig. 11 - Comparative pair: left trench void, right trench filled with Bentonite slurry (from Woods, et al, 1974).

In Fig. 10 it can be seen that all of the holes comprising the barrier do not appear to be circular. This is merely a consequence of the holographic interferometry technique. For some locations on the model, the angle to the hologram plate is such that parts of the interior walls of the holes are seen. Movement of the side walls of the holes also generates fringes and these are recorded in the interferograms making the holes appear out-of-round.

4.3 Slurry Filled Barriers

A fluid is capable of transmitting compressive strain but not shearing strain. The vertical component of the Rayleigh wave should, therefore, react to fluid filled barriers exactly in the same manner as to open barriers. The same geometrical parameters should apply. The holograms under both situations should show identical patterns because horizontal components are not recorded with the setup used in these tests.

A 5% Wyoming Bentonite slurry was tried in barriers consisting of trenches and cylindrical holes. Figure 11 shows a comparison between a void trench (left) and a bentonite filled trench immediately after filling (right). The results were as expected. Each pair of tests like that shown in Fig. 11 indicated that fluids do not transmit the shear component. However, caution must be exercised in interpreting these results for two reasons. First, field tests with bentonite slurry have shown little success because of solidification of the slurry, see Neumeuer (1963). Until this problem is solved, the practical application will not be feasible. Second, it must be remembered that the interferometer arrangement in the foregoing tests recorded only the vertical component of motion. The slurry was expected to be effective for the vertical component but not for the horizontal component. It is quite possible that most of the compressive component (horizontal component) would be efficiently transmitted through the barrier.

5 HORIZONTAL PARTICLE MOTION WITH HOLOGRAPHY

Because of the geometric and interferometric arrangements in all the tests described so far, the particle motion in the horizontal direction did not contribute to the generation of any fringe. But, as indicated in the previous section on slurry filled barriers, it is also necessary to determine the effectiveness of barriers in reducing horizontal particle movements.

As explained earlier, the direction of particle displacement which is recorded in an interferogram is controlled by the sensitivity vector. The sensitivity vector is totally determined by the geometry of the interferometer, that is, the optics, photographic plates and half-space model. By rearranging some components of the apparatus, a second sensitivity vector can be generated and a second direction of particle motion recorded. By combining two particle motion vectors in a common plane (vertical plane for example), two components of motion can be obtained.

400

Two arrangements were tried which permitted the generation of two sensitivity vectors. These are shown in Fig. 12. Figure 12a shows a single hologram arrangement. Two object beams, OB, and one reference beam, RB, impinge on the hologram plate to create two sets of fringes. The interpretation of these fringes is very difficult and while theoretically possible, it has not yet been done in these studies. Figure 12b shows a two hologram arrangement. Each pair of reference and object beams results in a sensitivity vector and each hologram records displacements in a single direction. Contours of motion associated with each sensitivity vector are recorded separately in the two holograms. By combining the displacements from both holograms, the particle motion in a vertical plane was obtained.

Interferograms of the two particle motions obtained by the two hologram techniques are shown in Fig. 13. The sensitivity vectors for points about 3 1/2 footing radii from the footing on the radial line shown in Fig. 13a were calculated. Using the sensitivity vectors and fringes from Fig. 13, the particle motion was found to be approximately ellipical as shown in

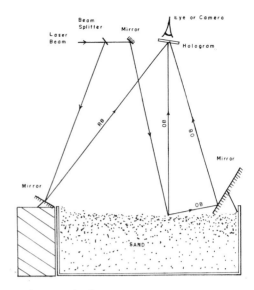

a) one hologram

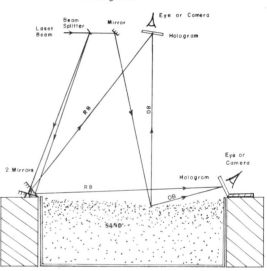

b) two holograms

Fig. 12 - Schematic of optics and plates for horizontal particle motion.

Fig. 14 as would be expected for a Rayleigh wave. In Fig. 14 the location of the ellipse along the x axis was chosen only for convenience, and the horizontal and vertical scales are in wavelengths of the laser light.

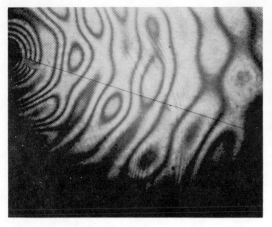

b) right nearly horizontal
   motion interferogram.

a) left vertical motion interferogram.

Fig. 13 - Interferograms for two hologram, horizontal particle
motion arrangements.

The two hologram method appears to provide a method of evaluating both horizontal and vertical components of particle
motion.   This technique will be used in the future to determine the effectiveness of barriers at reducing the horizontal
component of particle motion.

6   EMBEDDED FOOTINGS

Isolation tests described earlier in this paper as well as
those described by Woods (1968), Dolling (1970) and Barkan
(1962) have all considered footings on the surface of the half-
space.   The wave energy distribution between Rayleigh waves
and body waves would certainly be different for embedded

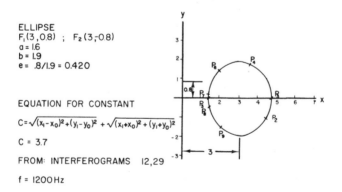

ELLIPSE
$F_1(3, 0.8)$; $F_2(3, -0.8)$
$a = 1.6$
$b = 1.9$
$e = .8/1.9 = 0.420$

EQUATION FOR CONSTANT

$C = \sqrt{(x_1-x_0)^2 + (y_1-y_0)^2} + \sqrt{(x_1+x_0)^2 + (y_1+y_0)^2}$

$C = 3.7$

FROM: INTERFEROGRAMS 12,29

$f = 1200\,Hz$

Fig. 14 - Particle motion at surface of sand, model half-space.

footings than for surface footings; consequently the effectiveness of barriers for screening waves generated by embedded footings was studied using holographic interferometry.

To get a qualitative idea of surface wave energy from an embedded circular footing, a series of tests was performed in which the energy on the surface at the half-space model in a 62° sector extending outward from a vibrating footing to a distance of $3\lambda_R$ was determined at embedment ratios from 0 to 1. The embedment ratio, ER, is the depth to the bottom of the footing, d, divided by the wavelength of the Rayleigh wave, λ_R. The energy on the surface (E_T) was determined by finding the area contained within the contours of each fringe, multiplying by the fringe number and then summing for the sector described above. This was a measure of the energy contained in the vertical particle motion only, because only vertical sensitivity vectors were used.

The energy on the surface vs. embedment ratio is shown in Fig. 15. The five test results presented in Fig. 15 represent Rayleigh Wavelengths from 8.0 cm to 9.0 cm.

It appears that there is a rapid reduction in the amount of energy transmitted by the vertical component of surface motion as the embedment increases from zero to 1/4, and a less rapid decrease for deeper footings. At an embedment ratio of 1, E_T is less than 1/2 as much as for an embedment ratio of 0.

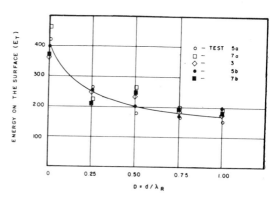

Fig. 15 - Energy in sector on surface of half-space.

To determine the effectiveness of an open trench barrier as the embedment ratio increased, the surface energy, E_{NB}, in a sector beyond the planned location of the barrier, but before excavation of the barrier, was determined first. Figure 16 shows E_{NB} vs. embedment ratio. The trends are identical to those shown in Fig. 15.

After determining the surface energy without a barrier, a trench was excavated to a depth of 1.4 λ_R and

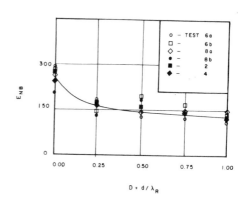

Fig. 16 - Energy in sector beyond imagined barrier.

403

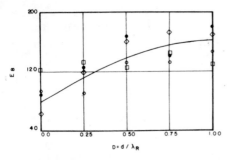
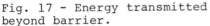

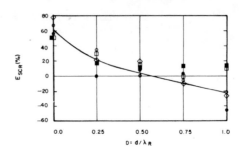

Fig. 17 - Energy transmitted beyond barrier.

Fig. 18 - Screening effect of open trench.

length of 2.5 λ_R, and the surface energy in the sector beyond the trench, E_B, was determined. Figure 17 shows E_B vs. embedment ratio.

Finally, the screening effect of an open trench on waves generated by an embedded footing, $E_{SCR}(\%)$, was defined as:

$$E_{SCR}(\%) = \frac{E_{NB} - E_B}{E_{NB}} \times 100.$$

Figure 18 shows E_{SCR} vs. embedment ratio. It is important to note that for an embedment ratio of about 1/2, a 1.4 λ_R deep barrier is of no value for screening the vertical component of surface motion. Furthermore for embedment ratios greater than about 1/2, the screening value of the trench is negative.

7 SUMMARY AND CONCLUSIONS

It has been demonstrated that double exposure, stroboscopic, holographic interferometry can be used to study the dynamic deformations of a sand half-space model as well as static deformations of model footings on a sand half-space. Using the holographic interferometry technique, the screening or isolation effectiveness of several types of barriers were studied. Also, an interferometric arrangement has been devised so that both horizontal and vertical particle motions can be observed and recorded. Finally, the effectiveness of trench barriers at screening waves generated by embedded footings was studied.

From the isolation or screening tests the following were observed:

1. A row of cylindrical holes can be an effective barrier if the diameter, spacing, and depth of the holes are properly selected, i.e., $H/\lambda_R \geq 1$, $D/\lambda_R \geq 1/6$ and $(S-D)/\lambda_R < 1/4$.

2. Double rows of cylindrical holes can be more effective than a single row, particularly for widely spaced holes.

3. Slurry filled barriers are effective as long as the slurry does not set or solidify.

4. Wave barriers are not as effective for vibrations generated by embedded footings as they are for surface footings. At an embedment ratio of $1/2$, a 1.4 λ_R deep trench has an effectiveness of zero.

8 ACKNOWLEDGEMENTS

The support of the National Science Foundation under Grant GK-16012 is gratefully acknowledged. The suggestions and encouragement of Professor F. E. Richart is also greatly appreciated. Professor N. E. Barnett collaborated on all matters concerning lasers and holography. The tests were performed by R. Sägesser, K. Karal, H. Papadapoulos, W. Haupt and D. Anderson all of whose help is gratefully acknowledged.

9 REFERENCES

Barkan, D. D. (1962), Dynamics of Bases and Foundations, McGraw-Hill Book Company (New York), 434 pp.

Brooks, R. E., Heflinger, L. O., and Wuerker, R. F. (1966), "Pulsed Laser Holograms", I.E.E.E. Jul. Quantum Electronics, QE-2, (8), p. 275.

Dolling, H. J. (1970), "Die Abschirumung von Erschü tterungen durch Bodenschlitze", Die Bautechnik, No. 5, p. 151.

Ennos, A. E. and Archbold, E. (1968), "Vibration Surface Viewed in Real-Time by Interference Holography", Laser Focus, Oct., p. 58.

Leith, E. N. and Upatnieks, J. (1965), "Photography by Laser", Scientific American, Vol. 212, No. 6, June, p. 24.

Neumeuer, H. (1963), "Untersuchungen über die Abschirmung eines bestehenden Gebäudes gegen Erschütterungen beim Bau und Betrieb einer U-Bahustrecke", Baumaschine und Bautechnik-10, Jahrgang, Neft 1, Jan.

Orr, L. W., Tehon, S. W. and Barnett, N. E. (1968), "Isophase Surfaces in Interference Holography", Applied Optics, Vol. 7, No. 1, Jan., pp. 202-203.

Powell, R. L. and Stetson, K. A. (1965), "Interferometric Vibration Analysis by Wave Front Reconstruction", J. Opt. Soc. America, Vol. 55, p. 593.

Sampson, R. C. (1970), "Holographic Interferometric Applications in Experimental Mechanics", Experimental Mechanics, Vol. 10, No. 8, August, p. 313.

Sampson, R. C. (1971), "Structural Measurements with Holographic Interferometry", Materials Researcn and Standards, Vol. 11, No. 9, Sept., p. 26.

Woods, R. D. (1968), "Screening of Surface Waves in Soils", Journal SMFD, ASCE, Vol. 94, No. SM4, p. 951.

Woods, R. D. and Sägesser, R. (1973), "Holographic Interferometry in Soil Dynamics", Proceedings of the Eighth International Conference on Soil Mechanics and Foundation Engineering, Vol. 1, Part 2, pp. 481-486.

Woods, R. D., Barnett, N. E., and Sägesser, R. (1974), "Holography - A New Tool for Soil Dynamics", Journal of the Geotechnical Engineering Division, ASCE, Vol. 100, No. GT11, Proc. Paper 10949, Nov., pp. 1231-1247.

10 NOTATION

D   Diameter of circular hole (cm)

d   Depth to bottom of footing (cm)

E_T Energy on surface of half-space

E_{NB} Energy beyond imagined barrier

E_B Energy beyond real barrier

E_{SCR} Screening effect of trench for embedded source (%)

F_A' Amplitude reduction factor, diameter and spacing variable

F_A'' Amplitude reduction factor, depth variable

g   Acceleration of gravity (cm)

H   Depth of wave barrier (cm)

L   Overall length of wave barrier (cm)

P(t)   Time varying force

R   Distance from vibration source to wave barrier (cm)

S   Spacing center to center of circular holes (cm)

Δ Settlement of footing (mm)

λ_R Rayleigh wavelength (cm)

Proceedings of DMSR 77 / Karlsruhe / 5-16 September 1977 / Volume 1

Measurement and evaluation of random vibrations

WERNER RÜCKER
Bundesanstalt für Materialprüfung, Berlin, Germany

SYNOPSIS

In many cases the input functions (excitation) of linear system
are statistical time functions. Very often these excitations ar
not accessible for direct measurement. It is demonstrated, that
for a continuous-input system a description of the excitation
signals can be obtained from measured field response data. This
description contains all the necessary informations in order to
statistically investigate the system. A numerical method is
described and its application is shown through means of traffic
vibrations.

1. INTRODUCTION

When investigating dynamically loaded structures it is
necessary to distinguish between two groups:
 1. structures which are loaded with deterministic excitation
such as unbalances of machinery, vibrations caused by bells, et
 2. structures which are loaded with stochastic loads such as
earthquakes, wind, seawaves and traffic vibrations.
 In investigations, the actual structure is normally represented by a mathematical model in which we can study by means
of variation of the system parameters the effects of changes
of individual structure values on the entire function of the
structure.
 In the case of deterministic excitation processes the responses of a structure can be indicated unambiguously. If, however, stochastic excitation processes are concerned, the vibration process of a structure is a stochastic process which cannot be expressed by deterministic time functions. Such excitation and response processes of linear structures can be
described only to a limited extent with the aid of the probability theory.
 The stochastic processes to be analysed in this study are
vibrations, generated by rail cars. In this case the dynamic
loading is not only produced by the moving mass of the train,
but also by imperfections on the surface of the rails. These
imperfections on the bearing surface produce an additional

dynamic loading, which is superposed on the original excitation.
Besides this factor there are a number of other influencing
factors which in general do not permit such excitation processes to be theoretically reconstructed.

It seems to be reasonable to carry out an estimate of these
excitation processes by means of measurements. In doing so two
problems have to be taken into consideration:

1. Since the excitation processes are random processes the
measurements have to be studied by statistical means. The problem now arises how such measurements are to be handled with
a view to statistics.

2. The second problem results from the fact that only the
responses of a structure can be measured. It must therefore be
established how the excitation processes can be determined on
the basis of the measured response processes.

2. INVESTIGATED SYSTEM

The system under study is shown in figure 1.

ANALYZED SYSTEM

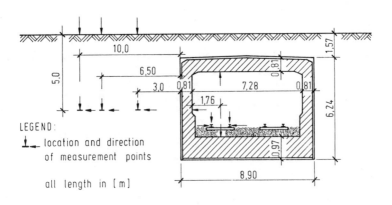

Fig. 1. Analyzed system(subway tunnel)with location of points
of measurement.

This system represents a continuum consisting of different materials, which is excited to mechanical vibrations by the running train. In order to study the propagation of vibrations and to determine the excitation functions, the vibrations generated by the running train have been measured at several points of the system.

At the tunnel itself the vibrations were measured in the horizontal and vertical directions at the rail, in the horizontal direction at the wall of the tunnel and in the vertical direction at the ceiling and the bed plate of tunnel at several cross-sections. In the surrounding soil measurements were made in both directions, horizontal and vertical at the soil surface and in boreholes drilled into the soil at several cross sections too.

All together more than 10.000 single signals were recorded. The speed of the trains varied from 30 km/h to 70 km/h. With respect to the result shown in this report the speed of the train was 70 km/h.

3. ELECTRICAL AND ELECTRONIC INSTRUMENTS

3.1 Transducers

The choice of the transducers, displacement gages, velocity gages or accelerometers depends on the frequency content of the observed signals.

For instance, signals with frequency content less than 1 Hz are usually detected with displacement transducers. Velocity transducers are employed in the frequency range 1 - 100 Hz and accelerometers are employed in the high frequency range >100 Hz

The frequency range of the signals generates by the running trains is usually between 1 - 100 Hz. Taking this into account geophones (velocity transducer) were used in all cases.

3.2 Monitoring system

For processes, with random characteristics a very great number of signals must be recorded in order to get a meaningful view inside the nature of the random process.

Another great difficulty in storage of such signals is related to the fact that one must deal simultaneously with 20 or more points of measurement.

Figure 2 illustrates in a simplified block diagram the storage and evaluation of vibration signals

First physical quantity of velocity is converted into electrical signals by the geophones. All electrical signals coming from the geophones then were amplified by a 24-channel-amplifie

After this, the amplified signals of all geophones pass through a PCM-System (PULSE CODE MODULATION). In this system the analog voltage signal is sampled at equivalent time intervals Δt and converted into a binary code.

The data words of each sample point and each channel are stored on a 4-track magnetic tape in a sequentia manner (8 channels on 1 track).

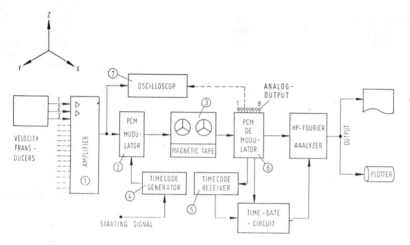

Fig. 2. Block diagram of storage and evaluation of vibrations.

Taking 30 ips for the speed of the magnetic tape, the sample frequency of each channel equals 1904 samples per second. In this working mode, signals with maximum frequency of about 450 Hz can be recorded.

To avoid the so-called aliasing phenomena, frequencies greater than approx.450 Hz were damped out by antialising filters incorporated in the PCM-System. For convenient analysis of stored data it is possible to store real time information generated by a time code generator on each track of the magnetic tape.

For a visual display of the signals a cathode-ray-oscilloscope may be connected to the system.

Figure 3 shows the whole monitoring and storage system.

3.3 Processing system

In the past years and until recently analysis of vibrations was done manually.

In the last year a digital computer analysis with necessary hardware and software has been developed by the Bundesanstalt für Materialprüfung, Berlin, in order to get a more sophisticated and quicker method to analyze determined or random vibrations.

By this method the digital data words stored on the magnetic tape first pass through a PCM-Demodulator and were then directly transcribed to a digital computer.

After some manipulation and rearrangement of the data words from sequential to parallel form the signals can be analyzed by standard methods, e.g. correlation in the time domain or Fourier analysis in the frequency domain. The frequency reso-

Fig. 3. Monitoring system.

lution Δf by Fourier transformation depends on the total time T of the record.

For instance, in the working mode of 30 ips of the magnetic tape the sampling rate is $\Delta t = {}^1/1904$ sec^{-1}. If N points define the signal (N can be chosen arbitrarily, but must be a integer) the total time T of the record is

$$T = N \cdot \Delta t \qquad (1)$$

Then the frequency resolution is

$$\Delta f = \frac{1}{T} = \frac{1}{N \Delta t} \qquad (2)$$

4. EVALUATION

4.1 Fourier-spectrum

Since the vibrations, generated by the running trains are random, the measured values must be averaged. In general such vibrations are transient. But often the whole transient time record can be subdivided into stationary segments. For the further analysis it is assumed, that the process is stationary but not necessary ergodic. Under this stationary condition any deterministic or random process can be regarded as superposition of harmonic vibrations of different frequencies. The amplitudes

of the harmonics contained in the random process are achieved
by applying a Fourier transform.

This yields a Fourier transform which will have positive
and negative real and imaginary values respectively, randomly
distributed across the spectrum. Thus, if the transform is
averaged, the results will average out to zero. It is therefore necessary to carry out a time synchronization, i.e., the
phase information contained in the spectrum due to the imaginary values must be eliminated. This is achieved by conjugate
complex multiplication the complex Fourier transform.

The result is a real positive value at each frequency, which
represents the energy content of this frequency.

Formally this procedure corresponds to establishing the
power spectral density G_{yy} (f)

$$G_{yy} (f) = \lim_{T \to \infty} \frac{|Y (f)|^2}{T} \tag{3}$$

where

$$Y (f) = \int_{-\infty}^{\infty} y (t) \cdot e^{-i\omega t} dt \tag{4}$$

y (t) represent the measured time dependent signal.

To normalize the power spectrum it is necessary to divide
the spectrum by the observed time duration T of the process.

In this way, measurements of random vibrations can be normalized to give equivalent reading independent of the sample
rate Δt and the frequency bandwidth Δf used in the Fourier
transform.

The actual value of T and Δf was T = 2.15 sec. and Δf = 1/T
= 0.46 Hz. The observed time duration of the whole process was
about 8 sec.

4.2 Statistics

For statistical investigations it is also necessary to determine the mean square value $\overline{y^2}$ of a random process. If a process
is Gaussian with zero mean value, the probability density p (y)
of the process is fully described by the variance $\sigma_y^2 = \overline{y^2}$.
This value is obtained by frequency integration over the power
spectral density G_{yy} (f).

$$\overline{y^2} = \int_{0}^{\infty} G_{yy} (f) df \tag{5}$$

4.3 Estimate of number of averages

For practical purposes it is very important to know how many
measurements must be averaged in order to get reliable data.

If the random process is a Gaussian process, then the real
and imaginary components a + ib of the Fourier transform are
also normally distributed.

When forming the power spectrum the term a + ib is multiplied by the conjugate complex. In this way a term $a^2 + b^2$ is
obtained the statistical distribution of which is a chi-square

distribution (χ^2) with two degrees of freedom.
By χ^2-distribution the variance of the expected value of
each spectral line of the power spectrum can be expressed as

$$\sigma = \frac{1}{\sqrt{n}} \cdot 100 \tag{6}$$

where σ standard deviation in percent
      n number of estimates.
For the special problem treated here it was found that after
approx. 70 averages there was no significant change in the
values of the spectral lines.

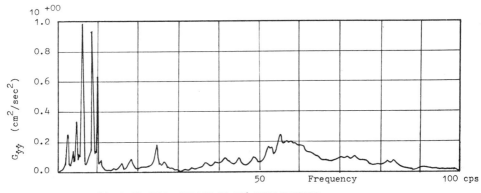

POWER-SPECTRA-AVERAGE OF 136 TIME RECORDS
MEAN SQUARE VALUE $\overline{y^2} = 746.10^{-2}$ cm^2/sec^2

Fig. 4. Vertical Component: at the rail.

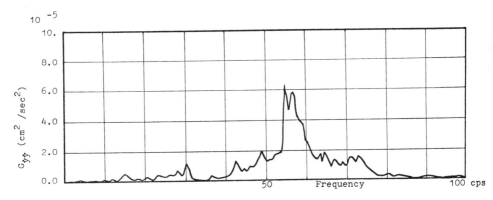

POWER-SPECTRA-AVERAGE OF 136 TIME RECORDS
MEAN SQUARE VALUE $\overline{y^2} = 786.10^{-6}$ cm^2/sec^2

Fig. 5. Vertical Component: on the bed plate foundation of the
tunnel.

413

5. DISCUSSION OF RESULTS

The results may be discussed here only briefly.

From figure 4 it is seen that the highest amplitudes occur at the rail in the frequency range 0 - 10 cps. These amplitudes with frequencies between 0 - 10 cps do not occur at any other point of measurement. At the rail there are also amplitudes with frequencies in the range of 50 - 75 cps.

The response of the bed plate of the tunnel (figure 5) and the response of the tunnel's ceiling (figure 6) lies mainly in one frequency, 55 cps and 15 cps, respectively.

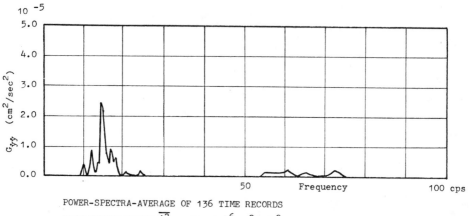

POWER-SPECTRA-AVERAGE OF 136 TIME RECORDS

MEAN SQUARE VALUE $\overline{y^2}$ = 102.10 $^{-6}$ cm^2/sec^2

Fig. 6. Vertical Component: tunnel ceiling.

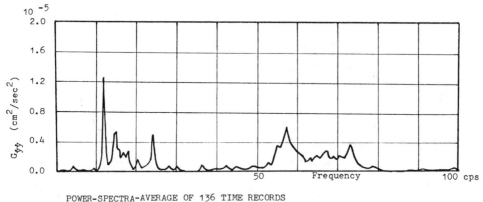

POWER-SPECTRA-AVERAGE OF 136 TIME RECORDS

MEAN SQUARE VALUE $\overline{y^2}$ = 111.10 $^{-6}$ cm^2/sec^2

Fig. 7. Horizontal Component: midside tunnel wall.

414

At the wall of the tunnel (figure 7) there are several peaks distributed across the spectrum.

The mean square values, which are a measure of the average energy contained in the processes, are approximately the same for the measured points of the wall and the ceiling.

For the surrounding soil (figures 8 -9), at a distance of 3 m from the tunnel and depth of 5 m the main response is in the frequency range between 50 - 74 cps.

The ratio of the mean square values for the vertical component at the depth of 5 m and a distance of 10 m compared with the vertical component at a distance of 3 m reduces to about 20 percent.

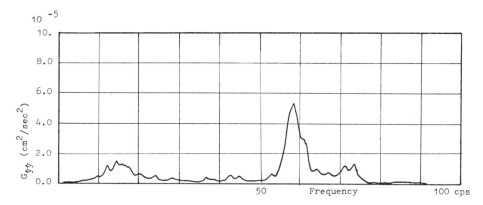

POWER-SPECTRA-AVERAGE OF 136 TIME RECORDS

MEAN SQUARE VALUE $\overline{\dot{y}^2}$ = 616.10^{-6} cm^2/sec^2

Fig. 8. Horizontal Component: 5 m under the soil surface, 3m from the tunnel.

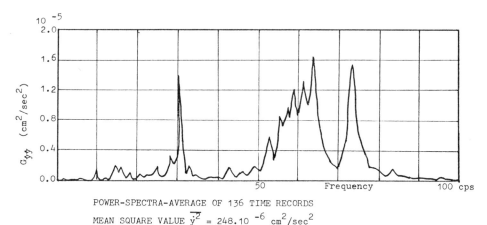

POWER-SPECTRA-AVERAGE OF 136 TIME RECORDS

MEAN SQUARE VALUE $\overline{\dot{y}^2}$ = 248.10^{-6} cm^2/sec^2

Fig. 9. Vertical Component: 5 m under the soil surface, 3 m from the tunnel.

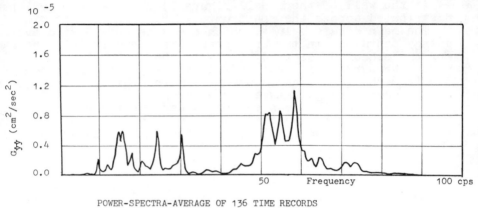

POWER-SPECTRA-AVERAGE OF 136 TIME RECORDS
MEAN SQUARE VALUE $\overline{y^2}$ = 151.10^{-6} cm^2/sec^2

Fig. 10. Vertical Component: on the soil surface 3 m from the tunnel.

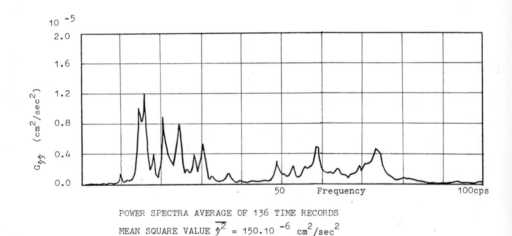

POWER SPECTRA AVERAGE OF 136 TIME RECORDS
MEAN SQUARE VALUE $\overline{y^2}$ = 150.10^{-6} cm^2/sec^2

Fig. 11. Vertical Component: on the soil surface, 6,5 m from the tunnel.

For the vertical component on the soil surface (figures 10-12) the response lies in 15, 25, 30 cps and between 50 - 60 cps.
The mean square values for the vertical components in a distance of 3 m and 6,5 m away from the tunnel are the same. But it must be noted that there is a change in amplitudes with increasing distance from the high frequency range at about 50 - 60 cps to the low frequency range at about 15 - 30 cps.

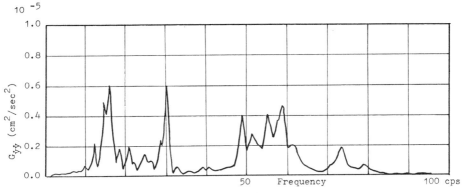

POWER-SPECTRA-AVERAGE OF 136 TIME RECORDS

MEAN SQUARE VALUE $\overline{y^2}$ = 100 .10^{-6} cm^2/sec^2

Fig. 12. Vertical component: on the soil surface, 10 m from the tunnel.

The ratio of the mean square values at a distance of 10 m and 6,5 m is about 67 percent.

6. DETERMINATION OF THE EXCITATION PROCESSES

Determining the excitation functions on the basis of the measured response processes the relationship between input x (t) and output y (t) of a linear system must be known.

This relationship can be obtained in the case of any deterministic or stochastic excitation by convolution by means of the Duhamel integral.

$$y \ (t) = x \ (t) \ _* \ h \ (t) = \int_{-\infty}^{\tau} x \ (t) \ h \ (\ \tau - t) \ dt \qquad (7)$$

h(t) is the response of the linear system to an unit impulse δ (t) (Dirac-impulse).

The star $_*$ in Eq. (7) symbolize convolution.

In the frequency domain this relationship becomes

$$Y \ (i\omega) = H \ (i\omega) \cdot X \ (i\omega) \qquad (8)$$

i.e. the spectral response Y (iω) is obtained as the product of the frequency response H (iω) due to unit excitation and the Fourier transform X (iω) of the excitation function.

For random processes the Fourier transform does not exist. But using the definition of power spectral densities Eq. (8) can be written as

$$G_{yy} \ (\omega) = \frac{Y \ (i\omega) \cdot Y \ (i\omega)*}{T} = |H \ (i\omega)|^2 \cdot G_{xx} \ (\omega) \qquad (9)$$

417

Y (iω)* is the conjugate complex of the Fourier transform Y(iω). This equation is valid for random processes.

For a continuous, discretized system subject to multiple inputs it is convenient to write equation (9) in matrix notation:

$$\lfloor G_{yy} \rfloor = [H]^* \lfloor G_{xx} \rfloor [H]^T \tag{10}$$

By inverting this matrix equation one obtains the desired power spectral density matrix $\lfloor G_{xx} \rfloor$ of the excitation process

$$\lfloor G_{xx} \rfloor = [H]^{*-1} \lfloor G_{yy} \rfloor [H]^{-T} \tag{11}$$

Although for the determination of power spectrum of the excitation processes x (t) the dynamic properties of the structure under consideration must be known, it is not necessary to know the dynamic properties of another structure which is loaded with the power spectrum determined. It must merely be ensured that the type and input geometry of the excitations are the same.

7. INPUT POWER SPECTRUM OF RUNNING TRAINS

The above procedure can best be demonstrated by considering the problem of exciting the structure seen in figure 1 through the running trains. For computation of the input spectrum G_{xx} the frequency responses at those points of the structure must be known where the response was measured.

In order to get the frequency responses the structure was analyzed by the finite element method. The structure was subdivided into 290 elements with 355 nodal points (Figure 13).

The greatest element size chosen was about 1/5 of the shortest wavelength of the shear wave.

The loading was a unit excitation on the surface of the ballast.

The necessary input parameters for the FE-model, modulus of elasticity in shear, G, and Poisson's ratio ν were computed from measured values of the compression-wave velocity v_p and shear wave velocity v_s.

This was done by cross-correlation of the signals of two measurement points at different locations .

The method of cross correlation yields the difference of travel time between two points of measurement for each wave type. Knowing this travel time and the distance L between two points one can calculate the values of velocity v_p and v_s.

Shear modulus G and Poisson's ratio ν are then obtained by the relations

$$G = \rho \cdot v_s^2 \tag{12}$$

$$\nu = (2 - v_p^2/v_s^2) \big/ (1 - 2\, v_p^2/v_s^2) \tag{13}$$

Mass density ρ was determined in the laboratory from soil samples. Damping was included in the model by a complex modulus

$$G^C = G_1 + iG_2 \qquad (14)$$

The ratio G_2/G_1 is defined by the loss angle δ_L

$$\tan\delta_L = G_2/G_1 \qquad (15)$$

which is related to the logarithmic decrement δ by

$$\delta = \Pi \cdot \tan\delta_L \qquad (16).$$

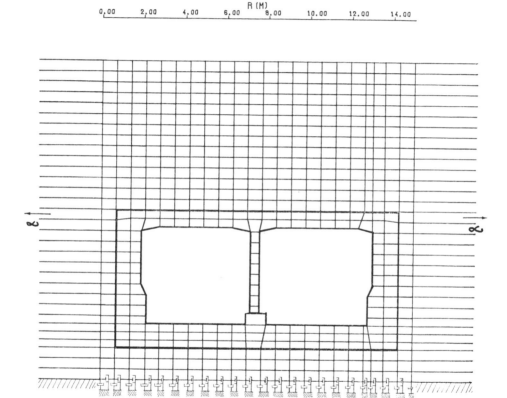

Fig. 13. FE-system of real system shown in figure 1.

The computed input power spectrum is shown in figure 14. It can be seen that there are small peaks at 10, 15, 25 and 30 cps. These frequencies are generated by static load repetitions of the running train. For the surrounding soil vibra-

419

tions with these frequencies are very important, because they do not decrease very rapidly with distance.

It can further be seen from figure 14 that there is a wide frequency range between 50 - 75 cps with very high spectral lines. But from the measured responses of the structure it can be seen that these frequencies correspond only to the bed plate of the tunnel in a significant manner. In the surrounding soil vibrations with these high frequencies are only important close by the tunnel. With increasing distance they decrease very rapidly.

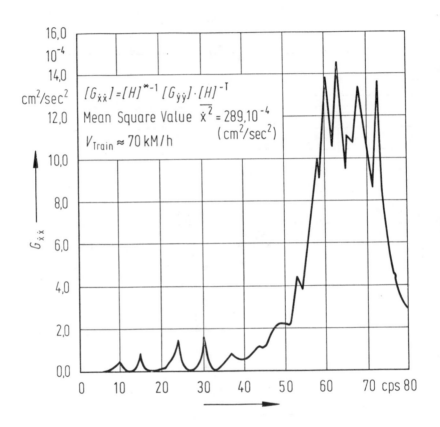

Fig.14. Powerspectrum $G_{\ddot{x}\ddot{x}}$ of the excitation process.

8. CONCLUSION

One possibility of analyzing random vibrations has been demonstrated. An upper limit for the choice of the number of time records regarding statistics was given. It has been shown how the excitation processes can be determined on the basis of measured response processes.

REFERENCES

Clough, R.W. and J. Penzien 1975, Dynamics of Structures,
    New York, McGraw-Hill.
Lysmer, J. and G. Waas 1972, Shear Waves in Plane Infinite
    Structures, J. Engng. Mech. Div.: 85 - 104.
Przemseniecke, J.S. 1971, Theorie of Matrix Structural Analysis,
    New York, McGraw-Hill Book Co.
Resch, F.J. and R. Abel 1975, Spectral Analysis Using Fourier
    Transform Technics, Int. Journal for Numerical Methods in
    Engineering, Vol. 9, pp 869 - 902.
Richarts, F.E., jr., J.R. Hall, jr. and R.D. Woods 1970,
    Vibrations of Soils and Foundations, New Jersey, Prentill-
    Wall, Inc. Englewoods Cliff.
Rücker, W., Untersuchung über die Auswirkungen von U-Bahn-Er-
    schütterungen auf geplante Gebäude (published shortly in
    "Die Bautechnik").
Vanmarcke, E.H. 1972, Properties of spectral moments with
    applications to random vibrations, J. Engng. Mech. Div.
    ASCE, 98. 425.

Excitation of uncoupled vibrations

B. PRANGE
University of Karlsruhe, Karlsruhe, Germany

Discussion Contribution to the Workshop:
"Ground Vibration Excitation and Measurement"

SYNOPSIS

By controlling amplitude and phase of the electrodynamic vibrators of an exciter system, uncoupled modes of excitation of
vibrations can be achieved.

INTRODUCTION

In many cases, wavefields are generated by a wave source consisting of an electro-dynamic vibrator with a circular contact
area between the vibrator mass and the surface of the subsoil.
Due to the fact that the centre of gravity of the vibrator and
the contact area in general do not coincide, only two of the
six possible degrees of freedom of excitation are uncoupled,
namely z translation and $\hat{z}$ rotation. In all the other degrees of
freedom, due to the height h of the c.g. coupling exists (e.g.
a translation x will result in a coupled rotation $\hat{y}$ etc., fig.1).

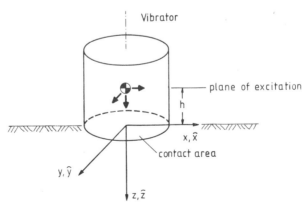

Figure 1. Vibrator on surface of subsoil

There are a number of soil-dynamical problems, however, which
demand the excitation of uncoupled vibrational modes:

423

Lumped Parameter Models

To check the validity of lumped parameter theories, experiments must be performed which come as close as possible to the theoretically assumed pure translations and rotations of the contact area (see e.g. Bycroft 1956, Richart, Hall and Woods 1970 e.al.). Coupled vibrations would overlay the experimental result with additional undesired parameters.

Wavefields

Some types of wavefields due to specific wave sources can be calculated analytically. To check the theoretical results by experiments, the wave source should perform only the specified translations or rotations without coupling effects to other vibration modes. In such cases, the resulting wavefields can be studied without considering the superposition of different wavefields from the different source types.

UNCOUPLED VIBRATOR

To overcome the beforementioned difficulties of coupled exciter vibrations, we provide in the case of translations x, y and rotations $\hat{x}$, $\hat{y}$ two vibrators on top of each other. By controlling amplitude and phase of each vibrator separately we are able to generate an additional moment or force in such a manner that the undesired coupling effects are cancelled.

The three governing parameters of vibrations are frequency, amplitude and phase. If the frequency for both vibrators in the same, we only have to control amplitude and phase of one vibrator, see fig.2.

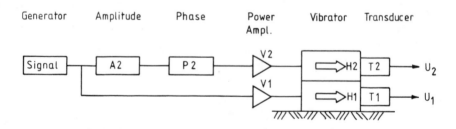

Figure 2. Vibrator with coupling compensation

The stationary signal from the signal generator is fed into power-amplifier V1 driving the lower vibrator H1. The same signal is controlled in amplitude by A2 and in phase by P2 and than fed into power-amplifier V2 driving the upper vibrator H2.

The resulting vibrations at the lower and upper level are measured by vibration transducers T1 and T2, respectively.

The two forces generated such are

$$H_1(t) = H_1 \cdot e^{i\omega t}$$

$$H_2(t) = H_2 \cdot e^{i(\omega t + \varepsilon)}$$

and the resulting translations

$$T_1(t) = T_1 \cdot e^{i(\omega t + \varphi_1)}$$

$$T_2(t) = T_2 \cdot e^{i(\omega t + \varphi_2)}$$

beeing transformed into the respective output voltages $U_1(t)$ and $U_2(t)$; ε, φ_1, φ_2 beeing the respective phase angles.

UNCOUPLED TRANSLATIONS X, Y

In order to achieve uncoupled translations in either the x or the y direction (rotating the uncoupled vibrator by 90° around $\hat{z}$) we control A2 and P2 such that $U_2(t)-U_1(t) = 0$ thereby cancelling any rotation around a horizontal axis. This can easily be done by feeding $U_1(t)$ and $U_2(t)$ into a differential amplifier and observing the output to be zero.

UNCOUPLED ROTATIONS $\hat{X}$, $\hat{Y}$

In this case we have to control A2 and P2 such that no translation at the contact area (Z = 0) occurs. This can easily be achieved by observing the output $U_1(t) = 0$ if T_1 is located at Z = 0 or $h_1 \cdot U_2(t) - h_2 \cdot U_1(t) = 0$ if the transducers are placed at the elevation $z_2 = -h_2$; $z_1 = -h_1$ respectively.
   Both cases can be controlled automatically by using the respective zero-conditions to control an automatic electronic feedback circuit which than provides the necessary amplitude and phase conditions. Furthermore, such an arrangement enables one to generate nonperiodical, nonstationary uncoupled signals at the wave source.

REFERENCES

Bycroft, G.N. 1956 Forced Vibrations of a Rigid Circular Plate on a Semiinfinite Elastic Space and on an Elastic Stratum. Phil.Trans.Royal Soc. London Ser.A Vol.248

Prange, B. 1978 Measurement of Vibrations. Proc.Dynamical Methods in Soil and Rock Mech. DMSR 77 Balkema, Rotterdam

Richart, F.E., Hall, J.R. and Woods, R.D. 1970 Vibrations of Soils and Foundations, Prentice Hall Inc., Englewood Cliffs, N.J.

List of contributors to volume 1

Gaul, L., Dr.-Ing.
Institute of Applied Mechanics
Department of Mechanical Engineering
The Technical University of Hannover
Hannover, Germany

Gonzalez, J.J.
INTECSA
Madrid, Spain

Haupt, W.A., Dr.-Ing.
Institute of Soil and Rock Mechanics
Department of Civil Engineering
The University of Karlsruhe
Karlsruhe, Germany

Holzlöhner, U., Dr.-Ing.
Bundesanstalt für Materialprüfung BAM
Berlin, Germany

Kausel, E., Dr.
Senior Structural Engineer
Stone & Webster Engineering Corporation
Boston, Massachusetts, USA

Novak, M., Professor
Department of Engineering Science
The University of Western Ontario
London, Canada

Prange, B., Dr.-Ing.
Privatdozent, Akademischer Direktor
Institute of Soil and Rock Mechanics
Department of Civil Engineering
The University of Karlsruhe
Karlsruhe, Germany

Richart, F.E.Jr.
W.J.Emmons Distinguished Professor
of Civil Engineering
Department of Civil Engineering
College of Engineering
The University of Michigan
Ann Arbor, Michigan, USA

Roesler, S.K., Dr.-Ing.
Institute of Soil and Rock Mechanics
Department of Civil Engineering
The University of Karlsruhe
Karlsruhe, Germany

Roesset, J.M., Professor
Department of Civil Engineering
Massachusetts Institute of Technology
Cambridge, Massachusetts, USA

Rücker, W., Dipl.-Ing.
Bundesanstalt für Materialprüfung BAM
Berlin, Germany

Savidis, S.A., Professor
Institute of Soil Mechanics
and Foundation Engineering
Department of Civil Engineering
The Technical University of Berlin
Berlin, Germany

Stokoe, K.H.II, Professor
College of Engineering
Department of Civil Engineering
The University of Texas
Austin, Texas, USA

Werkle, H., Dipl.-Ing.
Institute of Concrete Engineering
Department of Civil Engineering
The University of Karlsruhe
Karlsruhe, Germany

Woods, R.D., Professor
College of Engineering
Department of Civil Engineering
The University of Michigan
Ann Arbor, Michigan, USA